The Mechanics of
Piezoelectric Structures

T0338655

The Mechanics of
Piezoelectric Structures

Jiashi YANG

University of Nebraska-Lincoln, USA

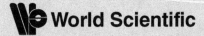

NEW JERSEY • LONDON • SINGAPORE • BEIJING • SHANGHAI • HONG KONG • TAIPEI • CHENNAI

Published by

World Scientific Publishing Co. Pte. Ltd.

5 Toh Tuck Link, Singapore 596224

USA office: 27 Warren Street, Suite 401-402, Hackensack, NJ 07601

UK office: 57 Shelton Street, Covent Garden, London WC2H 9HE

British Library Cataloguing-in-Publication Data
A catalogue record for this book is available from the British Library.

THE MECHANICS OF PIEZOELECTRIC STRUCTURES

ISBN-13 978-981-256-701-7
ISBN-10 981-256-701-1

Printed in Singapore

Preface

This book is a natural continuation of the author's previous book, *"An Introduction to the Theory of Piezoelectricity"* (Springer, New York, 2005), which discusses the three-dimensional theory of piezoelectricity. Three-dimensional theory presents complicated mathematical problems due to the anisotropy of piezoelectric crystals and electromechanical coupling. Very few problems in piezoelectric devices can be directly analyzed by the three-dimensional theory. To obtain results useful for device applications, usually numerical methods have to be used or structural theories have to be developed to simplify the problems so that theoretical analyses are possible. These two approaches are both very effective in the modeling and design of piezoelectric devices.

For piezoelectric devices, dynamic problems are frequently encountered. This is because many piezoelectric devices are resonant devices operating at a particular resonant frequency and mode of a structure. Both surface acoustic waves (SAW) and bulk acoustic waves (BAW) are used. In the analysis of resonant piezoelectric devices, usually vibration characteristics like frequency and wave speed are of primary interest, not the stress and strain for strength and failure consideration as in traditional structural engineering.

Another rather unique feature of the analysis of resonant piezoelectric devices is that BAW devices often operate with the so-called high-frequency modes. Take a plate as an example. The high frequency modes, e.g., thickness-shear and thickness-stretch, are modes whose frequencies are determined by the plate thickness, the smallest dimension. This is in contrast to the low frequency modes of extension and flexure in traditional structural engineering, whose frequencies depend strongly on the length and/or width of the plate. Another characteristic of the high frequency modes is that for long waves their frequencies do not go to zero but have finite cutoff frequencies. This has implications in certain unique behaviors of the high frequency modes such as the useful energy trapping phenomenon.

In applications to high-frequency, dynamic problems of piezoelectric devices, the accuracy of a structural theory is judged by its dispersion relation of the wave solution of the operating mode of a device in the frequency range and wave number range of interest. This is different from traditional structural engineering where, for example, the stress distribution over the cross section of a beam or plate is often of main interest.

The study of high frequency modes in piezoelectric plates by structural theories was initiated by R. D. Mindlin. Mindlin's effort in the shear deformation plate theory was mainly for the analysis of thickness-shear vibrations of crystal plates, a problem motivated by the study of piezoelectric resonators. Under the influence of the pioneering work of Cauchy, Poisson and Kirchhoff, Mindlin systematically derived equations for high-frequency vibrations of piezoelectric plates based on expansions and approximations in the variational formulation of the three-dimensional theory, and studied behaviors of the high-frequency modes using plate equations. A systematic treatment of high frequency vibrations of crystal plates was given by Mindlin in "An Introduction to the Mathematical Theory of Vibrations of Plates" (the U.S. Army Signal Corps Engineering Laboratories, Fort Monmouth, NJ, 1955), which was not formally published.

This book focuses on high-frequency, dynamic theories of piezoelectric structures for device applications. It emphasizes the development of theories and the determination of the frequency ranges and wave number ranges in which the theories are good approximations of the three-dimensional theory. Following a brief summary of the three-dimensional theories of electroelastic bodies in Chapter 1, the development of two-, one- and zero-dimensional theories for high-frequency vibrations of piezoelectric plates, shells, beams, rings and parallelepipeds is systematically presented in subsequent chapters. The range of applicability of the structural theories obtained is examined by comparing dispersion relations of simple wave solutions from the structural theories to the dispersion relations of the exact solutions of the same waves from the three-dimensional theories. In addition to linear piezoelectricity, certain nonlinear effects are also considered. As examples of applications, simple vibrations of piezoelectric plates, shells, beams and rings are analyzed. A few piezoelectric devices including resonators, actuators, a mass sensor, a fluid sensor, a transformer, a gyroscope and buckling of thin structures are also studied using structural theories.

The main purpose of the book is to present a procedure systemized by Mindlin for developing structural theories, rather than collecting all theories for piezoelectric structures. It is hoped that, having read a book like this, one can develop various structural theories needed when facing different device problems.

Due to the use of quite a few stress tensors and electric fields in nonlinear electroelasticity, a list of notation is provided in Appendix 1. Material constants of some common piezoelectric materials are given in Appendix 2.

I would like to take this opportunity to thank Ms. Deborah Derrick of the College of Engineering and Technology at UNL for editing assistance with the book, and Mr. Honggang Zhou, my graduate student, for plotting Figures 2.5.2 and 2.5.3.

JSY
Lincoln, NE
September, 2005

Contents

Chapter 1
Three-Dimensional Theories

In this chapter we summarize the three-dimensional equations of the nonlinear theory of electroelasticty for large deformations and strong fields [1,2], the linear theory of piezoelectricity for infinitesimal deformation and fields [3,4], the linear theory for small fields superposed on finite biasing or initial fields [5,6], and the theory for weak, cubic nonlinearity [7,8]. A systematic presentation of these theories can also be found in [9]. The structural theories of lower dimensions in later chapters will be derived from these three-dimensional theories. This chapter uses the two-point Cartesian tensor notation, the summation convention for repeated tensor indices, and the convention that a comma followed by an index denotes partial differentiation with respect to the coordinate associated with the index.

1.1 Nonlinear Electroelasticity for Strong Fields

Consider a deformable continuum which, in the reference configuration at time t_0, occupies a region V with a boundary surface S (see Figure 1.1.1). $\mathbf{N}$ is the unit exterior normal of S. In this state the body is free from deformation and fields. The position of a material point in this state is denoted by a position vector $\mathbf{X} = X_K\mathbf{I}_K$ in a rectangular coordinate system X_K. X_K denotes the reference or material coordinates of the material point. They are a continuous labeling of material particles so that they are identifiable. At time t, the body occupies a region v with a boundary surface s and an exterior normal $\mathbf{n}$. The current position of the material point associated with $\mathbf{X}$ is given by $\mathbf{y} = y_k\mathbf{i}_k$, which denotes the present or spatial coordinates of the material point.

Since the coordinate systems are othogonal,

$$\mathbf{i}_k \cdot \mathbf{i}_l = \delta_{kl}, \quad \mathbf{I}_K \cdot \mathbf{I}_L = \delta_{KL}, \tag{1.1.1}$$

where δ_{kl} and δ_{KL} are the Kronecker delta. In matrix notation,

1

$$[\delta_{kl}] = [\delta_{KL}] = \begin{bmatrix} 1 & 0 & 0 \\ 0 & 1 & 0 \\ 0 & 0 & 1 \end{bmatrix}. \tag{1.1.2}$$

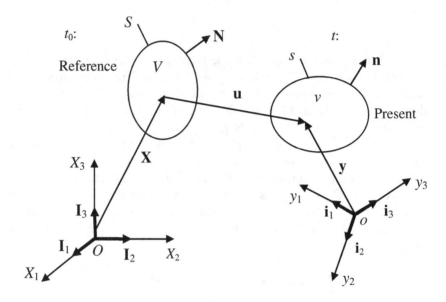

Figure 1.1.1. Motion of a continuum and coordinate systems.

In the rest of this book the two coordinate systems are chosen to be coincident, i.e.,

$$o = O, \quad \mathbf{i}_1 = \mathbf{I}_1, \quad \mathbf{i}_2 = \mathbf{I}_2, \quad \mathbf{i}_3 = \mathbf{I}_3. \tag{1.1.3}$$

The transformation coefficients (shifters) between the two coordinate systems are denoted by

$$\mathbf{i}_k \cdot \mathbf{I}_L = \delta_{kL}. \tag{1.1.4}$$

When the two coordinate systems are coincident, δ_{kL} is simply the Kronecker delta. It is still needed for notational homogeneity. A vector can be resolved into rectangular components in different coordinate systems. For example, we can also write

$$\mathbf{y} = y_K \mathbf{I}_K, \tag{1.1.5}$$

with

$$y_M = \delta_{Mi} y_i. \tag{1.1.6}$$

The motion of the body is described by $y_i = y_i(\mathbf{X}, t)$. The equations of motion and Gauss's equation of electrostatic (the charge equation) are

$$K_{Lj,L} + \rho_0 f_j = \rho_0 \ddot{y}_j,$$
$$\mathcal{D}_{K,K} = \rho_E, \tag{1.1.7}$$

where K_{Lj} is the two-point total stress tensor, ρ_0 is the reference mass density, f_j is the mechanical body force per unit mass, and $\mathcal{D}_K$ is the reference electric displacement vector. ρ_E, a scalar (E is not an index), is the free charge density per unit reference volume, and a superimposed dot represents the material time derivative

$$\ddot{y}_i = \frac{D^2 y_i}{Dt^2} = \left. \frac{\partial^2 y_i(\mathbf{X}, t)}{\partial t^2} \right|_{\mathbf{X}\ \text{fixed}}. \tag{1.1.8}$$

In Equation (1.1.7), K_{Lj} and $\mathcal{D}_K$ are given by:

$$K_{Lj} = F_{Lj} + M_{Lj},$$

$$F_{Lj} = y_{j,K} T_{KL}^S, \quad M_{Lj} = J X_{L,i} \varepsilon_0 (E_i E_j - \frac{1}{2} E_k E_k \delta_{ij}), \tag{1.1.9}$$

$$J = \det(y_{i,K}), \quad T_{KL}^S = \rho_0 \frac{\partial \psi}{\partial S_{KL}}, \quad E_i = -\phi_{,i},$$

and

$$\mathcal{D}_K = \varepsilon_0 J X_{K,i} D_i = \varepsilon_0 J C_{KL}^{-1} \mathcal{E}_L + \mathcal{P}_K,$$

$$C_{KL}^{-1} = X_{K,i} X_{L,i}, \tag{1.1.10}$$

$$\mathcal{E}_K = y_{i,K} E_i = -\phi_{,K}, \quad \mathcal{P}_K = J X_{K,i} P_i = -\rho_0 \frac{\partial \psi}{\partial \mathcal{E}_K},$$

where ε_0 is the electric permittivity of free space, E_i is the electric field, P_i is the electric polarization per unit present volume, and D_i is the electric displacement vector. $\mathcal{E}_K$ is the reference electric field vector, and $\mathcal{P}_K$ is the reference electric polarization vector. ϕ is the electric potential. C_{KL}^{-1} is the inverse of the deformation tensor. $\psi = \psi(S_{KL}, \mathcal{E}_K)$ is a free energy density per unit mass, which is a function of $\mathcal{E}_K$ and the following finite strain tensor:

$$S_{KL} = (y_{i,K} y_{i,L} - \delta_{KL})/2. \tag{1.1.11}$$

From Equations (1.1.9) and (1.1.10), we have

$$K_{Lj} = y_{j,K} \rho_0 \frac{\partial \psi}{\partial S_{KL}} + J X_{L,i} \varepsilon_0 (E_i E_j - \frac{1}{2} E_k E_k \delta_{ij}),$$

$$\mathcal{D}_K = \varepsilon_0 J C_{KL}^{-1} \mathcal{E}_L - \rho_0 \frac{\partial \psi}{\partial \mathcal{E}_K}. \tag{1.1.12}$$

With successive substitutions from Equations (1.1.9) through (1.1.11), Equation (1.1.7) can be written as four equations for the four unknowns $y_i(\mathbf{X},t)$ and $\phi(\mathbf{X},t)$.

The free energy ψ that determines the constitutive relations of nonlinear electroelastic materials may be written as

$$\rho_0 \psi(S_{KL}, \mathcal{E}_K)$$

$$= \frac{1}{2} \underset{2\,ABCD}{c} S_{AB} S_{CD} - e_{ABC} \mathcal{E}_A S_{BC} - \frac{1}{2} \underset{2\,AB}{\chi} \mathcal{E}_A \mathcal{E}_B$$

$$+ \frac{1}{6} \underset{3\,ABCDEF}{c} S_{AB} S_{CD} S_{EF} + \frac{1}{2} \underset{1\,ABCDE}{k} \mathcal{E}_A S_{BC} S_{DE}$$

$$- \frac{1}{2} b_{ABCD} \mathcal{E}_A \mathcal{E}_B S_{CD} - \frac{1}{6} \underset{3\,ABC}{\chi} \mathcal{E}_A \mathcal{E}_B \mathcal{E}_C$$

$$+ \frac{1}{24} \underset{4\,ABCDEFGH}{c} S_{AB} S_{CD} S_{EF} S_{GH} \tag{1.1.13}$$

$$+ \frac{1}{6} \underset{2\,ABCDEFG}{k} \mathcal{E}_A S_{BC} S_{DE} S_{FG}$$

$$+ \frac{1}{4} \underset{1\,ABCDEF}{a} \mathcal{E}_A \mathcal{E}_B S_{CD} S_{EF} + \frac{1}{6} \underset{3\,ABCDE}{k} \mathcal{E}_A \mathcal{E}_B \mathcal{E}_C S_{DE}$$

$$- \frac{1}{24} \underset{4\,ABCD}{\chi} \mathcal{E}_A \mathcal{E}_B \mathcal{E}_C \mathcal{E}_D + \cdots,$$

where the material constants

$$\underset{2\,ABCD}{c}, \quad e_{ABC}, \quad \underset{2\,AB}{\chi},$$

$$\underset{3\,ABCDEF}{c}, \quad \underset{1\,ABCDE}{k}, \quad b_{ABCD}, \quad \underset{3\,ABC}{\chi}, \tag{1.1.14}$$

$$\underset{4\,ABCDEFGH}{c}, \quad \underset{2\,ABCDEFG}{k}, \quad \underset{1\,ABCDEF}{a}, \quad \underset{3\,ABCDE}{k}, \quad \underset{4\,ABCD}{\chi}$$

are called the second-order elastic, piezoelectric, electric susceptibility, third-order elastic, first odd electroelastic, electrostrictive, third-order electric susceptibility, fourth-order elastic, second odd electroelastic, first even electroelastic, third odd electroelastic, and fourth-order electric susceptibility, respectively. The second-order constants are responsible for linear material behaviors. The third- and higher-order material constants are related to nonlinear behaviors of materials.

For mechanical boundary conditions S is partitioned into S_y and S_T, on which motion (or displacement) and traction are prescribed, respectively. Electrically S is partitioned into S_ϕ and S_D with prescribed electric potential and surface free charge, respectively, and

$$S_y \cup S_T = S_\phi \cup S_D = S,$$
$$S_y \cap S_T = S_\phi \cap S_D = 0. \tag{1.1.15}$$

The usual boundary value problem for an electroelastic body consists of Equation (1.1.7) and the following boundary conditions:

$$y_i = \bar{y}_i \quad \text{on} \quad S_y,$$
$$\phi = \bar{\phi} \quad \text{on} \quad S_\phi,$$
$$K_{Lk} N_L = \bar{T}_k \quad \text{on} \quad S_T, \tag{1.1.16}$$
$$\mathcal{D}_K N_K = -\bar{\sigma}_E \quad \text{on} \quad S_D,$$

where $\bar{y}_i$ and $\bar{\phi}$ are the prescribed boundary motion and potential, $\bar{T}_i$ is the surface traction per unit undeformed area, and $\bar{\sigma}_E$ is the surface free charge per unit undeformed area.

Consider the following variational functional:

$$\Pi(\mathbf{y}, \phi) = \int_{t_0}^{t_1} dt \int_V \left[\frac{1}{2} \rho_0 \dot{y}_i \dot{y}_i - \rho_0 \psi(S_{KL}, \mathcal{E}_K) \right.$$
$$\left. + \pi(S_{KL}, \mathcal{E}_K) + \rho_0 f_i y_i - \rho_E \phi \right] dV \tag{1.1.17}$$
$$+ \int_{t_0}^{t_1} dt \int_{S_T} \bar{T}_i y_i \, dS - \int_{t_0}^{t_1} dt \int_{S_D} \bar{\sigma}_E \phi \, dS,$$

where

$$\pi(S_{KL}, \mathcal{E}_K) = \frac{1}{2} \varepsilon_0 J E_k E_k = \frac{1}{2} \varepsilon_0 J C_{MN}^{-1} \mathcal{E}_M \mathcal{E}_N. \tag{1.1.18}$$

The admissible y_i and ϕ for Π satisfy the following initial and boundary conditions on S_y and S_ϕ:

$$\delta y_i\,|_{t=t_0} = 0, \quad \delta y_i\,|_{t=t_1} = 0 \quad \text{in} \quad V,$$

$$y_i = \bar{y}_i \quad \text{on} \quad S_y, \quad t_0 < t < t_1, \tag{1.1.19}$$

$$\phi = \bar{\phi} \quad \text{on} \quad S_\phi, \quad t_0 < t < t_1.$$

Then the first variation of Π is

$$
\begin{aligned}
\delta\Pi = &\int_{t_0}^{t_1} dt \int_V \, [(K_{Li,L} + \rho_0 f_i - \rho_0 \ddot{y}_i)\delta y_i \\
&+ (\mathcal{D}_{L,L} - \rho_E)\delta\phi]dV \\
&- \int_{t_0}^{t_1} dt \int_{S_T} (K_{Li} N_L - \bar{T}_i)\delta y_i \, dS \\
&- \int_{t_0}^{t_1} dt \int_{S_D} (\mathcal{D}_L N_L + \bar{\sigma}_E)\delta\phi \, dS.
\end{aligned}
\tag{1.1.20}
$$

Therefore the stationary condition of Π implies the following equations and natural boundary conditions:

$$
\begin{aligned}
K_{Lk,L} + \rho_0 f_k &= \rho_0 \ddot{y}_k \quad \text{in} \quad V, \\
\mathcal{D}_{K,K} &= \rho_E \quad \text{in} \quad V, \\
K_{Lk} N_L &= \bar{T}_k \quad \text{on} \quad S_T, \\
\mathcal{D}_K N_K &= -\bar{\sigma}_E \quad \text{on} \quad S_D.
\end{aligned}
\tag{1.1.21}
$$

Denoting

$$K_{LM} = K_{Lj}\delta_{jM}, \quad f_M = f_j\delta_{jM}, \quad \bar{T}_M = \bar{T}_i\delta_{iM}, \tag{1.1.22}$$

we can write Equation $(1.1.7)_1$ and Equation $(1.1.20)$ as

$$K_{LM,L} + \rho_0 f_M = \rho_0 \ddot{y}_M, \tag{1.1.23}$$

and

$$
\begin{aligned}
\delta\Pi = &\int_{t_0}^{t_1} dt \int_V \, [(K_{LM,L} + \rho_0 f_M - \rho_0 \ddot{y}_M)\delta y_M \\
&+ (\mathcal{D}_{L,L} - \rho_E)\delta\phi]dV \\
&- \int_{t_0}^{t_1} dt \int_{S_T} (K_{LM} N_L - \bar{T}_M)\delta y_M \, dS \\
&- \int_{t_0}^{t_1} dt \int_{S_D} (\mathcal{D}_L N_L + \bar{\sigma}_E)\delta\phi \, dS.
\end{aligned}
\tag{1.1.24}
$$

1.2 Linear Piezoelectricity for Weak Fields

In linear theory, we introduce the small displacement vector $\mathbf{u} = \mathbf{y} - \mathbf{X}$ and assume infinitesimal displacement gradient and electric potential gradient. The infinitesimal strain tensor is denoted by

$$S_{kl} = \frac{1}{2}(u_{l,k} + u_{k,l}). \qquad (1.2.1)$$

The material electric field becomes

$$\mathcal{E}_K = E_i y_{i,K} \cong E_i \delta_{iK} \rightarrow E_k. \qquad (1.2.2)$$

Similarly,

$$M_{Lj} \cong 0, \quad K_{Lj} \cong F_{Lj}, \quad \mathcal{P}_K \rightarrow P_k, \quad \mathcal{D}_K \rightarrow D_k. \qquad (1.2.3)$$

Since the various stress tensors are either approximately zero (quadratic or of higher order in the infinitesimal gradients) or about the same, we use T_{ij} to denote the stress tensor that is linear in the infinitesimal gradients. This notation follows the IEEE Standard on Piezoelectricity [3]. Our notation for the rest of the linear theory will also follow the IEEE Standard. Then

$$K_{Lj} \cong F_{Lj} \rightarrow T_{lj}, \quad T_{KL}^S \rightarrow T_{kl}. \qquad (1.2.4)$$

For small fields the free energy density can be approximated by

$$\begin{aligned}
&\rho_0 \psi(S_{KL}, \mathcal{E}_K) - \frac{1}{2}\varepsilon_0 J E_k E_k \\
&\cong \frac{1}{2}c_{2\,ABCD} S_{AB}S_{CD} - e_{ABC}\mathcal{E}_A S_{BC} \\
&\quad - \frac{1}{2}\chi_{2\,AB} \mathcal{E}_A \mathcal{E}_B - \frac{1}{2}\varepsilon_0 J E_k E_k \\
&\rightarrow \frac{1}{2}c_{ijkl}^E S_{ij}S_{kl} - e_{ijk}E_i S_{jk} - \frac{1}{2}\varepsilon_{ij}^S E_i E_j = H(S_{kl}, E_k),
\end{aligned} \qquad (1.2.5)$$

where

$$\varepsilon_{ij}^S = \chi_{2\,ij} + \varepsilon_0 \delta_{ij}. \qquad (1.2.6)$$

The superscript E in c_{ijkl}^E indicates that the independent electric constitutive variable is the electric field $\mathbf{E}$. The superscript S in ε_{ij}^S indicates that the mechanical constitutive variable is the strain tensor $\mathbf{S}$.

In Equation (1.2.5) we have also introduced the electric enthalpy H. The constitutive relations generated by H are:

$$T_{ij} = \frac{\partial H}{\partial S_{ij}} = c_{ijkl}^E S_{kl} - e_{kij} E_k,$$

$$D_i = -\frac{\partial H}{\partial E_i} = e_{ikl} S_{kl} + \varepsilon_{ik}^S E_k. \tag{1.2.7}$$

The material constants in Equation (1.2.7) have the following symmetries:

$$c_{ijkl}^E = c_{jikl}^E = c_{klij}^E,$$

$$e_{kij} = e_{kji}, \quad \varepsilon_{ij}^S = \varepsilon_{ji}^S. \tag{1.2.8}$$

We also assume that the elastic and dielectric material tensors are positive-definite in the following sense:

$$c_{ijkl}^E S_{ij} S_{kl} \geq 0 \quad \text{for any} \quad S_{ij} = S_{ji},$$

$$\text{and} \quad c_{ijkl}^E S_{ij} S_{kl} = 0 \quad \Rightarrow \quad S_{ij} = 0,$$

$$\varepsilon_{ij}^S E_i E_j \geq 0 \quad \text{for any} \quad E_i, \tag{1.2.9}$$

$$\text{and} \quad \varepsilon_{ij}^S E_i E_j = 0 \quad \Rightarrow \quad E_i = 0.$$

Similar to Equation (1.2.7), linear constitutive relations can also be written as [3]

$$T_{ij} = c_{ijkl}^D S_{kl} - h_{kij} D_k,$$

$$E_i = -h_{ikl} S_{kl} + \beta_{ik}^S D_k, \tag{1.2.10}$$

$$S_{ij} = s_{ijkl}^E T_{kl} + d_{kij} E_k,$$

$$D_i = d_{ikl} T_{kl} + \varepsilon_{ik}^T E_k, \tag{1.2.11}$$

and

$$S_{ij} = s_{ijkl}^D T_{kl} + g_{kij} D_k,$$

$$E_i = -g_{ikl} T_{kl} + \beta_{ik}^T D_k. \tag{1.2.12}$$

The equations of motion and the charge equation become

$$T_{ji,j} + \rho f_i = \rho \ddot{u}_i,$$

$$D_{i,i} = \rho_e, \tag{1.2.13}$$

where ρ is the present mass density, and ρ_e is the free charge density per unit present volume. The difference between ρ and ρ_0, and that between ρ_E and ρ_e are neglected in Equation (1.2.13).

In summary, the linear theory of piezoelectricity consists of the equations of motion and charge (1.2.13), the constitutive relations

$$T_{ij} = c_{ijkl}S_{kl} - e_{kij}E_k,$$
$$D_i = e_{ijk}S_{jk} + \varepsilon_{ij}E_j, \tag{1.2.14}$$

where the superscripts in the material constants in Equation (1.2.7) have been dropped, and the strain-displacement and electric field-potential relations

$$S_{ij} = (u_{i,j} + u_{j,i})/2,$$
$$E_i = -\phi_{,i}. \tag{1.2.15}$$

With successive substitutions from Equations (1.2.14) and (1.2.15), Equation (1.2.13) can be written as four equations for **u** and ϕ:

$$c_{ijkl}u_{k,lj} + e_{kij}\phi_{,kj} + \rho f_i = \rho\ddot{u}_i,$$
$$e_{ikl}u_{k,li} - \varepsilon_{ij}\phi_{,ij} = \rho_e. \tag{1.2.16}$$

Let the region occupied by the piezoelectric body be V and its boundary surface be S as shown in Figure 1.2.1. For linear piezoelectricity we use **x** as the independent spatial coordinates. Let the unit outward normal of S be **n**.

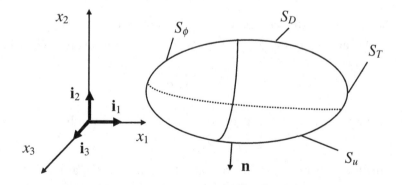

Figure 1.2.1. A piezoelectric body and partitions of its surface.

For boundary conditions we consider the following partitions of S:

$$S_u \cup S_T = S_\phi \cup S_D = S,$$
$$S_u \cap S_T = S_\phi \cap S_D = 0, \tag{1.2.17}$$

where S_u is the part of S on which the mechanical displacement is prescribed, and S_T is the part of S where the traction vector is prescribed. S_ϕ represents the part of S which is electroded where the electric potential is no more than a function of time, and S_D is the unelectroded part. We consider very thin electrodes whose mechanical effects can be neglected. For mechanical boundary conditions we have prescribed displacement $\bar{u}_i$

$$u_i = \bar{u}_i \quad \text{on} \quad S_u, \tag{1.2.18}$$

and prescribed traction $\bar{t}_j$

$$T_{ij} n_i = \bar{t}_j \quad \text{on} \quad S_T. \tag{1.2.19}$$

Electrically, on the electroded portion of S,

$$\phi = \bar{\phi} \quad \text{on} \quad S_\phi, \tag{1.2.20}$$

where $\bar{\phi}$ does not vary spatially. On the unelectroded part of S, the charge condition can be written as

$$D_j n_j = -\bar{\sigma}_e \quad \text{on} \quad S_D, \tag{1.2.21}$$

where $\bar{\sigma}_e$ is the free charge density per unit surface area.

On an electrode S_ϕ, the total free electric charge Q_e can be represented by

$$Q_e = \int_{S_\phi} -n_i D_i dS. \tag{1.2.22}$$

The electric current flowing out of the electrode is given by

$$I = -\dot{Q}_e. \tag{1.2.23}$$

Sometimes there are two (or more) electrodes on a body that are connected to an electric circuit. In this case, circuit equation(s) will need to be considered.

The equations and boundary conditions of linear piezoelectricity can be derived from a variational principle. Consider [4]

$$\Pi(\mathbf{u}, \phi) = \int_{t_0}^{t_1} dt \int_V \left[\frac{1}{2} \rho \dot{u}_i \dot{u}_i - H(\mathbf{S}, \mathbf{E}) + \rho f_i u_i - \rho_e \phi \right] dV$$

$$+ \int_{t_0}^{t_1} dt \int_{S_T} \bar{t}_i u_i dS - \int_{t_0}^{t_1} dt \int_{S_D} \bar{\sigma}_e \phi dS. \tag{1.2.24}$$

$\mathbf{u}$ and ϕ are variationally admissible if they are smooth enough and satisfy

$$\delta u_i \big|_{t_0} = \delta u_i \big|_{t_1} = 0 \quad \text{in} \quad V,$$

$$u_i = \bar{u}_i \quad \text{on} \quad S_u, \quad t_0 < t < t_1, \tag{1.2.25}$$

$$\phi = \bar{\phi} \quad \text{on} \quad S_\phi, \quad t_0 < t < t_1.$$

The first variation of Π is

$$\delta\Pi = \int_{t_0}^{t_1} dt \int_V \left[(T_{ji,j} + \rho f_i - \rho \ddot{u}_i) \delta u_i + (D_{i,i} - \rho_e) \delta\phi \right] dV$$

$$- \int_{t_0}^{t_1} dt \int_{S_T} (T_{ji} n_j - \bar{t}_i) \delta u_i dS - \int_{t_0}^{t_1} dt \int_{S_D} (D_i n_i + \bar{\sigma}_e) \delta\phi dS. \tag{1.2.26}$$

Therefore the stationary condition of Π is

$$T_{ji,j} + \rho f_i = \rho \ddot{u}_i \quad \text{in} \quad V, \quad t_0 < t < t_1,$$

$$D_{i,i} = \rho_e \quad \text{in} \quad V, \quad t_0 < t < t_1,$$

$$T_{ji} n_j = \bar{t}_i \quad \text{on} \quad S_T, \quad t_0 < t < t_1, \tag{1.2.27}$$

$$D_i n_i = -\bar{\sigma}_e \quad \text{on} \quad S_D, \quad t_0 < t < t_1.$$

We now introduce a compact matrix notation [3,4]. This notation consists of replacing pairs of indices ij or kl by single indices p or q, where i, j, k and l take the values of 1, 2, and 3, and p and q take the values of 1, 2, 3, 4, 5, and 6 according to

$$\begin{array}{ccccccc} ij \text{ or } kl: & 11 & 22 & 33 & 23 \text{ or } 32 & 31 \text{ or } 13 & 12 \text{ or } 21 \\ p \text{ or } q: & 1 & 2 & 3 & 4 & 5 & 6 \end{array}. \tag{1.2.28}$$

Thus

$$c_{ijkl} \to c_{pq}, \quad e_{ikl} \to e_{ip}, \quad T_{ij} \to T_p. \tag{1.2.29}$$

For the strain tensor, we introduce S_p such that

$$S_1 = S_{11}, \quad S_2 = S_{22}, \quad S_3 = S_{33},$$

$$S_4 = 2S_{23}, \quad S_5 = 2S_{31}, \quad S_6 = 2S_{12}. \tag{1.2.30}$$

The constitutive relations in Equation (1.2.7) can then be written as

$$T_p = c_{pq}^E S_q - e_{kp} E_k,$$
$$D_i = e_{iq} S_q + \varepsilon_{ik}^S E_k. \tag{1.2.31}$$

In matrix form, Equation (1.2.31) becomes

$$
\begin{Bmatrix} T_1 \\ T_2 \\ T_3 \\ T_4 \\ T_5 \\ T_6 \end{Bmatrix}
=
\begin{pmatrix}
c_{11}^E & c_{12}^E & c_{13}^E & c_{14}^E & c_{15}^E & c_{16}^E \\
c_{21}^E & c_{22}^E & c_{23}^E & c_{24}^E & c_{25}^E & c_{26}^E \\
c_{31}^E & c_{32}^E & c_{33}^E & c_{34}^E & c_{35}^E & c_{36}^E \\
c_{41}^E & c_{42}^E & c_{43}^E & c_{44}^E & c_{45}^E & c_{46}^E \\
c_{51}^E & c_{52}^E & c_{53}^E & c_{54}^E & c_{55}^E & c_{56}^E \\
c_{61}^E & c_{62}^E & c_{63}^E & c_{64}^E & c_{65}^E & c_{66}^E
\end{pmatrix}
\begin{Bmatrix} S_1 \\ S_2 \\ S_3 \\ S_4 \\ S_5 \\ S_6 \end{Bmatrix}
-
\begin{pmatrix}
e_{11} & e_{21} & e_{31} \\
e_{12} & e_{22} & e_{32} \\
e_{13} & e_{23} & e_{33} \\
e_{14} & e_{24} & e_{34} \\
e_{15} & e_{25} & e_{35} \\
e_{16} & e_{26} & e_{36}
\end{pmatrix}
\begin{Bmatrix} E_1 \\ E_2 \\ E_3 \end{Bmatrix},
$$

$$
\begin{Bmatrix} D_1 \\ D_2 \\ D_3 \end{Bmatrix}
=
\begin{bmatrix}
e_{11} & e_{12} & e_{13} & e_{14} & e_{15} & e_{16} \\
e_{21} & e_{22} & e_{23} & e_{24} & e_{25} & e_{26} \\
e_{31} & e_{32} & e_{33} & e_{34} & e_{35} & e_{36}
\end{bmatrix}
\begin{Bmatrix} S_1 \\ S_2 \\ S_3 \\ S_4 \\ S_5 \\ S_6 \end{Bmatrix}
+
\begin{pmatrix}
\varepsilon_{11}^S & \varepsilon_{12}^S & \varepsilon_{13}^S \\
\varepsilon_{21}^S & \varepsilon_{22}^S & \varepsilon_{22}^S \\
\varepsilon_{31}^S & \varepsilon_{32}^S & \varepsilon_{33}^S
\end{pmatrix}
\begin{Bmatrix} E_1 \\ E_2 \\ E_3 \end{Bmatrix}.
$$

$$\tag{1.2.32}$$

1.3 Linear Theory for Small Fields Superposed on a Finite Bias

The theory of linear piezoelectricity assumes infinitesimal deviations from an ideal reference state of the material in which there are no pre-existing mechanical and/or electrical fields (initial or biasing fields). The presence of biasing fields makes a material apparently behave like a different material, and renders the linear theory of piezoelectricity invalid. The behavior of electroelastic bodies under biasing fields can be described by the theory for infinitesimal incremental fields superposed on finite biasing fields [5,6], which is a consequence of the nonlinear theory of electroelasticity. This section presents the theory for small fields superposed on finite biasing fields in an electroelastic body.

Consider the following three states of an electroelastic body (see Figure 1.3.1).

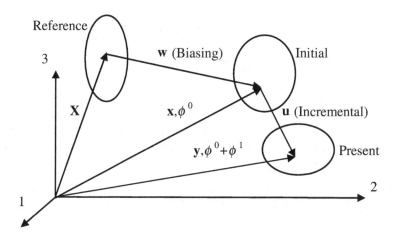

Figure 1.3.1. Reference, initial, and present configurations of an electroelastic body.

In the reference state the body is undeformed and free of electric fields. A generic point at this state is denoted by $\mathbf{X}$ with Cartesian coordinates X_K. The mass density is ρ_0.

In the initial state the body is deformed finitely and statically, and carries finite static electric fields. The body is under the action of body force f_α^0, body charge ρ_E^0, prescribed surface position $\bar{x}_\alpha$, surface traction $\bar{T}_\alpha^0$, surface potential $\bar{\phi}^0$ and surface charge $\bar{\sigma}_E^0$. The deformation and fields at this configuration are the initial or biasing fields. The position of the material point associated with $\mathbf{X}$ is given by $\mathbf{x} = \mathbf{x}(\mathbf{X})$ or $x_\gamma = x_\gamma(\mathbf{X})$, with strain S_{KL}^0. Greek indices are used for the initial configuration. The electric potential in this state is denoted by $\phi^0(\mathbf{X})$, with electric field E_α^0. $\mathbf{x}(\mathbf{X})$ and $\phi^0(\mathbf{X})$ satisfy the following static equations of nonlinear electroelasticity:

$$S_{KL}^0 = (x_{\alpha,K}\,x_{\alpha,L} - \delta_{KL})/2, \quad \mathcal{E}_K^0 = -\phi_{,K}^0, \quad E_\alpha^0 = -\phi_{,\alpha}^0,$$

$$T_{KL}^0 = \rho_0 \left.\frac{\partial \psi}{\partial S_{KL}}\right|_{S_{KL}^0,\,\mathcal{E}_K^0},$$

$$J^0 = \det(x_{\alpha,K}),$$

$$K^0_{K\alpha} = x_{\alpha,L}T^0_{KL} + M^0_{K\alpha}, \quad \mathcal{D}^0_K = \varepsilon_0 J^0 X_{K,\alpha} X_{L,\alpha} \mathcal{E}^0_L + \mathcal{P}^0_K,$$

$$M^0_{K\alpha} = J^0 X_{K,\beta} \varepsilon_0 (E^0_\beta E^0_\alpha - \frac{1}{2} E^0_\gamma E^0_\gamma \delta_{\beta\alpha}),$$

$$K^0_{K\alpha,K} + \rho_0 f^0_\alpha = 0, \quad \mathcal{D}^0_{K,K} = \rho^0_E.$$

(1.3.1)

In the present state, time-dependent, small, incremental deformations and electric fields are applied to the deformed body at the initial state. The body is under the action of f_i , ρ_E, $\bar{y}_i$, $\bar{T}_i$, $\bar{\phi}$ and $\bar{\sigma}_E$. The final position of $\mathbf{X}$ is given by $\mathbf{y} = \mathbf{y}(\mathbf{X},t)$, and the final electric potential is $\phi(\mathbf{X},t)$. $\mathbf{y}(\mathbf{X},t)$ and $\phi(\mathbf{X},t)$ satisfy the dynamic equations of nonlinear electroelasticity:

$$S_{KL} = (y_{i,K} y_{i,L} - \delta_{KL})/2, \quad \mathcal{E}_K = -\phi_{,K}, \quad E_i = -\phi_{,i}$$

$$T^S_{KL} = \rho_0 \frac{\partial \psi}{\partial S_{KL}}\bigg|_{S_{KL},\mathcal{E}_K}, \quad \mathcal{P}_K = -\rho_0 \frac{\partial \psi}{\partial \mathcal{E}_K}\bigg|_{S_{KL},\mathcal{E}_K},$$

$$K_{Lj} = y_{j,K} T^S_{KL} + M_{Lj}, \quad \mathcal{D}_K = \varepsilon_0 J C^{-1}_{KL} \mathcal{E}_L + \mathcal{P}_K,$$

(1.3.2)

$$M_{Lj} = J X_{L,i} \varepsilon_0 (E_i E_j - \frac{1}{2} E_k E_k \delta_{ij}),$$

$$K_{Lj,L} + \rho_0 f_j = \rho_0 \ddot{y}_j, \quad \mathcal{D}_{K,K} = \rho_E.$$

Let the incremental displacement be $\mathbf{u}(\mathbf{X},t)$ and the incremental potential be $\phi^1(\mathbf{X},t)$ (see Figure 1.3.1). $\mathbf{u}$ and ϕ^1 are assumed to be infinitesimal. We write $\mathbf{y}$ and ϕ as

$$y_i(\mathbf{X},t) = \delta_{i\alpha}[x_\alpha(\mathbf{X},t) + u_\alpha(\mathbf{X},t)],$$

$$\phi(\mathbf{X},t) = \phi^0(\mathbf{X},t) + \phi^1(\mathbf{X},t).$$

(1.3.3)

Then it can be shown that the equations governing the incremental fields $\mathbf{u}$ and ϕ^1 are

$$K^1_{K\alpha,K} + \rho_0 f^1_\alpha = \rho_0 \ddot{u}_\alpha,$$

$$\mathcal{D}^1_{K,K} = \rho^1_E,$$

(1.3.4)

where f_α^1 and ρ_E^1 are determined from

$$f_i = \delta_{i\alpha}(f_\alpha^0 + f_\alpha^1),$$
$$\rho_E = \rho_E^0 + \rho_E^1, \tag{1.3.5}$$

and the incremental stress tensor and electric displacement are given by the following constitutive relations:

$$K_{L\gamma}^1 = G_{L\gamma M\alpha}u_{\alpha,M} - R_{ML\gamma}\mathcal{E}_M^1,$$
$$\mathcal{D}_K^1 = R_{KL\gamma}u_{\gamma,L} + L_{KL}\mathcal{E}_L^1, \tag{1.3.6}$$

where $\mathcal{E}_K^1 = -\phi_{,K}^1$. Equation (1.3.6) shows that the incremental stress tensor and electric displacement vector depend linearly on the incremental displacement gradient and potential gradient. In Equation (1.3.6),

$$G_{K\alpha L\gamma} = x_{\alpha,M}\,\rho_0\,\frac{\partial^2\psi}{\partial S_{KM}\partial S_{LN}}\bigg|_{S_{KL}^0,\mathcal{E}_K^0}\,x_{\gamma,N}$$

$$+T_{KL}^0\delta_{\alpha\gamma} + g_{K\alpha L\gamma} = G_{L\gamma K\alpha},$$

$$R_{KL\gamma} = -\rho_0\,\frac{\partial^2\psi}{\partial\mathcal{E}_K\partial S_{ML}}\bigg|_{S_{KL}^0,\mathcal{E}_K^0}\,x_{\gamma,M} + r_{KL\gamma}, \tag{1.3.7}$$

$$L_{KL} = -\rho_0\,\frac{\partial^2\psi}{\partial\mathcal{E}_K\partial\mathcal{E}_L}\bigg|_{S_{KL}^0,\mathcal{E}_K^0} + l_{KL} = L_{LK}.$$

where

$$g_{K\alpha L\gamma} = \varepsilon_0 J^0 [E_\alpha^0 E_\beta^0(X_{K,\beta}X_{L,\gamma} - X_{K,\gamma}X_{L,\beta})$$

$$-E_\alpha^0 E_\gamma^0 X_{K,\beta}X_{L,\beta}$$

$$+E_\beta^0 E_\gamma^0(X_{K,\alpha}X_{L,\beta} - X_{K,\beta}X_{L,\alpha})$$

$$+\frac{1}{2}E_\beta^0 E_\beta^0(X_{K,\gamma}X_{L,\alpha} - X_{K,\alpha}X_{L,\gamma})], \tag{1.3.8}$$

$$r_{KL\gamma} = \varepsilon_0 J^0(E_\alpha^0 X_{K,\alpha}X_{L,\gamma} - E_\alpha^0 X_{K,\gamma}X_{L,\alpha} - E_\gamma^0 X_{K,\alpha}X_{L,\alpha}),$$

$$l_{KL} = \varepsilon_0 J^0 X_{K,\alpha}X_{L,\alpha}.$$

$G_{K\alpha L\gamma}$, $R_{KL\gamma}$, and L_{KL} are called the effective or apparent elastic, piezoelectric, and dielectric constants. They depend on the initial deformation $x_\alpha(\mathbf{X})$ and electric potential $\phi^0(\mathbf{X})$.

In summary, the boundary value problem for the incremental fields $\mathbf{u}$ and ϕ^1 consists of the following equations and boundary conditions:

$$K_{K\alpha,K}^1 + \rho_0 f_\alpha^1 = \rho_0 \ddot{u}_\alpha \quad \text{in} \quad V,$$

$$\mathcal{D}_{K,K}^1 = \rho_E^1 \quad \text{in} \quad V,$$

$$K_{L\gamma}^1 = G_{L\gamma M\alpha} u_{\alpha,M} + R_{ML\gamma}\phi_{,M}^1 \quad \text{in} \quad V,$$

$$\mathcal{D}_K^1 = R_{KL\gamma} u_{\gamma,L} - L_{KL}\phi_{,L}^1 \quad \text{in} \quad V,$$

$$u_\alpha = \bar{u}_\alpha \quad \text{on} \quad S_y, \tag{1.3.9}$$

$$\phi^1 = \bar{\phi}^1 \quad \text{on} \quad S_\phi,$$

$$K_{L\alpha}^1 N_L = \bar{T}_\alpha^1 \quad \text{on} \quad S_T,$$

$$\mathcal{D}_K^1 N_K = -\bar{\sigma}_E^1 \quad \text{on} \quad S_D.$$

Consider the following variational functional:

$$\Pi(\mathbf{u},\phi^1) = \int_{t_0}^{t_1} dt \int_V \; (\frac{1}{2}\rho_0 \dot{u}_\alpha \dot{u}_\alpha - \frac{1}{2}G_{K\alpha L\gamma} u_{K,\alpha} u_{L,\gamma}$$

$$- R_{KL\gamma}\phi_{,K}^1 u_{L,\gamma} + \frac{1}{2}L_{KL}\phi_{,K}^1 \phi_{,L}^1 + \rho_0 f_\alpha^1 u_\alpha - \rho_E^1 \phi^1)dV \tag{1.3.10}$$

$$+ \int_{t_0}^{t_1} dt \int_{S_T} \bar{T}_\alpha^1 u_\alpha dS - \int_{t_0}^{t_1} dt \int_{S_D} \bar{\sigma}_E^1 \phi^1 dS.$$

The admissible $\mathbf{u}$ and ϕ^1 must satisfy

$$\delta u_\alpha \big|_{t_0} = \delta u_\alpha \big|_{t_1} = 0 \quad \text{in} \quad V,$$

$$u_\alpha = \bar{u}_\alpha \quad \text{on} \quad S_u, \quad t_0 < t < t_1, \tag{1.3.11}$$

$$\phi^1 = \bar{\phi}^1 \quad \text{on} \quad S_\phi, \quad t_0 < t < t_1.$$

The first variation is found to be

$$
\begin{aligned}
\delta\Pi(\mathbf{u}, \phi^1) = &\int_{t_0}^{t_1} dt \int_V \ [(K_{L\alpha,L}^1 + \rho_0 f_\alpha^1 - \rho_0 \ddot{u}_\alpha)\delta u_\alpha \\
&+ (\mathcal{D}_{K,K}^1 - \rho_E^1)\delta\phi^1]dV \\
&- \int_{t_0}^{t_1} dt \int_{S_T} \ (K_{L\alpha}^1 N_L - \bar{T}_\alpha^1)\delta u_\alpha dS \\
&- \int_{t_0}^{t_1} dt \int_{S_D} \ (\mathcal{D}_K^1 N_K + \bar{\sigma}_E^1)\delta\phi^1 dS.
\end{aligned}
\tag{1.3.12}
$$

Therefore the stationary condition of the functional gives the following governing equations and boundary conditions:

$$
\begin{aligned}
K_{K\alpha,K}^1 + \rho_0 f_\alpha^1 &= \rho_0 \ddot{u}_\alpha \quad \text{in} \quad V, \\
\mathcal{D}_{K,K}^1 &= \rho_E^1 \quad \text{in} \quad V, \\
K_{L\alpha}^1 N_L &= \bar{T}_\alpha^1 \quad \text{on} \quad S_T, \\
\mathcal{D}_K^1 N_K &= -\bar{\sigma}_E^1 \quad \text{on} \quad S_D.
\end{aligned}
\tag{1.3.13}
$$

Denoting

$$
\begin{aligned}
K_{LM}^1 &= K_{L\alpha}^1 \delta_{\alpha M}, \quad f_M^1 = f_\alpha^1 \delta_{\alpha M}, \\
\bar{T}_M^1 &= \bar{T}_\alpha^1 \delta_{\alpha M}, \quad u_M = u_\alpha \delta_{\alpha M},
\end{aligned}
\tag{1.3.14}
$$

we can write Equation (1.3.12) as

$$
\begin{aligned}
\delta\Pi(\mathbf{u}, \phi^1) = &\int_{t_0}^{t_1} dt \int_V \ [(K_{LM,L}^1 + \rho_0 f_M^1 - \rho_0 \ddot{u}_M)\delta u_M \\
&+ (\mathcal{D}_{K,K}^1 - \rho_E^1)\delta\phi^1]dV \\
&- \int_{t_0}^{t_1} dt \int_{S_T} \ (K_{LM}^1 N_L - \bar{T}_M^1)\delta u_M dS \\
&- \int_{t_0}^{t_1} dt \int_{S_D} \ (\mathcal{D}_K^1 N_K + \bar{\sigma}_E^1)\delta\phi^1 dS.
\end{aligned}
\tag{1.3.15}
$$

In some applications, the biasing deformations and fields are also infinitesimal. In this case, usually only their first-order effects on the incremental fields need to be considered. Then the following energy density of a cubic polynomial is sufficient:

$$\rho_0\psi(S_{KL},\mathcal{E}_K) = \frac{1}{2}c_{ABCD}S_{AB}S_{CD} - e_{ABC}\mathcal{E}_A S_{BC} - \frac{1}{2}\chi_{AB}\mathcal{E}_A\mathcal{E}_B$$

$$+\frac{1}{6}c_{ABCDEF}S_{AB}S_{CD}S_{EF} + \frac{1}{2}k_{ABCDE}\mathcal{E}_A S_{BC}S_{DE} \qquad (1.3.16)$$

$$-\frac{1}{2}b_{ABCD}\mathcal{E}_A\mathcal{E}_B S_{CD} - \frac{1}{6}\chi_{ABC}\mathcal{E}_A\mathcal{E}_B\mathcal{E}_C,$$

where the subscripts indicating the orders of the material constants have been dropped. For small biasing fields it is convenient to introduce the small displacement vector **w** of the initial deformation (see Figure 1.3.1), given as

$$x_\alpha = \delta_{\alpha K}X_K + w_\alpha. \qquad (1.3.17)$$

Then, neglecting the quadratic terms of the gradients of **w** and ϕ^0, the effective material constants take the following form [5,6]:

$$G_{K\alpha L\gamma} = c_{K\alpha L\gamma} + \hat{c}_{K\alpha L\gamma},$$

$$R_{KL\gamma} = e_{KL\gamma} + \hat{e}_{KL\gamma}, \qquad (1.3.18)$$

$$L_{KL} = \varepsilon_{KL} + \hat{\varepsilon}_{KL},$$

where

$$\hat{c}_{K\alpha L\gamma} = T^0_{KL}\delta_{\alpha\gamma} + c_{K\alpha LN}w_{\gamma,N} + c_{KNL\gamma}w_{\alpha,N}$$

$$+ c_{K\alpha L\gamma AB}S^0_{AB} + k_{AK\alpha L\gamma}\mathcal{E}^0_A,$$

$$\hat{e}_{KL\gamma} = e_{KLM}w_{\gamma,M} - k_{KL\gamma AB}S^0_{AB} + b_{AKL\gamma}\mathcal{E}^0_A$$

$$+ \varepsilon_0(\mathcal{E}^0_K\delta_{L\gamma} - \mathcal{E}^0_L\delta_{K\gamma} - \mathcal{E}^0_M\delta_{M\gamma}\delta_{KL}),$$

$$\hat{\varepsilon}_{KL} = b_{KLAB}S^0_{AB} + \chi_{KLA}\mathcal{E}^0_A + \varepsilon_0(S^0_{MM}\delta_{KL} - 2S^0_{KL}), \qquad (1.3.19)$$

$$T^0_{KL} = c_{KLAB}S^0_{AB} - e_{AKL}\mathcal{E}^0_A,$$

$$S^0_{AB} \equiv (w_{A,B} + w_{B,A})/2,$$

$$\mathcal{E}^0_K = -\phi^0_{,K}.$$

In certain applications, e.g., buckling of thin structures, consideration of initial stresses without initial deformations is sufficient. Such a theory is called the initial stress theory in elasticity. It can be reduced from the theory for small fields superposed on a bias. First we set **x** = **X**. Furthermore, for buckling analysis, a quadratic expression of ψ with second-order material constants only and the corresponding linear constitutive relations are sufficient. The biasing fields can be treated as infinitesimal fields. Then the effective material constants sufficient for describing the buckling phenomenon take the following simple form:

$$G_{K\alpha L\gamma} = c_{K\alpha L\gamma} + T_{KL}^0 \delta_{\alpha\gamma},$$

$$R_{KL\gamma} = e_{KL\gamma} + \varepsilon_0 (\mathcal{E}_K^0 \delta_{L\gamma} - \mathcal{E}_L^0 \delta_{K\gamma} - \mathcal{E}_M^0 \delta_{M\gamma} \delta_{KL}), \qquad (1.3.20)$$

$$L_{KL} = \varepsilon_{KL},$$

where T_{KL}^0 is the initial stress and $\mathcal{E}_K^0$ is the initial electric field.

1.4 Cubic Theory for Weak Nonlinearity

By cubic theory we mean that effects of all terms up to the third power of the displacement and potential gradients or their products are included [7]. Cubic theory is an approximate theory for relatively weak nonlinearities, and can be obtained by expansions and truncations from the nonlinear theory in the first section of this chapter. The resulting equations are:

$$
\begin{aligned}
F_{Lj} \cong \delta_{jM} \Bigg[& c_{2LMAB} u_{A,B} + e_{ALM} \phi_{,A} + \frac{1}{2} c_{2LMAB} u_{K,A} u_{K,B} \\
& + c_{2LKAB} u_{M,K} u_{A,B} + \frac{1}{2} c_{3LMABCD} u_{A,B} u_{CD} \\
& + e_{ALK} u_{M,K} \phi_{,A} - d_{1ABCLM} u_{B,C} \phi_{,A} - \frac{1}{2} b_{ABLM} \phi_{,A} \phi_{,B} \\
& + \frac{1}{2} c_{2LRAB} u_{M,R} u_{K,A} u_{K,B} + \frac{1}{2} c_{3LKABCD} u_{M,K} u_{A,B} u_{CD} \\
& + \frac{1}{2} c_{3LMABcD} u_{A,B} u_{K,C} u_{K,D} + \frac{1}{6} c_{4LMABCDEF} u_{A,B} u_{CD} u_{E,F} \\
& - d_{1ABCLK} u_{B,C} u_{M,K} \phi_{,A} - \frac{1}{2} d_{1ABCLM} u_{K,B} u_{K,C} \phi_{,A} \\
& - \frac{1}{2} d_{2ABCDELM} u_{B,C} u_{D,E} \phi_{,A} - \frac{1}{2} b_{ABLK} u_{M,K} \phi_{,A} \phi_{,B} \\
& + \frac{1}{2} a_{1ABCDLM} u_{c,D} \phi_{,A} \phi_{,B} + \frac{1}{6} d_{3ABCLM} \phi_{,A} \phi_{,B} \phi_{,C} \Bigg],
\end{aligned}
\qquad (1.4.1)
$$

$$\mathcal{P}_L \cong e_{LBC} u_{B,C} - \chi_{2\,AL} \phi_{,A} + \frac{1}{2} e_{LBC} u_{K,B} u_{K,C}$$

$$-\frac{1}{2} d_{1\,LBCDE} u_{B,C} u_{D,E} - b_{ALCD} u_{C,D} \phi_{,A}$$

$$+\frac{1}{2}\chi_{3\,ABL}\phi_{,A}\phi_{,B} - \frac{1}{2}d_{1\,LBCDE} u_{B,C} u_{K,D} u_{K,E}$$

$$-\frac{1}{6}d_{2\,LBCDEFG} u_{B,C} u_{D,E} u_{F,G} - \frac{1}{2}b_{ALCD} u_{K,C} u_{K,D}\phi_{,A}$$ (1.4.2)

$$+\frac{1}{2}a_{1\,ALCDEF} u_{C,D} u_{E,F}\phi_{,A} + \frac{1}{2}d_{3\,ABLDE} u_{D,E}\phi_{,A}\phi_{,B}$$

$$-\frac{1}{6}\chi_{4\,ABCL}\phi_{,A}\phi_{,B}\phi_{,C},$$

$$M_{Lj} \cong \varepsilon_0 \delta_{jM}\left[\phi_{,L}\phi_{,M} - \frac{1}{2}\phi_{,K}\phi_{,K}\delta_{LM} - \phi_{,K}\phi_{,M} u_{K,L}\right.$$

$$-\phi_{,K}\phi_{,M} u_{L,K} + \phi_{,L}\phi_{,M} u_{K,K} - \phi_{,L}\phi_{,K} u_{K,M}$$ (1.4.3)

$$\left.+\phi_{,K}\phi_{,R} u_{R,K}\delta_{LM} + \frac{1}{2}\phi_{,K}\phi_{,K} u_{L,M} - \frac{1}{2}\phi_{,R}\phi_{,R} u_{K,K}\delta_{LM}\right],$$

$$\varepsilon_0 J C_{KL}^{-1}\mathcal{E}_K \cong \varepsilon_0\left[-\phi_{,L} + \phi_{,K} u_{L,K} - \phi_{,L} u_{K,K} + \phi_{,K} u_{K,L}\right.$$

$$-\phi_{,M} u_{L,K} u_{K,M} + \phi_{,K} u_{M,M} u_{L,K} - \frac{1}{2}\phi_{,L} u_{K,K} u_{M,M}$$ (1.4.4)

$$+\frac{1}{2}\phi_{,L} u_{K,M} u_{M,K} - \phi_{,M} u_{L,K} u_{M,K}$$

$$\left.+\phi_{,M} u_{M,L} u_{K,K} - \phi_{,M} u_{M,K} u_{K,L}\right].$$

A special case of cubic theory is the case of relatively large mechanical deformations and weak electric fields [8]. In this case all electrical nonlinearities can be neglected. The following energy density is sufficient:

$$\rho_0\psi = \frac{1}{2}c_{ABCD}S_{AB}S_{CD} - e_{ABC}\mathcal{E}_A S_{BC} - \frac{1}{2}\chi_{AB}\mathcal{E}_A\mathcal{E}_B$$

$$+\frac{1}{6}c_{ABCDEF}S_{AB}S_{CD}S_{EF} + \frac{1}{24}c_{ABCDEFGH}S_{AB}S_{CD}S_{EF}S_{GH}.$$ (1.4.5)

Keeping the linear terms of the electric potential gradient and up to cubic terms of the displacement gradient, we obtain

$$K_{LM} = c_{LMRS} u_{R,S} + e_{KLM} \phi_{,K}$$
$$+ c^e_{LMRSKN} u_{R,S} u_{K,N} + c^e_{LMRSKNIJ} u_{R,S} u_{K,N} u_{I,J},$$
$$\mathcal{D}_K = e_{KRS} u_{R,S} - \varepsilon_{KL} \phi_{,L}, \tag{1.4.6}$$

where

$$c^e_{LMRSKN} = \frac{1}{2} (c_{LMRSKN} + c_{LMNS} \delta_{KR} + c_{LNRS} \delta_{KM}),$$

$$c^e_{LMRSKNIJ} = \frac{1}{6} c_{LMRSKNIJ} \tag{1.4.7}$$

$$+ \frac{1}{2} (c_{LMKNSJ} \delta_{RI} + c_{LNSJ} \delta_{MK} \delta_{RI} + c_{LNRSIJ} \delta_{MK}).$$

Chapter 2
Piezoelectric Plates

In this chapter we derive two-dimensional equations for a piezoelectric plate. First we examine a few exact solutions of vibration modes and propagating waves in plates from three-dimensional equations. They provide guidance in developing two-dimensional theories and serve as criteria for determining the accuracy of two-dimensional theories. Then two-dimensional plate equations are systematically derived.

2.1 Exact Modes in a Plate

The specific three-dimensional problems to be examined are the thickness-shear vibration of a quartz plate [10] and waves propagating in a plate of polarized ferroelectric ceramics [11,12].

2.1.1 Thickness-shear vibration of a quartz plate

Quartz of crystal class 32 is probably the most widely used piezoelectric crystal. Plates of rotated Y-cut quartz [4] are particularly useful for thickness-shear resonators, filters, and sensors because of the existence of pure thickness-shear modes and their frequency stability. Langasite and some of its isomorphs (langanite and langatate) are emerging piezoelectric crystals which have stronger piezoelectric coupling than quartz and also belong to crystal class 32. Rotated Y-cut quartz exhibits monoclinic symmetry of class 2 (or C_2) in a coordinate system (x_1, x_2) in and normal to the plane of the plate. Consider an unbounded, rotated Y-cut quartz plate (see Figure 2.1.1). The two major surfaces are traction-free and are electroded, with a driving voltage $V \exp(i\omega t)$ across the thickness.

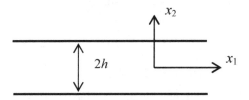

Figure 2.1.1. An electroded quartz plate.

2.1.1.1 Boundary value problem

The boundary value problem is:

$$T_{ji,j} = \rho \ddot{u}_i, \quad D_{i,i} = 0, \quad |x_2| < h,$$

$$T_{ij} = c_{ijkl} S_{kl} - e_{kij} E_k, \quad D_i = e_{ikl} S_{kl} + \varepsilon_{ik} E_k, \quad |x_2| < h,$$

$$S_{ij} = (u_{i,j} + u_{j,i})/2, \quad E_i = -\phi_{,i}, \quad |x_2| < h, \qquad (2.1.1)$$

$$T_{2j} = 0, \quad x_2 = \pm h,$$

$$\phi(x_2 = h) - \phi(x_2 = -h) = V \exp(i\omega t).$$

For monoclinic crystals, the material tensors c_{ijkl}^E, e_{ijk} and ε_{ij}^S can be represented by the following matrices under the compact matrix notation:

$$\begin{pmatrix} c_{11} & c_{12} & c_{13} & c_{14} & 0 & 0 \\ c_{21} & c_{22} & c_{23} & c_{24} & 0 & 0 \\ c_{31} & c_{32} & c_{33} & c_{34} & 0 & 0 \\ c_{41} & c_{42} & c_{43} & c_{44} & 0 & 0 \\ 0 & 0 & 0 & 0 & c_{55} & c_{56} \\ 0 & 0 & 0 & 0 & c_{65} & c_{66} \end{pmatrix},$$

$$\begin{pmatrix} e_{11} & e_{12} & e_{13} & e_{14} & 0 & 0 \\ 0 & 0 & 0 & 0 & e_{25} & e_{26} \\ 0 & 0 & 0 & 0 & e_{35} & e_{36} \end{pmatrix}, \begin{pmatrix} \varepsilon_{11} & 0 & 0 \\ 0 & \varepsilon_{22} & \varepsilon_{23} \\ 0 & \varepsilon_{32} & \varepsilon_{33} \end{pmatrix}. \qquad (2.1.2)$$

Consider the possibility of the following displacement and potential fields:

$$u_1 = u_1(x_2)\exp(i\omega t), \quad u_2 = u_3 = 0,$$
$$\phi = \phi(x_2)\exp(i\omega t). \tag{2.1.3}$$

The nontrivial components of strain, electric field, stress, and electric displacement are

$$2S_{12} = u_{1,2}, \quad E_2 = -\phi_{,2}, \tag{2.1.4}$$

and

$$T_{31} = c_{56}u_{1,2} + e_{25}\phi_{,2}, \quad T_{12} = c_{66}u_{1,2} + e_{26}\phi_{,2},$$
$$D_2 = e_{26}u_{1,2} - \varepsilon_{22}\phi_{,2}, \quad D_3 = e_{36}u_{1,2} - \varepsilon_{23}\phi_{,2}, \tag{2.1.5}$$

where the time-harmonic factor has been dropped. The equation of motion and the charge equation require that

$$T_{21,2} = c_{66}u_{1,22} + e_{26}\phi_{,22} = -\rho\omega^2 u_1,$$
$$D_{2,2} = e_{26}u_{1,22} - \varepsilon_{22}\phi_{,22} = 0. \tag{2.1.6}$$

Equation (2.1.6)$_2$ can be integrated to yield

$$\phi = \frac{e_{26}}{\varepsilon_{22}}u_1 + B_1 x_2 + B_2, \tag{2.1.7}$$

where B_1 and B_2 are integration constants and B_2 is immaterial. Substituting Equation (2.1.7) into the expressions for T_{21}, D_2, and Equation (2.1.6)$_1$, we obtain

$$T_{21} = \bar{c}_{66}u_{1,2} + e_{26}B_1, \quad D_2 = -\varepsilon_{22}B_1, \tag{2.1.8}$$

$$\bar{c}_{66}u_{1,22} = -\rho\omega^2 u_1, \tag{2.1.9}$$

where

$$\bar{c}_{66} = c_{66}(1 + k_{26}^2), \quad k_{26}^2 = \frac{e_{26}^2}{\varepsilon_{22}c_{66}}. \tag{2.1.10}$$

The general solution to Equation (2.1.9) and the corresponding expression for the electric potential are

$$u_1 = A_1 \sin \xi x_2 + A_2 \cos \xi x_2,$$

$$\phi = \frac{e_{26}}{\varepsilon_{22}} (A_1 \sin \xi x_2 + A_2 \cos \xi x_2) + B_1 x_2 + B_2, \quad (2.1.11)$$

where A_1 and A_2 are integration constants, and

$$\xi^2 = \frac{\rho}{\bar{c}_{66}} \omega^2 . \quad (2.1.12)$$

Then the expression for the stress component relevant to boundary conditions is

$$T_{21} = \bar{c}_{66} (A_1 \xi \cos \xi x_2 - A_2 \xi \sin \xi x_2) + e_{26} B_1 . \quad (2.1.13)$$

The boundary conditions require that

$$\bar{c}_{66} A_1 \xi \cos \xi h - \bar{c}_{66} A_2 \xi \sin \xi h + e_{26} B_1 = 0,$$

$$\bar{c}_{66} A_1 \xi \cos \xi h + \bar{c}_{66} A_2 \xi \sin \xi h + e_{26} B_1 = 0, \quad (2.1.14)$$

$$2 \frac{e_{26}}{\varepsilon_{22}} A_1 \sin \xi h + 2 B_1 h = V.$$

We can also add the first two, and subtract the first two from each other:

$$\bar{c}_{66} A_1 \xi \cos \xi h + e_{26} B_1 = 0,$$

$$\bar{c}_{66} A_2 \xi \sin \xi h = 0, \quad (2.1.15)$$

$$2 \frac{e_{26}}{\varepsilon_{22}} A_1 \sin \xi h + 2 B_1 h = V.$$

2.1.1.2 Free vibration solution

First, consider free vibrations with $V = 0$. Equation (2.1.15) decouples into two sets of equations. For symmetric modes,

$$\bar{c}_{66} A_2 \xi \sin \xi h = 0 . \quad (2.1.16)$$

Nontrivial solutions may exist if

$$\sin \xi h = 0, \quad (2.1.17)$$

or

$$\xi^{(n)}h = \frac{n\pi}{2}, \quad n = 0, 2, 4, 6, \cdots, \tag{2.1.18}$$

which determines the following resonant frequencies:

$$\omega^{(n)} = \frac{n\pi}{2h}\sqrt{\frac{\bar{c}_{66}}{\rho}}, \quad n = 0, 2, 4, 6, \cdots. \tag{2.1.19}$$

Equation (2.1.17) implies that $B_1 = 0$ and $A_1 = 0$. The corresponding modes are

$$u_1 = \cos\xi^{(n)}x_2, \quad \phi = \frac{e_{26}}{\varepsilon_{22}}\cos\xi^{(n)}x_2, \tag{2.1.20}$$

where $n = 0$ represents a rigid body mode. For anti-symmetric modes,

$$\bar{c}_{66}A_1\xi\cos\xi h + e_{26}B_1 = 0,$$
$$2\frac{e_{26}}{\varepsilon_{22}}A_1\sin\xi h + 2B_1h = 0. \tag{2.1.21}$$

The resonance frequencies are determined by

$$\begin{vmatrix} \bar{c}_{66}\xi\cos\xi h & e_{26} \\ \dfrac{e_{26}}{\varepsilon_{22}}\sin\xi h & h \end{vmatrix} = \bar{c}_{66}\xi h\cos\xi h - \frac{e_{26}^2}{\varepsilon_{22}}\sin\xi h = 0, \tag{2.1.22}$$

or

$$\tan\xi h = \frac{\xi h}{\bar{k}_{26}^2}, \tag{2.1.23}$$

where

$$\bar{k}_{26}^2 = \frac{e_{26}^2}{\varepsilon_{22}\bar{c}_{66}} = \frac{e_{26}^2}{\varepsilon_{22}c_{66}(1+k_{26}^2)} = \frac{k_{26}^2}{1+k_{26}^2}. \tag{2.1.24}$$

Equations (2.1.23) and (2.1.21) determine the resonant frequencies and modes. If the small piezoelectric coupling for quartz is neglected in Equation (2.1.23), a set of frequencies similar to Equation (2.1.19) with n equals odd numbers can be determined for a set of modes with sine dependence on the thickness coordinate. Static thickness-shear deformation and the first few thickness-shear modes in a plate are shown in Figure 2.1.2.

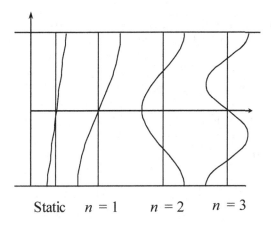

Figure 2.1.2. Thickness-shear deformation and modes in a plate.

2.1.1.3 Forced vibration solution

For forced vibration we have $A_2 = 0$ and

$$A_1 = \frac{\begin{vmatrix} 0 & e_{26} \\ V & 2h \end{vmatrix}}{\begin{vmatrix} \bar{c}_{66}\xi\cos\xi h & e_{26} \\ 2\dfrac{e_{26}}{\varepsilon_{22}}\sin\xi h & 2h \end{vmatrix}} = \frac{-e_{26}V}{2\bar{c}_{66}\xi h\cos\xi h - 2\dfrac{e_{26}^2}{\varepsilon_{22}}\sin\xi h}, \qquad (2.1.25)$$

$$B_1 = \frac{\begin{vmatrix} \bar{c}_{66}\xi\cos\xi h & 0 \\ 2\dfrac{e_{26}}{\varepsilon_{22}}\sin\xi h & V \end{vmatrix}}{\begin{vmatrix} \bar{c}_{66}\xi\cos\xi h & e_{26} \\ 2\dfrac{e_{26}}{\varepsilon_{22}}\sin\xi h & 2h \end{vmatrix}} = \frac{V\bar{c}_{66}\xi\cos\xi h}{2\bar{c}_{66}\xi h\cos\xi h - 2\dfrac{e_{26}^2}{\varepsilon_{22}}\sin\xi h}. \qquad (2.1.26)$$

Hence

$$D_2 = -\varepsilon_{22}B_1 = -\varepsilon_{22}\frac{V}{2h}\frac{\xi h}{\xi h - \bar{k}_{26}^2 \tan \xi h} = -\sigma_e, \qquad (2.1.27)$$

where σ_e is the surface free charge per unit area on the electrode at $x_2 = h$. The capacitance per unit area is then

$$C = \frac{\sigma_e}{V} = \frac{\varepsilon_{22}}{2h}\frac{\xi h}{\xi h - \bar{k}_{26}^2 \tan \xi h}. \qquad (2.1.28)$$

Note the following limits:

$$\lim_{e_{26} \to 0} C = \frac{\varepsilon_{22}}{2h},$$

$$\lim_{\omega \to 0} C = \frac{\varepsilon_{22}}{2h}\frac{1}{1 - \dfrac{k_{26}^2}{1 + k_{26}^2}} = \frac{\varepsilon_{22}}{2h}(1 + k_{26}^2). \qquad (2.1.29)$$

2.1.2 Propagating waves in a plate

Consider an unbounded piezoelectric plate of thickness $2h$ as schematically illustrated in Figure 2.1.3. The major surfaces of the plate are traction-free and are electroded. The electrodes are shorted.

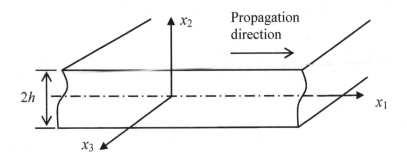

Figure 2.1.3. Propagating waves in a piezoelectric plate.

2.1.2.1 Eigenvalue value problem

We study straight-crested waves without x_3 dependence. Then the homogeneous form of Equation (1.2.16) takes the following form:

$$c_{11}u_{1,11} + c_{12}u_{2,21} + c_{14}u_{3,21} + c_{15}u_{3,11} + c_{16}(u_{1,21} + u_{2,11})$$
$$+ e_{11}\phi_{,11} + e_{21}\phi_{,21}$$
$$+ c_{16}u_{1,12} + c_{26}u_{2,22} + c_{46}u_{3,22} + c_{56}u_{3,12} + c_{66}(u_{1,22} + u_{2,12})$$
$$+ e_{16}\phi_{,12} + e_{26}\phi_{,22} = \rho\ddot{u}_1,$$

$$c_{16}u_{1,11} + c_{26}u_{2,21} + c_{46}u_{3,21} + c_{56}u_{3,11} + c_{66}(u_{1,21} + u_{2,11})$$
$$+ e_{16}\phi_{,11} + e_{26}\phi_{,21}$$
$$+ c_{12}u_{1,12} + c_{22}u_{2,22} + c_{24}u_{3,22} + c_{25}u_{3,12} + c_{26}(u_{1,22} + u_{2,12})$$
$$+ e_{12}\phi_{,12} + e_{22}\phi_{,22} = \rho\ddot{u}_2,$$

$$c_{15}u_{1,11} + c_{25}u_{2,21} + c_{45}u_{3,21} + c_{55}u_{3,11} + c_{56}(u_{1,21} + u_{2,11})$$
$$+ e_{15}\phi_{,11} + e_{25}\phi_{,21}$$
$$+ c_{14}u_{1,12} + c_{24}u_{2,22} + c_{44}u_{3,22} + c_{45}u_{3,12} + c_{46}(u_{1,22} + u_{2,12})$$
$$+ e_{14}\phi_{,12} + e_{24}\phi_{,22} = \rho\ddot{u}_3,$$

$$e_{11}u_{1,11} + e_{12}u_{2,21} + e_{14}u_{3,21} + e_{15}u_{3,11} + e_{16}(u_{1,21} + u_{2,11})$$
$$- \varepsilon_{11}\phi_{,11} - \varepsilon_{12}\phi_{,21}$$
$$+ e_{21}u_{1,12} + e_{22}u_{2,22} + e_{24}u_{3,22} + e_{25}u_{3,12} + e_{26}(u_{1,22} + u_{2,12})$$
$$- \varepsilon_{12}\phi_{,12} - \varepsilon_{22}\phi_{,22} = 0. \tag{2.1.30}$$

We seek solutions representing waves propagating in the x_1 direction:

$$u_j(\mathbf{x},t) = A_j \exp(k\eta x_2)\exp[i(kx_1 - \omega t)],$$
$$\phi(\mathbf{x},t) = A_4 \exp(k\eta x_2)\exp[i(kx_1 - \omega t)], \tag{2.1.31}$$

where k and ω are the wave number in the x_1 direction and the frequency, respectively. ηk is related to the wave number in the x_2 direction. A_j ($j = 1, 2, 3$) and A_4 are complex constants, representing the wave amplitude.

Substitution of Equation (2.1.31) into Equation (2.1.30) leads to the following four linear algebraic equations for A_j and A_4:

$$[\rho\omega^2 + k^2(\eta^2 c_{66} + i2\eta c_{16} - c_{11})]A_1$$
$$+ k^2(\eta^2 c_{26} + i\eta c_{66} + i\eta c_{12} - c_{16})A_2$$
$$+ k^2(\eta^2 c_{46} + i\eta c_{56} + i\eta c_{14} - c_{15})A_3$$
$$+ k^2(\eta^2 e_{26} + i\eta e_{21} + i\eta e_{16} - e_{11})A_4 = 0,$$
$$k^2(\eta^2 c_{26} + i\eta c_{21} + i\eta c_{66} - c_{16})A_1$$
$$+ [\rho\omega^2 + k^2(\eta^2 c_{22} + i2\eta c_{26} - c_{66})]A_2$$
$$+ k^2(\eta^2 c_{24} + i\eta c_{25} + i\eta c_{46} - c_{56})A_3$$
$$+ k^2(\eta^2 e_{22} + i\eta e_{26} + i\eta e_{12} - e_{16})A_4 = 0$$
$$k^2(\eta^2 c_{46} + i\eta c_{14} + i\eta c_{56} - c_{15})A_1$$
$$+ k^2(\eta^2 c_{24} + i\eta c_{46} + i\eta c_{25} - c_{56})A_2$$
$$+ [\rho\omega^2 + k^2(\eta^2 c_{44} + i2\eta c_{45} - c_{55})]A_3$$
$$+ k^2(\eta^2 e_{24} + i\eta e_{25} + i\eta e_{14} - e_{15})A_4 = 0,$$
$$(\eta^2 e_{26} + i\eta e_{21} + i\eta e_{16} - e_{11})A_1$$
$$+ (\eta^2 e_{22} + i\eta e_{26} + i\eta e_{12} - e_{16})A_2$$
$$+ (\eta^2 e_{24} + i\eta e_{25} + i\eta e_{14} - e_{15})A_3 \qquad (2.1.32)$$
$$- (\eta^2 \varepsilon_{22} + i2\eta \varepsilon_{12} - \varepsilon_{11})A_4 = 0.$$

For nontrivial solutions of A_j and/or A_4, the determinant of the coefficient matrix of the above equations must vanish. This leads to a polynomial equation of degree eight for η. We denote the eight roots of the equation by $\eta_{(m)}$, and the corresponding eigenvectors by $(\overline{A}_j^{(m)}, \overline{A}_4^{(m)})$, $m = 1, 2,$ ..., 8. Thus, the general wave solution to Equation (2.1.30) in the form of Equation (2.1.31) can be written as

$$u_i = \sum_{m=1}^{8} C_{(m)} \overline{A}_i^{(m)} \exp(k\eta_{(m)} x_2) \exp[i(kx_1 - \omega t)],$$
$$\qquad (2.1.33)$$
$$\phi = \sum_{m=1}^{8} C_{(m)} \overline{A}_4^{(m)} \exp(k\eta_{(m)} x_2) \exp[i(kx_1 - \omega t)],$$

where the constants $C_{(m)}$ ($m = 1, 2, \ldots, 8$) are to be determined. Substituting Equation (2.1.33) into

$$T_{2j}(x_2 = \pm h) = [c_{2jkl}u_{k,l} + e_{k2j}\phi_{,k}]_{x_2 = \pm h} = 0 \qquad (2.1.34)$$

and

$$\phi(x_2 = \pm h) = 0 \qquad (2.1.35)$$

yields the following eight linear algebraic equations for $C_{(m)}$:

$$\sum_{m=1}^{8}(c_{12}i\overline{A}_1^{(m)} + c_{22}\eta_{(m)}\overline{A}_2^{(m)} + c_{24}\eta_{(m)}\overline{A}_3^{(m)}$$
$$+ c_{25}i\overline{A}_3^{(m)} + c_{26}\eta_{(m)}\overline{A}_1^{(m)} + c_{26}i\overline{A}_2^{(m)}$$
$$+ e_{12}i\overline{A}_4^{(m)} + e_{22}\eta_{(m)}\overline{A}_4^{(m)})e^{\pm k\eta_{(m)}h}C_{(m)} = 0,$$

$$\sum_{m=1}^{8}(c_{16}i\overline{A}_1^{(m)} + c_{26}\eta_{(m)}\overline{A}_2^{(m)} + c_{46}\eta_{(m)}\overline{A}_3^{(m)}$$
$$+ c_{56}i\overline{A}_3^{(m)} + c_{66}\eta_{(m)}\overline{A}_1^{(m)} + c_{66}i\overline{A}_2^{(m)}$$
$$+ e_{16}i\overline{A}_4^{(m)} + e_{26}\eta_{(m)}\overline{A}_4^{(m)})e^{\pm k\eta_{(m)}h}C_{(m)} = 0,$$

$$\sum_{m=1}^{8}(c_{14}i\overline{A}_1^{(m)} + c_{24}\eta_{(m)}\overline{A}_2^{(m)} + c_{44}\eta_{(m)}\overline{A}_3^{(m)}$$
$$+ c_{45}i\overline{A}_3^{(m)} + c_{46}\eta_{(m)}\overline{A}_1^{(m)} + c_{46}i\overline{A}_2^{(m)}$$
$$+ e_{14}i\overline{A}_4^{(m)} + e_{24}\eta_{(m)}\overline{A}_4^{(m)})e^{\pm k\eta_{(m)}h}C_{(m)} = 0,$$

$$\sum_{j=1}^{8}\overline{A}_4^{(m)}e^{\pm k\eta_{(m)}h}C_{(m)} = 0. \qquad (2.1.36)$$

For nontrivial solutions of $C_{(m)}$, the determinant of the coefficient matrix of Equation (2.1.36) has to vanish, which yields the frequency equation that contains ω and k. The above derivation is for materials with general anisotropy.

2.1.2.2 Numerical example

As a numerical example, consider a plate of polarized ceramics poled in the x_3 direction with

$$
\begin{pmatrix}
c_{11} & c_{12} & c_{13} & 0 & 0 & 0 \\
c_{12} & c_{11} & c_{13} & 0 & 0 & 0 \\
c_{13} & c_{13} & c_{33} & 0 & 0 & 0 \\
0 & 0 & 0 & c_{44} & 0 & 0 \\
0 & 0 & 0 & 0 & c_{44} & 0 \\
0 & 0 & 0 & 0 & 0 & c_{66}
\end{pmatrix},
$$

$$
\begin{bmatrix}
0 & 0 & 0 & 0 & e_{15} & 0 \\
0 & 0 & 0 & e_{15} & 0 & 0 \\
e_{31} & e_{31} & e_{33} & 0 & 0 & 0
\end{bmatrix},
\begin{pmatrix}
\varepsilon_{11} & 0 & 0 \\
0 & \varepsilon_{11} & 0 \\
0 & 0 & \varepsilon_{33}
\end{pmatrix},
\qquad (2.1.37)
$$

where $c_{66} = (c_{11} - c_{12})/2$. Polarized ceramics are transversely isotropic. Their linear behavior described by Equation (2.1.37) is the same as crystals of 6mm symmetry. For the above straight-crested waves, when

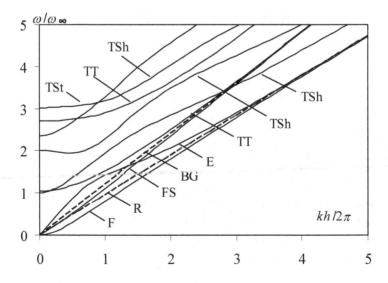

Figure 2.1.4. Dispersion relations of waves in an electroded ceramic plate poled along the x_3 direction.

the poling direction is along the x_3 axis, u_1 is coupled with u_2, and u_3 is coupled with ϕ, but the two groups do not couple to each other. The dispersion relations (ω versus k) for PZT-5H are plotted in Figure 2.1.4.

There exist an infinite number of branches of dispersion relations. Only the first nine branches are shown. The wave frequencies are normalized by the lowest thickness-shear frequency

$$\omega_\infty = \frac{\pi}{2h}\sqrt{\frac{c_{44}}{\rho}}. \qquad (2.1.38)$$

The dispersion relations are labeled as follows, along with the dominant displacement component at small wave numbers:

E = extension (u_1),

F = flexure (u_2),

FS = face-shear (u_3),

TSh = thickness-shear (u_1),

TSt = thickness-stretch (u_2),

TT = thickness-twist (u_3),

R = Rayleigh surface wave (u_1 and u_2),

BG = Bleustein-Gulyaev surface wave (u_3).

In the figure, the three branches passing the origin are the so-called low frequency branches. They represent the extensional, flexural, and face-shear waves. The other six branches are high-frequency branches, which have finite intercepts with the ω axis. These intercepts are called cutoff frequencies, below which the corresponding waves cannot propagate. Cutoff frequencies are in fact the frequencies of pure thickness modes. The six high frequency branches shown represent three thickness-shear waves, one thickness-stretch wave, and two thickness-twist waves. One of the two dotted lines in the figure is the well-known Rayleigh surface wave, which can propagate over an elastic half-space and is not dispersive. The other dotted line is the well known Bleustein-Gulyaev surface wave which has only one displacement component u_3 and can propagate over a piezoelectric half-space but does not have an elastic counterpart. These two surface waves are included as references. It is seen that for short waves with larger k, the frequencies of the extensional and flexural waves approach that of the Rayleigh surface wave. Similarly, for short waves, the frequencies of the face-shear wave approach that of the Bleustein-Gulyaev wave.

2.2 Power Series Expansion

Consider a piezoelectric plate (see Figure 2.2.1). The plate is assumed to be thin in the sense that its thickness is much smaller than the in-plane dimensions or the in-plane wavelength we are interested in. The key to the derivation of two-dimensional plate equations is the approximation of the variation of the fields through the plate thickness (see Figure 2.1.2) by some known and simple functions. Then plate equations can be derived systematically by inserting the approximate fields into the variational formulation of the three-dimensional theory. This procedure can be traced back to Cauchy, Poisson and Kirchhoff according to Mindlin [13]. The resulting two-dimensional plate equations are much simpler than the three-dimensional equations, and therefore often allow analytical solutions.

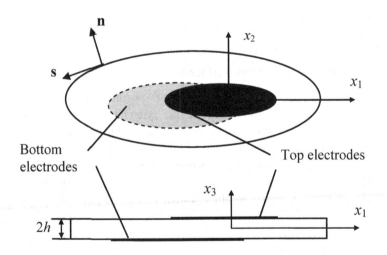

Figure 2.2.1. A thin piezoelectric plate.

2.2.1 Expansions of displacement and potential

2.2.1.1 Polynomial approximation of thickness modes

The exact thickness modes (except the static one) are sinusoidal along the plate thickness (see Figure 2.1.2). Since power series are simple to differentiate and integrate, early approximations of the thickness variations of the fields were in terms of power series. This approximation can be very accurate because trigonometric functions can be well approximated by power series. For example, in Figure 2.2.2, the first and the second thickness-shear modes in Figure 2.1.2 are compared with two simple polynomials, where

$$\text{Black triangle: } y = \sin \pi x / 2,$$

$$\text{Cross: } y = \frac{3}{2}(x - \frac{x^3}{3}),$$

$$\text{Square: } y = \cos \pi x,$$

$$\text{White triangle: } y = 2(x^2 - 1)^2 - 1.$$

(2.2.1)

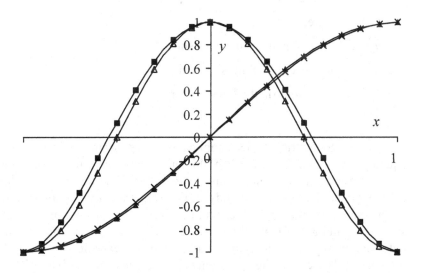

Figure 2.2.2. Thickness-shear modes and their polynomial approximations.

2.2.1.2 Polynomial expansions

In a series of papers [14-17], Mindlin developed theories for high frequency vibrations of isotropic, anisotropic, and piezoelectric plates by power series expansions in the plate thickness coordinate. We now derive two-dimensional equations for a piezoelectric plate in the manner of [17]. First we expand the mechanical displacement and the electric potential into power series in x_3

$$u_i(x_1, x_2, x_3, t) = \sum_{n=0}^{\infty} x_3^n u_i^{(n)}(x_1, x_2, t),$$

$$\phi(x_1, x_2, x_3, t) = \sum_{n=0}^{\infty} x_3^n \phi^{(n)}(x_1, x_2, t).$$

(2.2.2)

Our goal is to obtain two-dimensional equations for $u_i^{(n)}$ and $\phi^{(n)}$. The lower order two-dimensional displacements can describe the following deformations:

$u_1^{(0)}, u_2^{(0)}$ – extension,

$u_3^{(0)}$ – flexure,

$u_1^{(1)}, u_2^{(1)}$ – fundamental thickness-shear,

$u_3^{(1)}$ – fundamental thickness-stretch,

$u_1^{(2)}, u_2^{(2)}$ – symmetric thickness-shear,

....

For piezoelectric device applications we want to derive plate equations that can describe the thickness-shear and thickness-stretch deformations well. It is important to note that these deformations have different behaviors in static and dynamic problems. For example, the first or fundamental thickness-shear mode of $n = 1$ in Figure 2.1.2 is sinusoidal along the plate thickness. However, the static thicken-shear deformation shown in the same figure is linearly varying along the plate thickness. Therefore the fundamental thickness-shear has different distributions along the plate thickness in static and dynamic problems. One simple expression can only approximate either the static or the dynamic deformation well, but not both.

We are mainly interested in dynamic problems. Obviously the sinusoidal variation of the fundamental thickness-shear deformation can be well approximated by a cubic polynomial. However, we are not interested in accurately describing the field variation along the plate thickness. Our main goal is to develop plate theories than can predict the frequencies and dispersion relations of these modes accurately. For this purpose, following Mindlin, we use a linear function $x_3 u_a^{(1)}(x_1, x_2, t)$, $a =$ 1, 2, to approximately describe the fundamental thicken-shear motion. The main advantage of using a linear function is its simplicity. The error due to the approximation in resonant frequencies can be reduced or removed by introducing some correction factors. Overall this is a simple and accurate approach.

Shear correction factors are essentially not needed when using sinusoidal or higher-degree polynomial approximations for thickness-shear modes in the sense that the thickness-shear frequencies can be predicted exactly or accurately. This approach involves more algebra. In addition, the static thickness-shear behavior cannot be described well by a sine function or a cubic or higher-degree polynomial. Hence corrections may be needed for low frequency behaviors.

2.2.2 Strains and electric fields

From Equation (2.2.2) and the strain-displacement relation and the electric field-potential relation in Equation (1.2.15), we obtain

$$S_{ij} = \sum_n x_3^n S_{ij}^{(n)}, \quad E_i = \sum_n x_3^n E_i^{(n)}, \tag{2.2.3}$$

where

$$S_{ij}^{(n)} = \frac{1}{2}[u_{j,i}^{(n)} + u_{i,j}^{(n)} + (n+1)(\delta_{i3} u_j^{(n+1)} + \delta_{3j} u_i^{(n+1)})],$$
$$E_i^{(n)} = -\phi_{,i}^{(n)} - (n+1)\delta_{3i}\phi^{(n+1)}. \tag{2.2.4}$$

The first few orders of strains and electric fields have the following form:

$$S_1^{(0)} = u_{1,1}^{(0)}, \quad S_2^{(0)} = u_{2,2}^{(0)}, \quad S_3^{(0)} = u_3^{(1)},$$
$$S_4^{(0)} = u_{3,2}^{(0)} + u_2^{(1)}, \quad S_5^{(0)} = u_{3,1}^{(0)} + u_1^{(1)}, \quad S_6^{(0)} = u_{1,2}^{(0)} + u_{2,1}^{(0)}, \tag{2.2.5}$$

$$S_1^{(1)} = u_{1,1}^{(1)}, \quad S_2^{(1)} = u_{2,2}^{(1)}, \quad S_3^{(1)} = 2u_3^{(2)},$$
$$S_4^{(1)} = u_{3,2}^{(1)} + 2u_2^{(2)}, \quad S_5^{(1)} = u_{3,1}^{(1)} + 2u_1^{(2)}, \quad S_6^{(1)} = u_{1,2}^{(1)} + u_{2,1}^{(1)}, \tag{2.2.6}$$

$$S_1^{(2)} = u_{1,1}^{(2)}, \quad S_2^{(2)} = u_{2,2}^{(2)}, \quad S_3^{(2)} = 3u_3^{(3)},$$
$$S_4^{(2)} = u_{3,2}^{(2)} + 3u_2^{(3)}, \quad S_5^{(2)} = u_{3,1}^{(2)} + 3u_1^{(3)}, \quad S_6^{(2)} = u_{1,2}^{(2)} + u_{2,1}^{(2)}, \tag{2.2.7}$$

$$E_1^{(0)} = -\phi_{,1}^{(0)}, \quad E_2^{(0)} = -\phi_{,2}^{(0)}, \quad E_3^{(0)} = -\phi^{(1)}, \tag{2.2.8}$$

$$E_1^{(1)} = -\phi_{,1}^{(1)}, \quad E_2^{(1)} = -\phi_{,2}^{(1)}, \quad E_3^{(1)} = -2\phi^{(2)}, \tag{2.2.9}$$

$$E_1^{(2)} = -\phi_{,1}^{(2)}, \quad E_2^{(2)} = -\phi_{,2}^{(2)}, \quad E_3^{(2)} = -3\phi^{(3)}. \tag{2.2.10}$$

The zero-order plate strains in Equation (2.2.5) describe homogeneous deformations of a plate element. The first-order strains in Equation (2.2.6) represent higher-order deformations of a plate element like curvature and twist, etc. Pictures of zero- and first-order deformed plate elements can be found in [13].

2.2.3 Constitutive relations

The plate resultants of various orders are defined by

$$T_{ij}^{(n)} = \int_{-h}^{h} T_{ij} x_3^n dx_3, \quad D_i^{(n)} = \int_{-h}^{h} D_i x_3^n dx_3. \tag{2.2.11}$$

Substituting the three-dimensional constitutive relations from Equation (1.2.14) into Equation (2.2.11), we obtain the plate constitutive relations as:

$$T_{ij}^{(n)} = \sum_m B_{mn} (c_{ijkl} S_{kl}^{(m)} - e_{kij} E_k^{(m)}),$$
$$D_i^{(n)} = \sum_m B_{mn} (e_{ijk} S_{jk}^{(m)} + \varepsilon_{ij} E_j^{(m)}), \tag{2.2.12}$$

where

$$B_{mn} = \int_{-h}^{h} x_3^m x_3^n dx_3 = \begin{cases} 2h^{m+n+1}/(m+n+1), & m+n \quad \text{even}, \\ 0, & m+n \quad \text{odd}. \end{cases} \tag{2.2.13}$$

Pictures of zero- and first-order plate resultants are given in [13]. They represent extensional and shear forces, and bending and twisting moments, etc.

2.2.4 Equations of motion and charge

We use the variational formulation in Equation (1.2.26) to derive the plate equations of motion and charge. For convenience we introduce a convention in which subscripts a and b assume 1 and 2 only but not 3. Let A be a two-dimensional area in the x_1-x_2 plane with a boundary curve C. Then Equation (1.2.26) can be written as

$$
\delta\Pi = \int_{t_0}^{t_1} dt \int_A dA \int_{-h}^{h} dx_3 \big[(T_{aj,a} + \rho f_j - \rho \ddot{u}_j)\delta u_j + T_{3j,3}\delta u_j\big]
$$
$$
+ \int_{t_0}^{t_1} dt \int_A dA \int_{-h}^{h} dx_3 \big[(D_{a,a} - \rho_e)\delta\phi + D_{3,3}\delta\phi\big]
$$
$$
- \int_{t_0}^{t_1} dt \int_{C_T} dl \int_{-h}^{h} dx_3 (T_{aj}n_a - \bar{t}_j)\delta u_i
$$
$$
- \int_{t_0}^{t_1} dt \int_{C_D} dl \int_{-h}^{h} dx_3 (D_a n_a + \bar{\sigma}_e)\delta\phi,
$$

(2.2.14)

where C_T has prescribed traction and C_D has prescribed surface charge. Substituting Equation (2.2.2) into Equation (2.2.14), with integration by parts with respect to x_3 and time, we obtain

$$
\delta\Pi = \sum_n \int_{t_0}^{t_1} dt \int_A \left(T_{aj,a}^{(n)} - nT_{3j}^{(n-1)} + F_j^{(n)} - \rho\sum_m B_{mn}\ddot{u}_j^{(m)} \right)\delta u_j^{(n)} dA
$$
$$
+ \sum_n \int_{t_0}^{t_1} dt \int_A \left(D_{a,a}^{(n)} - nD_3^{(n-1)} + D^{(n)} \right)\delta\phi^{(n)} dA
$$
$$
- \sum_n \int_{t_0}^{t_1} dt \int_{C_T} (T_{aj}^{(n)}n_a - \bar{t}_j^{(n)})\delta u_j^{(n)} dl
$$
$$
- \sum_n \int_{t_0}^{t_1} dt \int_{C_D} (D_a^{(n)}n_a + \bar{\sigma}_e^{(n)})\delta\phi^{(n)} dl,
$$

(2.2.15)

where the body and surface loads of various orders are defined by

$$F_j^{(n)} = [x_3^n T_{3j}]_{-h}^h + \int_{-h}^h \rho f_j x_3^n dx_3,$$

$$D^{(n)} = [x_3^n D_3]_{-h}^h - \int_{-h}^h \rho_e x_3^n dx_3, \qquad (2.2.16)$$

$$\bar{t}_j^{(n)} = \int_{-h}^h \bar{t}_j x_3^n dx_3, \quad \bar{\sigma}_e^{(n)} = \int_{-h}^h \bar{\sigma}_e x_3^n dx_3.$$

For independent variations of $\delta u_i^{(n)}$ and $\delta \phi^{(n)}$, we obtain

$$T_{aj,a}^{(n)} - n T_{3j}^{(n-1)} + F_j^{(n)} = \rho \sum_m B_{mn} \ddot{u}_j^{(m)} \quad \text{in} \quad A,$$

$$D_{a,a}^{(n)} - n D_3^{(n-1)} + D^{(n)} = 0 \quad \text{in} \quad A,$$

$$T_{aj}^{(n)} n_a = \bar{t}_j^{(n)} \quad \text{on} \quad C_T, \qquad (2.2.17)$$

$$D_a^{(n)} n_a = -\bar{\sigma}_e^{(n)} \quad \text{on} \quad C_D.$$

We note that electrodes in fact impose constraints on $\delta \phi^{(n)}$ [18]. This will be discussed in the next section.

In addition to power series expansions, it was pointed out in [13] that trigonometric series could also be used. Two-dimensional equations obtained from trigonometric expansions were given in [19-21]. More references on various expansions can be found in a review article [22]. In this book we mainly focus on power series expansions. We note that the following polynomial expansion of the electric potential [23]:

$$\phi = \phi^{(0)} + x_3 \phi^{(1)} + (x_3^2 - h^2)(\phi^{(2)} + x_3 \phi^{(3)} + \cdots) \qquad (2.2.18)$$

has an important feature, i.e., only the first two terms do not vanish at $x_3 = \pm h$. Therefore only $\phi^{(0)}$ and $\phi^{(1)}$ are responsible for the voltage across the plate thickness. Then in the plate equations of electrostatics only the zero- and first-order equations have surface charge terms. These will make it convenient for an electroded plate, especially for higher order equations.

2.3 Zero-Order Theory for Extension

In this section we develop equations for extensional motions of thin plates. The propagation of extensional and face-shear waves is examined and compared with the three-dimensional solutions so that the range of applicability of the plate equations can be established. The equations are specialized to plates of polarized ceramics. As an example, the equations are used to study radial vibrations of a circular ceramic disk.

2.3.1 Equations for zero-order theory

By a zero-order theory we mean a theory for extensional motions of a plate with $u_1^{(0)}$ and $u_2^{(0)}$ as the major displacements. For the electrical behavior of the plate we are interested in what is governed by $\phi^{(0)}$ and $\phi^{(1)}$. The approximate displacement and potential fields are

$$u_i \cong u_i^{(0)} + x_3 u_i^{(1)},$$
$$\phi \cong \phi^{(0)} + x_3 \phi^{(1)}. \tag{2.3.1}$$

Although we are mainly interested in $u_1^{(0)}$ and $u_2^{(0)}$, we have included a few other displacement components in Equation (2.3.1). Among these additional displacements, $u_3^{(1)}$ represents the thickness stretch or contraction accompanying extension due to Poisson's effect, and must be included. $u_3^{(0)}$ describes flexure. $u_1^{(1)}$ and $u_2^{(1)}$ represent thickness-shear. From Equation (2.2.5) it can be seen that $u_3^{(0)}$ together with $u_1^{(1)}$ and $u_2^{(1)}$ contribute to thickness-shear deformations $S_4^{(0)}$ and $S_5^{(0)}$, which may couple to extension due to anisotropy and should be allowed. The two-dimensional plate equations we will obtain are for the extensional displacements $u_1^{(0)}$ and $u_2^{(0)}$. Other displacements will be eliminated through a stress relaxation procedure. Within the approximation in Equation (2.3.1), the strains and electric fields in Equations (2.2.5), (2.2.8) and (2.2.9) become:

$$S_1^{(0)} = u_{1,1}^{(0)}, \quad S_2^{(0)} = u_{2,2}^{(0)}, \quad S_3^{(0)} = u_3^{(1)},$$
$$S_4^{(0)} = u_{3,2}^{(0)} + u_2^{(1)}, \quad S_5^{(0)} = u_{3,1}^{(0)} + u_1^{(1)}, \quad S_6^{(0)} = u_{1,2}^{(0)} + u_{2,1}^{(0)}, \tag{2.3.2}$$

$$E_1^{(0)} = -\phi_{,1}^{(0)}, \quad E_2^{(0)} = -\phi_{,2}^{(0)}, \quad E_3^{(0)} = -\phi^{(1)}, \tag{2.3.3}$$

$$E_1^{(1)} = -\phi_{,1}^{(1)}, \quad E_2^{(1)} = -\phi_{,2}^{(1)}, \quad E_3^{(1)} = 0. \tag{2.3.4}$$

Higher order strains $S_{ij}^{(1)}$ are neglected.

From Equation $(2.2.17)_1$, for $j = 1, 2$ and $n = 0$, we obtain

$$T_{ab,a}^{(0)} + F_b^{(0)} = 2h\rho\ddot{u}_b^{(0)}, \quad a,b = 1,2, \tag{2.3.5}$$

where we have truncated the right hand side by keeping the B_{00} term only in the summation.

For the electrostatic equations we discuss several cases below. In the theory of piezoelectricity the electric potential is at most a function of time on an electrode. For example, if both the top and bottom surfaces of the plate are electroded, we can write

$$\phi = V_1(t), \quad x_3 = h,$$
$$\phi = V_2(t), \quad x_3 = -h. \tag{2.3.6}$$

Since

$$\phi \cong \phi^{(0)} + x_3\phi^{(1)}, \tag{2.3.7}$$

we have

$$\phi^{(0)} + h\phi^{(1)} = V_1(t),$$
$$\phi^{(0)} - h\phi^{(1)} = V_2(t). \tag{2.3.8}$$

Equation (2.3.8) imposes constraints in the variational formulation in Equation (2.2.15). These constraints can be systematically treated by the method of Lagrange multipliers [18]. In our case with $\phi^{(0)}$ and $\phi^{(1)}$ only, the constraints are relatively simple so we will proceed in the manner of [24,25]. We discuss four possibilities separately.

(i) An unelectroded plate

In this case $\delta\phi^{(0)}$ and $\delta\phi^{(1)}$ are independent functions of x_1, x_2 and t. We have the following two-dimensional equations of electrostatics:

$$D_{a,a}^{(0)} + D^{(0)} = 0,$$
$$D_{a,a}^{(1)} - D_3^{(0)} + D^{(1)} = 0. \tag{2.3.9}$$

(ii) A symmetrically electroded plate

In this case $x_3 = \pm h$ are both electroded. $\phi^{(0)}$ and $\phi^{(1)}$ are directly determined by Equation (2.3.8) as functions of time. No differential equations for $\phi^{(0)}$ and $\phi^{(1)}$ result from Equation (2.2.15) and no differential equations are needed for determining $\phi^{(0)}$ and $\phi^{(1)}$.

(iii) A plate with the upper surface electroded

From Equation (2.3.8)$_1$

$$\delta\phi^{(0)} + h\delta\phi^{(1)} = 0. \tag{2.3.10}$$

Substituting Equation (2.3.10) into Equation (2.2.15) we obtain

$$-h(D_{a,a}^{(0)} + D^{(0)}) + D_{a,a}^{(1)} - D_3^{(0)} + D^{(1)} = 0. \tag{2.3.11}$$

Since $\phi^{(0)}$ and $\phi^{(1)}$ are related by (2.3.8)$_1$, we only need one differential equation (2.3.11) to determine one of them. When the upper surface is electroded, $D_3(h)$ is unknown. Equation (2.3.11) can also be obtained by eliminating $D_3(h)$ between Equations (2.3.9)$_1$ and (2.3.9)$_2$. This procedure can also be used to treat prescribed displacement on a major surface of a plate when using two-dimensional equations [26].

(iv) A plate with the lower surface electroded

From Equation (2.3.8)$_2$

$$\delta\phi^{(0)} - h\delta\phi^{(1)} = 0. \tag{2.3.12}$$

Substituting Equation (2.3.12) into Equation (2.2.15) we obtain

$$h(D_{a,a}^{(0)} + D^{(0)}) + D_{a,a}^{(1)} - D_3^{(0)} + D^{(1)} = 0, \tag{2.3.13}$$

which is equivalent to eliminating $D_3(-h)$ between Equations (2.3.9)$_1$ and (2.3.9)$_2$.

For plate constitutive relations, we truncate Equation (2.2.12) by keeping the B_{00} and B_{11} terms only in the summation:

$$T_{ij}^{(0)} = 2h(c_{ijkl}S_{kl}^{(0)} - e_{kij}E_k^{(0)}),$$

$$D_i^{(0)} = 2h(e_{ijk}S_{jk}^{(0)} + \varepsilon_{ij}E_j^{(0)}), \tag{2.3.14}$$

$$D_i^{(1)} = \frac{2h^3}{3}\varepsilon_{ij}E_j^{(1)},$$

where $S_{ij}^{(1)}$ have been neglected. Since $S_{ij}^{(0)}$ contains $u_3^{(0)}$ and $u_i^{(1)}$, Equations (2.3.14) are not yet ready to be used for a theory of extension. To obtain the proper constitutive relations for extension we proceed as follows [13]. Since the plate is assumed to be very thin and for extension the dominating stress components are T_{11}, T_{22} and T_{12}, we take the following to be approximately true:

$$T_{3j}^{(0)} = 0. \tag{2.3.15}$$

Equation (2.3.15) is called stress relaxation. According to the compact matrix notation, with the range of p, q as 1, 2, ... and 6, Equation (2.3.15) can be written as

$$T_q^{(0)} = 0, \quad q = 3,4,5. \tag{2.3.16}$$

For convenience we introduce another index convention in which subscripts u, v, w take the values 3, 4, and 5 while subscripts r and s take the remaining values 1, 2, and 6. Then Equation (2.3.14)$_{1,2}$ can be written as

$$T_r^{(0)} = 2h(c_{rs}S_s^{(0)} + c_{ru}S_u^{(0)} - e_{kr}E_k^{(0)}),$$

$$T_v^{(0)} = 2h(c_{vs}S_s^{(0)} + c_{vw}S_w^{(0)} - e_{kv}E_k^{(0)}) = 0, \tag{2.3.17}$$

$$D_i^{(0)} = 2h(e_{is}S_s^{(0)} + e_{iu}S_u^{(0)} + \varepsilon_{ij}E_j^{(0)}),$$

where Equation (2.3.16) has been used. From Equations (2.3.17)$_2$ we have

$$S_u^{(0)} = -c_{uv}^{-1}c_{vs}S_s^{(0)} + c_{uv}^{-1}e_{kv}E_k^{(0)}. \tag{2.3.18}$$

Substitution of Equation (2.3.18) into Equations (2.3.17)$_{1,3}$ gives

$$T_r^{(0)} = 2h(\gamma_{rs}S_s^{(0)} - \psi_{kr}E_k^{(0)}),$$
$$D_i^{(0)} = 2h(\psi_{is}S_s^{(0)} + \zeta_{ij}E_j^{(0)}),$$

(2.3.19)

where the material constants relaxed for thin plates are

$$\gamma_{rs} = c_{rs} - c_{rv}c_{vw}^{-1}c_{ws}, \quad r,s = 1,2,6,$$
$$\psi_{ks} = e_{ks} - e_{kw}c_{wv}^{-1}c_{vs}, \quad v,w = 3,4,5,$$
$$\zeta_{kj} = \varepsilon_{kj} + e_{kv}c_{vw}^{-1}e_{jw}. \quad j,k = 1,2,3.$$

(2.3.20)

In summary, in the case of an unelectroded plate, we have obtained

$$T_{ab,a}^{(0)} + F_b^{(0)} = 2h\rho\ddot{u}_b^{(0)}, \quad a,b = 1,2,$$
$$D_{a,a}^{(0)} + D^{(0)} = 0,$$
$$D_{a,a}^{(1)} - D_3^{(0)} + D^{(1)} = 0,$$
$$T_r^{(0)} = 2h(\gamma_{rs}S_s^{(0)} - \psi_{kr}E_k^{(0)}), \quad r,s = 1,2,6,$$
$$D_i^{(0)} = 2h(\psi_{is}S_s^{(0)} + \zeta_{ij}E_j^{(0)}),$$

(2.3.21)

$$D_i^{(1)} = \frac{2h^3}{3}\varepsilon_{ij}E_j^{(1)},$$
$$S_1^{(0)} = u_{1,1}^{(0)}, \quad S_2^{(0)} = u_{2,2}^{(0)}, \quad S_6^{(0)} = u_{1,2}^{(0)} + u_{2,1}^{(0)},$$
$$E_1^{(0)} = -\phi_{,1}^{(0)}, \quad E_2^{(0)} = -\phi_{,2}^{(0)}, \quad E_3^{(0)} = -\phi^{(1)},$$
$$E_1^{(1)} = -\phi_{,1}^{(1)}, \quad E_2^{(1)} = -\phi_{,2}^{(1)}, \quad E_3^{(1)} = 0.$$

With successive substitutions from Equations (2.3.21)$_{4\text{-}9}$, Equations (2.3.21)$_{1\text{-}3}$ can be written as four equations for $u_1^{(0)}$, $u_2^{(0)}$, $\phi^{(0)}$ and $\phi^{(1)}$. At the boundary of a plate with an in-plane unit exterior normal **n** and an in-plane unit tangent **s** (see Figure 2.2.1), we may prescribe

$$T_{nn}^{(0)} \text{ or } u_n^{(0)}, \quad T_{ns}^{(0)} \text{ or } u_s^{(0)},$$
$$D_n^{(0)} \text{ or } \phi^{(0)}, \quad D_n^{(1)} \text{ or } \phi^{(1)}.$$

(2.3.22)

When electrodes are present the differential equations are fewer, and so are the boundary conditions.

We note that in Equation (2.3.21) only $S_s^{(0)}$ or $u_1^{(0)}$ and $u_2^{(0)}$ are involved. The other three strain components $S_u^{(0)}$ ($u = 3, 4, 5$) are due to Poisson's effect for couplings among extensions in different directions, and couplings between extensions and shears in anisotropic materials. $S_u^{(0)}$ and the related displacement components $u_3^{(0)}$ and $u_i^{(1)}$ can be determined approximately from the stress relaxation condition in Equation (2.3.18) once the solution to Equation (2.3.21) has been obtained.

Compared to Poisson's effect for couplings among extensions in different directions, couplings between extensions and shears in anisotropic materials are usually weaker and do not even exist in certain crystal classes. If couplings between extensions and shears are neglected as an approximation, a simpler stress relaxation can be performed. From Equation (2.3.14)$_1$, by setting $i = j = 3$, we require

$$T_{33}^{(0)} = 2h(c_{33kl}S_{kl}^{(0)} - e_{k33}E_k^{(0)}) = 0 . \tag{2.3.23}$$

This implies the following expression for $S_{33}^{(0)}$:

$$S_{33}^{(0)} = -\frac{1}{c_{3333}}(c_{33kl}S_{kl}^{(0)} - c_{3333}S_{33}^{(0)} - e_{k33}E_k^{(0)}) . \tag{2.3.24}$$

In Equation (2.3.24), $S_{33}^{(0)}$ has been eliminated on the right hand side because when $i = j = 3$, the two terms containing $S_{33}^{(0)}$ cancel with each other. From Equation (2.3.24) the thickness expansion or contraction accompanying the extension of the plate due to Poisson's effect can be found if interested. Substituting Equation (2.3.24) back into Equations (2.3.14)$_{1,2}$, we obtain the following constitutive relations relaxed for thin plates:

$$T_{ij}^{(0)} = 2h(\bar{c}_{ijkl}S_{kl}^{(0)} - \bar{e}_{kij}E_k^{(0)}),$$
$$D_i^{(0)} = 2h(\bar{e}_{ikl}S_{kl}^{(0)} + \bar{\varepsilon}_{ij}E_j^{(0)}), \tag{2.3.25}$$

where the relaxed material constants are defined by

$$\bar{c}_{ijkl} = c_{ijkl} - c_{ij33}c_{33kl} / c_{3333},$$
$$\bar{e}_{kij} = e_{kij} - e_{k33}c_{33ij} / c_{3333}, \tag{2.3.26}$$
$$\bar{\varepsilon}_{ij} = \varepsilon_{ij} + e_{i33}e_{j33} / c_{3333} .$$

We note that the right hand sides of Equation (2.3.25) do not contain $S_{33}^{(0)}$ and $T_{33}^{(0)} = 0$ is automatically satisfied by Equation (2.3.25). When using Equation (2.3.25) for extension, $S_{31}^{(0)}$ and $S_{32}^{(0)}$ are taken to be zero. The structures of the $\bar{c}_{pq}$ and $\bar{e}_{iq}$ matrices are

$$[\bar{c}_{pq}] = \begin{pmatrix} \bar{c}_{11} & \bar{c}_{12} & 0 & \bar{c}_{14} & \bar{c}_{15} & \bar{c}_{16} \\ \bar{c}_{21} & \bar{c}_{22} & 0 & \bar{c}_{24} & \bar{c}_{25} & \bar{c}_{26} \\ 0 & 0 & 0 & 0 & 0 & 0 \\ \bar{c}_{41} & \bar{c}_{42} & 0 & \bar{c}_{44} & \bar{c}_{45} & \bar{c}_{46} \\ \bar{c}_{51} & \bar{c}_{52} & 0 & \bar{c}_{54} & \bar{c}_{55} & \bar{c}_{56} \\ \bar{c}_{61} & \bar{c}_{62} & 0 & \bar{c}_{64} & \bar{c}_{65} & \bar{c}_{66} \end{pmatrix}, \tag{2.3.27}$$

$$[\bar{e}_{iq}] = \begin{bmatrix} \bar{e}_{11} & \bar{e}_{12} & 0 & \bar{e}_{14} & \bar{e}_{15} & \bar{e}_{16} \\ \bar{e}_{21} & \bar{e}_{22} & 0 & \bar{e}_{24} & \bar{e}_{25} & \bar{e}_{26} \\ \bar{e}_{31} & \bar{e}_{32} & 0 & \bar{e}_{34} & \bar{e}_{35} & \bar{e}_{36} \end{bmatrix}. \tag{2.3.28}$$

2.3.2 Extensional and face-shear waves

To examine the basic behavior of the two-dimensional equations obtained, consider the propagation of the following extensional wave in a plate with traction-free surfaces which are electroded and the electrodes are shorted:

$$u_1^{(0)} = \exp i(\xi x_1 - \omega t), \quad u_2^{(0)} = 0,$$
$$\phi^{(0)} = 0, \quad \phi^{(1)} = 0. \tag{2.3.29}$$

In this case Equation (2.3.21) reduces to

$$\gamma_{11} u_{1,11}^{(0)} = \rho \ddot{u}_1^{(0)}. \tag{2.3.30}$$

Substitution of Equation (2.3.29) into Equation (2.3.30), we obtain the dispersion relation of the wave as

$$\omega = \sqrt{\frac{\gamma_{11}}{\rho}} \, \xi. \tag{2.3.31}$$

Similarly, if we consider the propagation of the following wave:

$$u_1^{(0)} = 0, \quad u_2^{(0)} = \exp i(\xi x_1 - \omega t),$$
$$\phi^{(0)} = 0, \quad \phi^{(1)} = 0,$$

(2.3.32)

which is called a face-shear wave, Equation (2.3.21) reduces to

$$\gamma_{66} u_{2,11}^{(0)} = \rho \ddot{u}_2^{(0)},$$

(2.3.33)

and the corresponding dispersion relation is

$$\omega = \sqrt{\frac{\gamma_{66}}{\rho}} \, \xi.$$

(2.3.34)

We plot the dispersion relations in Equations (2.3.31) and (2.3.34) qualitatively in Figure 2.3.1. Face-shear waves usually have a lower wave speed (slope) than extensional waves.

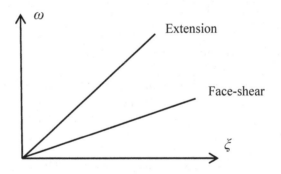

Figure 2.3.1. Dispersion relations of extensional and face-shear waves.

We note that the exact dispersion relations of extensional and face-shear waves in Figure 2.1.4 are curved, representing dispersive waves. The approximate dispersion relations in Figure 2.3.1 are straight lines for nondispersive waves. These straight lines correspond to the tangents of the corresponding curves in Figure 2.1.4 at the origin where the frequency is low and the wave number is small. Therefore the approximate two-dimensional equations we have obtained are low frequency, long wave approximations of the three-dimensional theory. They are valid when $kh/2\pi \ll 1$, or the wave length is much larger than the plate thickness. This gives a dynamic criterion of whether a plate is thin or not. It is relative to the wave length we are considering.

2.3.3 Equations for ceramic plates

We consider two cases of ceramic plates with thickness and in-plane poling.

2.3.3.1 Thickness poling

First consider a ceramic plate with thickness poling (see Figure 2.3.2).

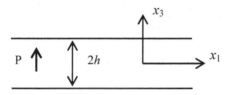

Figure 2.3.2. A ceramic plate with thickness poling.

The material tensors c_{ijkl}^{E}, e_{ijk} and ε_{ij}^{S} are given by the matrices in Equation (2.1.37). For these materials there is no coupling between extension and shear. The stress relaxation in Equation (2.3.15) yields the following results:

$$[\gamma_{rs}] = \begin{bmatrix} c_{11}^{P} & c_{12}^{P} & 0 \\ c_{12}^{P} & c_{11}^{P} & 0 \\ 0 & 0 & c_{66} \end{bmatrix}, \quad [\psi_{ks}] = \begin{bmatrix} 0 & 0 & 0 \\ 0 & 0 & 0 \\ e_{31}^{P} & e_{31}^{P} & 0 \end{bmatrix},$$

$$[\varsigma_{kj}] = \begin{bmatrix} \varepsilon_{11}^{P} & 0 & 0 \\ 0 & \varepsilon_{11}^{P} & 0 \\ 0 & 0 & \varepsilon_{33}^{P} \end{bmatrix}, \quad r,s = 1,2,6, \quad k,j = 1,2,3,$$

(2.3.35)

where

$$c_{11}^p = c_{11}^E - (c_{13}^E)^2 / c_{33}^E = s_{11}^E / [(s_{11}^E)^2 - (s_{12}^E)^2],$$

$$c_{12}^p = c_{12}^E - (c_{13}^E)^2 / c_{33}^E = -s_{12}^E / [(s_{11}^E)^2 - (s_{12}^E)^2],$$

$$c_{66} = c_{66}^E = 1/s_{66}^E = (c_{11}^p - c_{12}^p)/2,$$

$$e_{31}^p = e_{31} - e_{33}c_{13}^E / c_{33}^E = d_{31}/(s_{11}^E + s_{12}^E),$$ (2.3.36)

$$\varepsilon_{11}^p = \varepsilon_{11}^S + e_{15}^2 / c_{44}^E,$$

$$\varepsilon_{33}^p = \varepsilon_{33}^S + e_{33}^2 / c_{33}^E = \varepsilon_{33}^T - 2d_{31}e_{31}^p.$$

c_{pq}^p, e_{iq}^p and ε_{ij}^p are common notations for these plate material constants [3] which clearly indicate that the constants are for a plate. The constitutive relations then take the following form:

$$T_{11}^{(0)} = 2h(c_{11}^p u_{1,1}^{(0)} + c_{12}^p u_{2,2}^{(0)} - e_{31}^p E_3^{(0)}),$$

$$T_{22}^{(0)} = 2h(c_{12}^p u_{1,1}^{(0)} + c_{11}^p u_{2,2}^{(0)} - e_{31}^p E_3^{(0)}),$$ (2.3.37)

$$T_{12}^{(0)} = 2hc_{66}(u_{1,2}^{(0)} + u_{2,1}^{(0)}),$$

$$D_a^{(0)} = 2h\varepsilon_{11}^p E_a^{(0)},$$

$$D_3^{(0)} = 2h(e_{31}^p u_{a,a}^{(0)} + \varepsilon_{33}^p E_3^{(0)}),$$ (2.3.38)

$$D_a^{(1)} = -\frac{2}{3}h^3 \varepsilon_{11}\phi_{,a}^{(1)}.$$ (2.3.39)

Substitution of the above into the equations of motion and charge gives:

$$c_{11}^p u_{1,11}^{(0)} + c_{66}u_{1,22}^{(0)} + (c_{12}^p + c_{66})u_{2,21}^{(0)} + e_{31}^p \phi_{,1}^{(1)} + \frac{1}{2h}F_1^{(0)} = \rho \ddot{u}_1^{(0)},$$

$$(c_{12}^p + c_{66})u_{1,12}^{(0)} + c_{66}u_{2,11}^{(0)} + c_{11}^p u_{2,22}^{(0)} + e_{31}^p \phi_{,2}^{(1)} + \frac{1}{2h}F_2^{(0)} = \rho \ddot{u}_2^{(0)},$$ (2.3.40)

$$-\varepsilon_{11}^p \phi_{,aa}^{(0)} + \frac{1}{2h}D^{(0)} = 0,$$

$$-\varepsilon_{11}\phi_{,aa}^{(1)} + 3h^{-2}\varepsilon_{33}^p \phi^{(1)} - 3h^{-2}e_{31}^p u_{a,a}^{(0)} + \frac{3}{2h^3}D^{(1)} = 0.$$ (2.3.41)

The stress relaxation in Equation (2.3.23) yields:

$$[\bar{c}_{pq}] = \begin{pmatrix} c_{11}^P & c_{12}^P & 0 & 0 & 0 & 0 \\ c_{21}^P & c_{11}^P & 0 & 0 & 0 & 0 \\ 0 & 0 & 0 & 0 & 0 & 0 \\ 0 & 0 & 0 & c_{44} & 0 & 0 \\ 0 & 0 & 0 & 0 & c_{44} & 0 \\ 0 & 0 & 0 & 0 & 0 & c_{66} \end{pmatrix}, \qquad (2.3.42)$$

$$[\bar{e}_{iq}] = \begin{bmatrix} 0 & 0 & 0 & 0 & e_{15} & 0 \\ 0 & 0 & 0 & e_{15} & 0 & 0 \\ e_{31}^P & e_{31}^P & 0 & 0 & 0 & 0 \end{bmatrix}, \qquad (2.3.43)$$

$$[\bar{\varepsilon}_{kl}] = \begin{bmatrix} \varepsilon_{11} & 0 & 0 \\ 0 & \varepsilon_{11} & 0 \\ 0 & 0 & \varepsilon_{33}^P \end{bmatrix}, \quad \varepsilon_{11} = \varepsilon_{11}^S = \varepsilon_{11}^T - d_{15}^2 / s_{55}^E. \qquad (2.3.44)$$

2.3.3.2 In-plane poling

Next consider a ceramic plate with in-plane poling along the x_1 direction (see Figure 2.3.3).

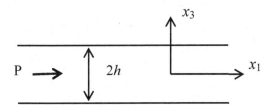

Figure 2.3.3. A ceramic plate with in-plane poling.

In this case the material matrices can be obtained by tensor transformation or reordering rows and columns of the matrices in Equation (2.1.37) properly, with the result [9]:

$$
\begin{pmatrix}
c_{33} & c_{13} & c_{13} & 0 & 0 & 0 \\
c_{13} & c_{11} & c_{12} & 0 & 0 & 0 \\
c_{13} & c_{12} & c_{11} & 0 & 0 & 0 \\
0 & 0 & 0 & c_{66} & 0 & 0 \\
0 & 0 & 0 & 0 & c_{44} & 0 \\
0 & 0 & 0 & 0 & 0 & c_{44}
\end{pmatrix},
$$

$$
\begin{bmatrix}
e_{33} & e_{31} & e_{31} & 0 & 0 & 0 \\
0 & 0 & 0 & 0 & 0 & e_{15} \\
0 & 0 & 0 & 0 & e_{15} & 0
\end{bmatrix},
\begin{pmatrix}
\varepsilon_{33} & 0 & 0 \\
0 & \varepsilon_{11} & 0 \\
0 & 0 & \varepsilon_{11}
\end{pmatrix}, \qquad (2.3.45)
$$

where the elements of the matrices have the same meaning as those in Equation (2.1.37). For example, c_{33} is always the stiffness in the poling direction. The indices of the elements of the matrices in Equation (2.1.37) represent the positions of the elements in the matrices, but the indices of the elements of the matrices in Equation (2.1.45) do not. The stress relaxation in Equation (2.3.15) yields the following results:

$$
[\gamma_{rs}] =
\begin{bmatrix}
\gamma_{11} & \gamma_{12} & 0 \\
\gamma_{12} & \gamma_{22} & 0 \\
0 & 0 & c_{44}
\end{bmatrix}, \quad
[\psi_{ks}] =
\begin{bmatrix}
\psi_{11} & \psi_{12} & 0 \\
0 & 0 & e_{15} \\
0 & 0 & 0
\end{bmatrix},
$$

$$
[\varsigma_{kj}] =
\begin{bmatrix}
\varsigma_{11} & 0 & 0 \\
0 & \varepsilon_{11} & 0 \\
0 & 0 & \varsigma_{33}
\end{bmatrix}, \quad r,s = 1,2,6, \quad k,j = 1,2,3,
\qquad (2.3.46)
$$

where

$$
\gamma_{11} = c_{33}^{E} - (c_{13}^{E})^{2} / c_{11}^{E}, \quad \gamma_{12} = c_{13}^{E} - c_{12}^{E} c_{13}^{E} / c_{11}^{E},
$$
$$
\gamma_{22} = c_{11}^{E} - (c_{12}^{E})^{2} / c_{11}^{E}, \quad c_{44} = c_{44}^{E} = 1 / s_{44}^{E},
$$
$$
\psi_{11} = e_{33} - e_{31} c_{13}^{E} / c_{11}^{E}, \quad \psi_{12} = e_{31} - e_{31} c_{12}^{E} / c_{11}^{E},
$$
$$
\varsigma_{11} = \varepsilon_{33}^{S} + e_{31}^{2} / c_{11}^{E}, \quad \varepsilon_{11} = \varepsilon_{11}^{S}, \quad \varsigma_{33} = \varepsilon_{11}^{S} + e_{15}^{2} / c_{44}^{E}.
\qquad (2.3.47)
$$

The constitutive relations then take the following form:

$$T_{11}^{(0)} = 2h(\gamma_{11}u_{1,1}^{(0)} + \gamma_{12}u_{2,2}^{(0)} - \psi_{11}E_1^{(0)}),$$
$$T_{22}^{(0)} = 2h(\gamma_{12}u_{1,1}^{(0)} + \gamma_{22}u_{2,2}^{(0)} - \psi_{12}E_1^{(0)}), \qquad (2.3.48)$$
$$T_{12}^{(0)} = 2h[c_{44}(u_{1,2}^{(0)} + u_{2,1}^{(0)}) - e_{15}E_2^{(0)}],$$

$$D_1^{(0)} = 2h[\psi_{11}u_{1,1}^{(0)} + \psi_{12}u_{2,2}^{(0)} + \zeta_{11}E_1^{(0)}],$$
$$D_2^{(0)} = 2h[e_{15}(u_{1,2}^{(0)} + u_{2,1}^{(0)}) + \varepsilon_{11}E_2^{(0)}], \qquad (2.3.49)$$
$$D_3^{(0)} = 2h\zeta_{33}E_3^{(0)},$$

$$D_1^{(1)} = -\frac{2}{3}h^3\varepsilon_{33}\phi_{,1}^{(1)},$$
$$\qquad (2.3.50)$$
$$D_2^{(1)} = -\frac{2}{3}h^3\varepsilon_{11}\phi_{,2}^{(1)}.$$

Substitution of the above into the equations of motion and charge gives:

$$\gamma_{11}u_{1,11}^{(0)} + c_{44}u_{1,22}^{(0)} + (\gamma_{12} + c_{44})u_{2,21}^{(0)}$$
$$+ \psi_{11}\phi_{,11}^{(0)} + e_{15}\phi_{,22}^{(0)} + \frac{1}{2h}F_1^{(0)} = \rho\ddot{u}_1^{(0)},$$
$$(\gamma_{12} + c_{44})u_{1,12}^{(0)} + c_{44}u_{2,11}^{(0)} + \gamma_{22}u_{2,22}^{(0)} \qquad (2.3.51)$$
$$+ (e_{15} + \psi_{12})\phi_{,12}^{(0)} + \frac{1}{2h}F_2^{(0)} = \rho\ddot{u}_2^{(0)},$$

$$-\zeta_{11}\phi_{,11}^{(0)} - \varepsilon_{11}\phi_{,22}^{(0)}$$
$$+ \psi_{11}u_{1,11}^{(0)} + e_{15}u_{1,22}^{(0)} + (\psi_{12} + e_{15})u_{2,12}^{(0)} + \frac{1}{2h}D^{(0)} = 0, \qquad (2.3.52)$$
$$-\varepsilon_{33}\phi_{,11}^{(1)} - \varepsilon_{11}\phi_{,22}^{(1)} + 3h^{-2}\zeta_{33}\phi^{(1)} + \frac{3}{2h^3}D^{(1)} = 0.$$

The stress relaxation in Equation (2.3.23) yields:

$$[\bar{c}_{pq}] = \begin{pmatrix} \bar{c}_{11} & \bar{c}_{12} & 0 & 0 & 0 & 0 \\ \bar{c}_{21} & \bar{c}_{22} & 0 & 0 & 0 & 0 \\ 0 & 0 & 0 & 0 & 0 & 0 \\ 0 & 0 & 0 & c_{66} & 0 & 0 \\ 0 & 0 & 0 & 0 & c_{44} & 0 \\ 0 & 0 & 0 & 0 & 0 & c_{44} \end{pmatrix}, \tag{2.3.53}$$

$$[\bar{e}_{iq}] = \begin{bmatrix} \bar{e}_{11} & \bar{e}_{12} & 0 & 0 & 0 & 0 \\ 0 & 0 & 0 & 0 & 0 & e_{15} \\ 0 & 0 & 0 & 0 & e_{15} & 0 \end{bmatrix}, \tag{2.3.54}$$

$$[\bar{\varepsilon}_{kl}] = \begin{bmatrix} \bar{\varepsilon}_{11} & 0 & 0 \\ 0 & \varepsilon_{11} & 0 \\ 0 & 0 & \varepsilon_{11} \end{bmatrix}, \tag{2.3.55}$$

where

$$\bar{c}_{11} = c_{33} - c_{13}^2 / c_{11}, \quad \bar{c}_{12} = c_{13} - c_{12}c_{13} / c_{11},$$
$$\bar{c}_{22} = c_{11} - c_{12}^2 / c_{11}, \quad \bar{e}_{11} = e_{33} - e_{31}c_{13} / c_{11}, \tag{2.3.56}$$
$$\bar{e}_{12} = e_{31} - e_{31}c_{12} / c_{11}, \quad \bar{\varepsilon}_{11} = \varepsilon_{33}^S + e_{31}^2 / c_{11}.$$

2.3.4 Radial vibration of a circular ceramic disk

Consider a circular disk of a piezoelectric ceramic poled in the thickness direction positioned in a coordinate system as shown in Figure 2.3.4. The faces of the disk are traction-free and are completely coated with electrodes. A voltage $V\exp(i\omega t)$ is applied across the electrodes. We consider axi-symmetric radial vibrations [3].

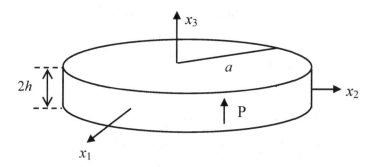

Figure 2.3.4. A circular ceramic plate with thickness poling.

The plate electric potentials for the applied voltage are

$$\phi^{(0)} = 0, \quad \phi^{(1)} = \frac{V}{2h}\exp(i\omega t).$$ (2.3.57)

Introduce the following vector notation:

$$\mathbf{u}^{(0)} = u_1^{(0)}\mathbf{i}_1 + u_2^{(0)}\mathbf{i}_2,$$
$$\nabla = \mathbf{i}_1\partial_1 + \mathbf{i}_2\partial_2, \quad \nabla^2 = \partial_1^2 + \partial_2^2.$$ (2.3.58)

Then Equation (2.3.40) can be written into coordinate independent form as

$$(c_{11}^p - c_{12}^p)h\nabla^2\mathbf{u}^{(0)} + (c_{11}^p + c_{12}^p)h\nabla(\nabla\cdot\mathbf{u}^{(0)})$$
$$+ 2he_{31}^p\nabla\phi^{(1)} + \left[T_{31}\mathbf{e}_1 + T_{32}\mathbf{e}_2\right]_{-h}^h = 2\rho h\ddot{\mathbf{u}}^{(0)}.$$ (2.3.59)

In polar coordinates

$$(\nabla^2\mathbf{u}^{(0)})_r = \frac{\partial^2 u_r^{(0)}}{\partial r^2} + \frac{1}{r}\frac{\partial u_r^{(0)}}{\partial r} - \frac{u_r^{(0)}}{r^2},$$
$$[\nabla(\nabla\cdot\mathbf{u}^{(0)})]_r = \frac{\partial^2 u_r^{(0)}}{\partial r^2} + \frac{1}{r}\frac{\partial u_r^{(0)}}{\partial r} - \frac{u_r^{(0)}}{r^2}.$$ (2.3.60)

Then for axi-symmetric motions Equation (2.3.59) becomes

$$c_{11}^P \left(\frac{\partial^2 u_r^{(0)}}{\partial r^2} + \frac{1}{r} \frac{\partial u_r^{(0)}}{\partial r} - \frac{u_r^{(0)}}{r^2} \right) = \rho \ddot{u}_r^{(0)}. \qquad (2.3.61)$$

The relevant resultants are

$$\begin{aligned}
T_{rr}^{(0)} &= 2h(c_{11}^P S_{rr}^{(0)} + c_{12}^P S_{\theta\theta}^{(0)} - e_{31}^P E_3^{(0)}) \\
&= 2h(c_{11}^P u_{r,r}^{(0)} + c_{12}^P u_r^{(0)} / r) + e_{31}^P V \exp(i\omega t), \qquad (2.3.62) \\
D_3^{(0)} &= 2h e_{31}^P (u_{r,r}^{(0)} + u_r^{(0)} / r) - \varepsilon_{33}^P V \exp(i\omega t).
\end{aligned}$$

For steady state motions, Equation (2.3.61) reduces to

$$u_{r,rr}^{(0)} + \frac{u_{r,r}^{(0)}}{r} + \left(\xi^2 - \frac{1}{r^2} \right) u_r^{(0)} = 0, \qquad (2.3.63)$$

where

$$\xi^2 = \frac{\omega^2}{(v^P)^2}, \quad (v^P)^2 = c_{11}^P / \rho. \qquad (2.3.64)$$

Equation (2.3.63) can be written as Bessel's equation of order one. For a solid disk, the motion at the origin is zero and the general solution is

$$u_r^{(0)} = B J_1(\xi r) \exp(i\omega t), \qquad (2.3.65)$$

where J_1 is the first kind Bessel function of the first order. Equation (2.3.65) is subject to the boundary condition

$$T_{rr}^{(0)} = 0, \quad r = a, \qquad (2.3.66)$$

hence Equation (2.3.66) requires that

$$c_{11}^P B \frac{dJ_1}{dr} \bigg|_{r=a} + c_{12}^P B \frac{J_1}{a} = -e_{31}^P \frac{V}{2h}, \qquad (2.3.67)$$

where, for convenience, the argument of the Bessel function is not written. From Equation (2.3.67) B can be expressed in terms of V as follows:

$$B = \left[(1 - \sigma^P) \frac{J_1(\xi a)}{a} - \xi J_0(\xi a) \right]^{-1} \frac{e_{31}^P}{c_{11}^P} \frac{V}{2h}, \qquad (2.3.68)$$

where

$$\frac{dJ_1(x)}{dx} = J_0(x) - \frac{J_1(x)}{x} \qquad (2.3.69)$$

has been used and

$$\sigma^P = c_{12}^P / c_{11}^P, \qquad (2.3.70)$$

which may be interpreted as a planar Poisson's ratio, since the material is isotropic in a plane with a normal along x_3. The total charge on the electrode at the bottom of the plate is given by

$$Q_e = \int_A \frac{1}{2h} D_3^{(0)} dA = 2\pi \int_0^a \frac{1}{2h} D_3^{(0)} r \, dr . \qquad (2.3.71)$$

Substitution of Equation (2.3.65) into Equation (2.3.62)$_2$ and then into Equation (2.3.71) yields

$$Q_e = 2\pi e_{31}^P a B J_1 (\xi a) - \pi \varepsilon_{33}^P V a^2 / 2h . \qquad (2.3.72)$$

Hence we obtain for the current that flows to the resonator

$$I = \frac{dQ_e}{dt} = i\omega \left[\frac{2(k_{31}^P)^2 J_1(\xi a)}{(1 - \sigma^P) J_1(\xi a) - \xi a J_0(\xi a)} - 1 \right] \frac{\varepsilon_{33}^P \pi a^2 V}{2h} , \qquad (2.3.73)$$

where

$$(k_{31}^P)^2 = \frac{(e_{31}^P)^2}{\varepsilon_{33}^P c_{11}^P} . \qquad (2.3.74)$$

At mechanical resonance, the applied voltage can be zero, and from Equation (2.3.67),

$$\left. \frac{dJ_1}{dr} \right|_{r=a} + \sigma^P \frac{J_1}{a} = 0 . \qquad (2.3.75)$$

Or, at the resonance frequency, the current goes to infinity. This condition is determined by setting the square bracketed factor in the denominator of Equation (2.3.68) equal to zero. The resulting equation is

$$\frac{\xi\, a J_0(\xi a)}{J_1(\xi a)} = 1 - \sigma^P, \tag{2.3.76}$$

which can be brought into the same form as (2.3.75). The antiresonance frequency results when the current goes to zero. The resulting equation is

$$\frac{\xi\, a J_0(\xi a)}{J_1(\xi a)} = 1 - \sigma^P - 2(k_{31}^p)^2. \tag{2.3.77}$$

The above results from plate equations are the same as the results in [3,9] obtained by directly making approximations in the three-dimensional equations.

2.4 First-Order Theory

In this section the first-order terms and equations are examined. A set of equations governing extension and flexure with shear deformations are obtained. These equations are probably the most widely used plate equations. The equations are specialized to the cases of ceramic and quartz plates.

2.4.1 Coupled extension, flexure and thickness-shear

By a first-order theory we mean a theory for coupled extensional ($u_a^{(0)}$, $a = 1, 2$), flexural ($u_3^{(0)}$) and thickness-shear ($u_a^{(1)}$) motions of a plate. For the electrical behavior of the plate we are interested in what is governed by $\phi^{(0)}$ and $\phi^{(1)}$. The approximate displacement and potential fields are

$$u_i \cong u_i^{(0)} + x_3 u_i^{(1)} + x_3^2 u_i^{(2)},$$
$$\phi \cong \phi^{(0)} + x_3 \phi^{(1)}. \tag{2.4.1}$$

Although we are mainly interested in $u_i^{(0)}$ and $u_a^{(1)}$, we have included a few additional displacement components in Equation (2.4.1). Among these additional displacements, $u_3^{(1)}$ and $u_3^{(2)}$ represent the thickness stretch or contraction accompanying extension and flexure due to Poisson's effect, and must be included. From Equation (2.2.6) it can be seen that $u_3^{(1)}$ together with $u_1^{(2)}$ and $u_2^{(2)}$ contribute to symmetric thickness-shear deformations $S_4^{(1)}$ and $S_5^{(1)}$, which may couple to the other first-order strains due to anisotropy and should be allowed. The two-dimensional plate equations we will obtain are for $u_a^{(0)}$, $u_3^{(0)}$ and $u_a^{(1)}$ only. Other displacements will be eliminated through stress relaxation procedures. Within the approximation in Equation (2.4.1), the strains and electric fields in Equations (2.2.5), (2.2.6), (2.2.8) and (2.2.9) become:

$$S_1^{(0)} = u_{1,1}^{(0)}, \quad S_2^{(0)} = u_{2,2}^{(0)}, \quad S_3^{(0)} = u_3^{(1)},$$
$$S_4^{(0)} = u_{3,2}^{(0)} + u_2^{(1)}, \quad S_5^{(0)} = u_{3,1}^{(0)} + u_1^{(1)}, \quad S_6^{(0)} = u_{1,2}^{(0)} + u_{2,1}^{(0)}, \tag{2.4.2}$$

$$S_1^{(1)} = u_{1,1}^{(1)}, \quad S_2^{(1)} = u_{2,2}^{(1)}, \quad S_3^{(1)} = 2u_3^{(2)},$$
$$S_4^{(1)} = u_{3,2}^{(1)} + 2u_2^{(2)}, \quad S_5^{(1)} = u_{3,1}^{(1)} + 2u_1^{(2)}, \quad S_6^{(1)} = u_{1,2}^{(1)} + u_{2,1}^{(1)}, \tag{2.4.3}$$

$$E_1^{(0)} = -\phi_{,1}^{(0)}, \quad E_2^{(0)} = -\phi_{,2}^{(0)}, \quad E_3^{(0)} = -\phi^{(1)}, \tag{2.4.4}$$

$$E_1^{(1)} = -\phi_{,1}^{(1)}, \quad E_2^{(1)} = -\phi_{,2}^{(1)}, \quad E_3^{(1)} = 0. \tag{2.4.5}$$

Higher order strains $S_{ij}^{(2)}$ are neglected. From Equation (2.2.17)$_1$, for $j = 1, 2, 3$ and $n = 0$, and for $j = 1, 2$ and $n = 1$, we obtain

$$T_{ab,a}^{(0)} + F_b^{(0)} = 2h\rho\ddot{u}_b^{(0)}, \quad a,b = 1,2,$$
$$T_{a3,a}^{(0)} + F_3^{(0)} = 2h\rho\ddot{u}_3^{(0)}, \tag{2.4.6}$$
$$T_{ab,a}^{(1)} - T_{3b}^{(0)} + F_b^{(1)} = \frac{2h^3}{3}\rho\ddot{u}_b^{(1)},$$

where we have truncated the right hand side by keeping the B_{00} term only in the summation in the zero-order equations, and the B_{11} term only in the summation in the first-order equations. Equation (2.4.6)$_1$ is for extension, Equation (2.4.6)$_2$ is for flexure, and Equation (2.4.6)$_3$ is for thickness-shear. The electrostatic equations are the same as those in the zero-order theory in the previous section. We have, for an unelectroded plate:

$$D_{a,a}^{(0)} + D^{(0)} = 0,$$
$$D_{a,a}^{(1)} - D_3^{(0)} + D^{(1)} = 0,$$

(2.4.7)

for a symmetrically electroded plate:

$$\phi^{(0)} + h\phi^{(1)} = V_1(t),$$
$$\phi^{(0)} - h\phi^{(1)} = V_2(t),$$

(2.4.8)

for a plate electroded at the upper surface:

$$\phi^{(0)} + h\phi^{(1)} = V_1(t),$$
$$-h(D_{a,a}^{(0)} + D^{(0)}) + D_{a,a}^{(1)} - D_3^{(0)} + D^{(1)} = 0,$$

(2.4.9)

and for a plate electroded at the lower surface:

$$h(D_{a,a}^{(0)} + D^{(0)}) + D_{a,a}^{(1)} - D_3^{(0)} + D^{(1)} = 0,$$
$$\phi^{(0)} - h\phi^{(1)} = V_2(t).$$

(2.4.10)

For plate constitutive relations, we truncate Equation (2.2.12) by keeping the B_{00} and B_{11} terms only in the summation:

$$T_{ij}^{(0)} = 2h(c_{ijkl}S_{kl}^{(0)} - e_{kij}E_k^{(0)}),$$
$$D_i^{(0)} = 2h(e_{ijk}S_{jk}^{(0)} + \varepsilon_{ij}E_j^{(0)}),$$

(2.4.11)

$$T_{ij}^{(1)} = \frac{2h^3}{3}(c_{ijkl}S_{kl}^{(1)} - e_{kij}E_k^{(1)}),$$
$$D_i^{(1)} = \frac{2h^3}{3}(e_{ijk}S_{jk}^{(1)} + \varepsilon_{ij}E_j^{(1)}),$$

(2.4.12)

where $S_{ij}^{(2)}$ have been neglected. Since $S_{ij}^{(0)}$ contains $u_3^{(1)}$ and $S_{ij}^{(1)}$ contains $u_3^{(1)}$ and $u_i^{(2)}$, Equations (2.4.11) and (2.4.12) are not yet ready to be used for a theory of extension, flexure and thickness-shear. To obtain the proper constitutive relations we proceed as follows [13,17]. For the zero-order constitutive relations we take the following stress relaxation to be approximately true:

$$T_{33}^{(0)} = 0.$$

(2.4.13)

Then, in the manner of the derivation of Equations (2.3.23) through (2.3.26), we obtain the relaxed zero-order constitutive relations:

$$T_{ij}^{(0)} = 2h(\bar{c}_{ijkl} S_{kl}^{(0)} - \bar{e}_{kij} E_k^{(0)}),$$
$$D_i^{(0)} = 2h(\bar{e}_{ikl} S_{kl}^{(0)} + \bar{\varepsilon}_{ij} E_j^{(0)}),$$

(2.4.14)

where

$$\bar{c}_{ijkl} = c_{ijkl} - c_{ij33} c_{33kl} / c_{3333},$$
$$\bar{e}_{kij} = e_{kij} - e_{k33} c_{33ij} / c_{3333},$$
$$\bar{\varepsilon}_{ij} = \varepsilon_{ij} + e_{i33} e_{j33} / c_{3333}.$$

(2.4.15)

Since we use a linear function $x_3 u_a^{(1)}$ as an approximation for the fundamental thickness-shear mode in this first-order theory, which is approximate in dynamic problems where the fundamental thickness-shear displacement has a sinusoidal variation through the plate thickness, following [13,17], we introduce two shear correction factors κ_1 and κ_2 as follows. First a two-dimensional electric enthalpy can be defined based on Equation (2.4.14). Then we replace the following zero-order strains in the two-dimensional enthalpy function:

$$S_{31}^{(0)} \to \kappa_1 S_{31}^{(0)}, \quad S_{32}^{(0)} \to \kappa_2 S_{32}^{(0)}.$$

(2.4.16)

This modifies $\bar{c}_{ijkl}$ and $\bar{e}_{ijk}$ into $\bar{c}_{ijkl}'$ and $\bar{e}_{ijk}'$ in the zero-order constitutive relations. Then Equation (2.4.14) takes the following form:

$$T_p^{(0)} = 2h(\bar{c}_{pq}' S_q^{(0)} - \bar{e}_{kp}' E_k^{(0)}),$$
$$D_i^{(0)} = 2h(\bar{e}_{iq}' S_q^{(0)} + \bar{\varepsilon}_{ij} E_j^{(0)}).$$

(2.4.17)

The structures of the $\bar{c}_{pq}'$ and $\bar{e}_{iq}'$ matrices are

$$[\bar{c}_{pq}'] = \begin{pmatrix} \bar{c}_{11} & \bar{c}_{12} & 0 & \kappa_2 \bar{c}_{14} & \kappa_1 \bar{c}_{15} & \bar{c}_{16} \\ \bar{c}_{21} & \bar{c}_{22} & 0 & \kappa_2 \bar{c}_{24} & \kappa_1 \bar{c}_{25} & \bar{c}_{26} \\ 0 & 0 & 0 & 0 & 0 & 0 \\ \kappa_2 \bar{c}_{41} & \kappa_2 \bar{c}_{42} & 0 & \kappa_2 \kappa_2 \bar{c}_{44} & \kappa_1 \kappa_2 \bar{c}_{45} & \kappa_2 \bar{c}_{46} \\ \kappa_1 \bar{c}_{51} & \kappa_1 \bar{c}_{52} & 0 & \kappa_1 \kappa_2 \bar{c}_{54} & \kappa_1 \kappa_1 \bar{c}_{55} & \kappa_1 \bar{c}_{56} \\ \bar{c}_{61} & \bar{c}_{62} & 0 & \kappa_2 \bar{c}_{64} & \kappa_1 \bar{c}_{65} & \bar{c}_{66} \end{pmatrix},$$

(2.4.18)

$$[\bar{e}'_{iq}] = \begin{bmatrix} \bar{e}_{11} & \bar{e}_{12} & 0 & \kappa_2\bar{e}_{14} & \kappa_1\bar{e}_{15} & \bar{e}_{16} \\ \bar{e}_{21} & \bar{e}_{22} & 0 & \kappa_2\bar{e}_{24} & \kappa_1\bar{e}_{25} & \bar{e}_{26} \\ \bar{e}_{31} & \bar{e}_{32} & 0 & \kappa_2\bar{e}_{34} & \kappa_1\bar{e}_{35} & \bar{e}_{36} \end{bmatrix}. \tag{2.4.19}$$

The two correction factors should be determined by requiring the two fundamental thickness-shear resonant frequencies obtained from the two-dimensional plate equations to be equal to the corresponding exact frequencies predicted by the three-dimensional equations. With shear correction factors thus determined, the two-dimensional plate equations and the exact three-dimensional equations yield the same frequencies for a particular motion, i.e., the thickness-shear vibration of a plate in the two fundamental thickness-shear modes. This is particularly important to the analysis of piezoelectric devices operating with thickness-shear modes.

For the first-order constitutive relations, we set [13,17]

$$T_q^{(1)} = 0, \quad q = 3,4,5. \tag{2.4.20}$$

Then, similar to the derivation of Equations (2.3.17) through (2.3.20), we obtain the relaxed first-order constitutive relations:

$$
\begin{aligned}
T_r^{(1)} &= \frac{2h^3}{3}(\gamma_{rs}S_s^{(1)} - \psi_{kr}E_k^{(1)}), \\
D_i^{(1)} &= \frac{2h^3}{3}(\psi_{is}S_s^{(1)} + \varsigma_{ij}E_j^{(1)}),
\end{aligned}
\tag{2.4.21}
$$

where

$$
\begin{aligned}
\gamma_{rs} &= c_{rs} - c_{rv}c_{vw}^{-1}c_{ws}, & r,s &= 1,2,6, \\
\psi_{ks} &= e_{ks} - e_{kw}c_{wv}^{-1}c_{vs}, & v,w &= 3,4,5, \\
\varsigma_{kj} &= \varepsilon_{kj} + e_{kv}c_{vw}^{-1}e_{jw}, & j,k &= 1,2,3.
\end{aligned}
\tag{2.4.22}
$$

In summary, in the case of an unelectroded plate, we have obtained

$$T_{ab,a}^{(0)} + F_b^{(0)} = 2h\rho\ddot{u}_b^{(0)}, \quad a,b = 1,2,$$

$$T_{a3,a}^{(0)} + F_3^{(0)} = 2h\rho\ddot{u}_3^{(0)},$$

$$T_{ab,a}^{(1)} - T_{3b}^{(0)} + F_b^{(1)} = \frac{2h^3}{3}\rho\ddot{u}_b^{(1)}, \qquad (2.4.23)$$

$$D_{a,a}^{(0)} + D^{(0)} = 0,$$

$$D_{a,a}^{(1)} - D_3^{(0)} + D^{(1)} = 0,$$

$$T_{ij}^{(0)} = 2h(\bar{c}_{ijkl}'S_{kl}^{(0)} - \bar{e}_{kij}'E_k^{(0)}),$$

$$D_i^{(0)} = 2h(\bar{e}_{ikl}'S_{kl}^{(0)} + \bar{\varepsilon}_{ij}E_j^{(0)}),$$

$$T_r^{(1)} = \frac{2h^3}{3}(\gamma_{rs}S_s^{(1)} - \psi_{kr}E_k^{(1)}), \quad r,s = 1,2,6, \qquad (2.4.24)$$

$$D_i^{(1)} = \frac{2h^3}{3}(\psi_{is}S_s^{(1)} + \zeta_{ij}E_j^{(1)}),$$

$$S_1^{(0)} = u_{1,1}^{(0)}, \quad S_2^{(0)} = u_{2,2}^{(0)},$$

$$S_4^{(0)} = u_{3,2}^{(0)} + u_2^{(1)}, \quad S_5^{(0)} = u_{3,1}^{(0)} + u_1^{(1)}, \quad S_6^{(0)} = u_{1,2}^{(0)} + u_{2,1}^{(0)},$$

$$S_1^{(1)} = u_{1,1}^{(1)}, \quad S_2^{(1)} = u_{2,2}^{(1)}, \quad S_6^{(1)} = u_{1,2}^{(1)} + u_{2,1}^{(1)}, \qquad (2.4.25)$$

$$E_1^{(0)} = -\phi_{,1}^{(0)}, \quad E_2^{(0)} = -\phi_{,2}^{(0)}, \quad E_3^{(0)} = -\phi^{(1)},$$

$$E_1^{(1)} = -\phi_{,1}^{(1)}, \quad E_2^{(1)} = -\phi_{,2}^{(1)}, \quad E_3^{(1)} = 0.$$

With successive substitutions from Equations (2.4.24) and (2.4.25), Equations (2.4.23) can be written as seven equations for $u_i^{(0)}$, $u_a^{(1)}$, $\phi^{(0)}$ and $\phi^{(1)}$. At the boundary of a plate with an in-plane unit exterior normal **n** and an in-plane unit tangent **s** (see Figure 2.2.1), we may prescribe

$$T_{nn}^{(0)} \text{ or } u_n^{(0)}, \quad T_{ns}^{(0)} \text{ or } u_s^{(0)}, \quad T_{n3}^{(0)} \text{ or } u_3^{(0)},$$

$$T_{nn}^{(1)} \text{ or } u_n^{(1)}, \quad T_{ns}^{(1)} \text{ or } u_s^{(1)}, \qquad (2.4.26)$$

$$D_n^{(0)} \text{ or } \phi^{(0)}, \quad D_n^{(1)} \text{ or } \phi^{(1)}.$$

When electrodes are present the differential equations are fewer, and so are the boundary conditions. We note that Equations (2.4.23) through (2.4.25) do not contain $S_3^{(0)}$ and $S_w^{(1)}$ (w = 3, 4, 5), and the related displacement components $u_3^{(1)}$ and $u_i^{(2)}$. These strain and displacements can be determined approximately from the stress relaxation conditions in Equations (2.4.13) and (2.4.20) once the solution to Equation (2.4.23) through (2.4.25) has been obtained.

2.4.2 Flexural and thickness-shear waves

To examine the basic behavior of the equations obtained, we study the special case of an isotropic, elastic plate for which

$$[c_{pq}] = \begin{pmatrix} c_{11} & c_{12} & c_{12} & 0 & 0 & 0 \\ c_{12} & c_{11} & c_{12} & 0 & 0 & 0 \\ c_{12} & c_{12} & c_{11} & 0 & 0 & 0 \\ 0 & 0 & 0 & c_{44} & 0 & 0 \\ 0 & 0 & 0 & 0 & c_{44} & 0 \\ 0 & 0 & 0 & 0 & 0 & c_{44} \end{pmatrix},$$

$$c_{11} = \lambda + 2\mu,$$
$$c_{12} = \lambda,$$
$$c_{44} = (c_{11} - c_{12})/2,$$
$$[e_{ip}] = 0,$$

$$(2.4.27)$$

where λ and μ are Lamè constants. In this case extension is not coupled to flexure and thickness-shear. Consider coupled flexure $u_3^{(0)}(x_1,t)$ and thickness-shear in the x_1 direction $u_1^{(1)}(x_1,t)$. The relevant equations reduce to

$$2h\mu\kappa^2(u_{3,11}^{(0)} + u_{1,1}^{(1)}) = 2h\rho\ddot{u}_3^{(0)},$$

$$Du_{1,11}^{(1)} - 2h\mu\kappa^2(u_{3,1}^{(0)} + u_1^{(1)}) = \frac{2h^3}{3}\rho\ddot{u}_1^{(1)},$$

$$(2.4.28)$$

where [13]

$$D = \frac{2h^3 E}{3(1 - v^2)}, \quad \kappa = \kappa_1 = \kappa_2 = \frac{\pi}{\sqrt{12}}. \quad (2.4.29)$$

D is the bending stiffness of the plate. E is Young's modulus. v is Poisson's ratio. We look for simple wave solutions

$$\begin{aligned}
u_1^{(1)} &= A_1 \exp i(\xi x_1 - \omega t), \\
u_3^{(0)} &= A_3 \exp i(\xi x_1 - \omega t),
\end{aligned} \quad (2.4.30)$$

where A_1 and A_3 are constants. Substitution of Equation (2.4.30) into Equation (2.4.28) results in two linear equations for A_1 and A_3. For nontrivial solutions the determinant of the coefficient matrix of the linear equations has to vanish. This yields the following equation that determines the dispersion relations for coupled flexural and thickness-shear waves:

$$3\Omega^4 - 3\Omega^2 - 3\Omega^2 \hat{\gamma}_{11} X^2 - X^2 \Omega^2 + \hat{\gamma}_{11} X^4 = 0, \quad (2.4.31)$$

where

$$X = \xi h, \quad \Omega = \omega / \omega_\infty,$$

$$\omega_\infty^2 = \frac{\pi^2}{4h^2} \frac{\mu}{\rho}, \quad \hat{\gamma}_{11} = \frac{8}{\pi^2} \frac{1}{1 - v}. \quad (2.4.32)$$

X is a dimensionless wave number. Ω is a dimensionless frequency, normalized by the exact fundamental thickness-shear frequency ω_∞. There are two branches of dispersion relations. For short waves with a small X, the asymptotic expressions of the two dispersion relations are

$$\Omega^2 \cong \frac{\hat{\gamma}_{11}}{3} X^4, \quad \text{flexure}, \quad (2.4.33)$$

and

$$\Omega^2 \approx 1 + (\hat{\gamma}_{11} + \frac{1}{3}) X^2, \quad \text{thickness-shear}. \quad (2.4.34)$$

The two dispersion curves are qualitatively shown in Figure 2.4.1. With the correction factor the plate equations predict the exact cutoff frequency at $\Omega = 1$.

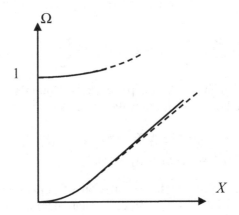

Figure 2.4.1. Dispersion curves for coupled flexural and thickness-shear
waves. Dotted lines: from plate equations. Solid lines: exact.

2.4.3 Reduction to classical flexure

For many applications in which the flexural motion is dominant, the
Kirchhoff classical theory of flexure without shear deformation is
sufficient.

2.4.3.1 Elimination of thickness-shear

To reduce the above first-order theory for flexure with shear
deformation to the classical theory of flexure, we take the plate
thickness-shear strains $S_{a3}^{(0)}$ to vanish [13], and then from Equation
(2.4.25)

$$u_a^{(1)} = -u_{3,a}^{(0)} \quad \Rightarrow \quad S_{ab}^{(1)} = -u_{3,ab}^{(0)}. \qquad (2.4.35)$$

Equation (2.4.35) enables us to eliminate $u_a^{(1)}$ in Equation (2.4.25).
Another approximation we need to make to obtain the classical theory is
to neglect the rotatory inertia $\rho 2h^3/3$ in Equation (2.4.23)$_3$. Then
Equation (2.4.23)$_3$ takes the following form:

$$T_{ab,a}^{(1)} - T_{3b}^{(0)} + F_b^{(1)} = 0, \qquad (2.4.36)$$

from which we can solve for $T_{3b}^{(0)}$ and substitute the result into Equation (2.4.23)$_2$ to obtain the equation for classical flexure

$$T_{ab,ab}^{(1)} + F_{b,b}^{(1)} + F_3^{(0)} = 2h\rho\ddot{u}_3^{(0)}. \tag{2.4.37}$$

In summary, the equations for extension and classical flexure are:

$$
\begin{aligned}
&T_{ab,a}^{(0)} + F_b^{(0)} = 2h\rho\ddot{u}_b^{(0)}, \quad a,b = 1,2, \\
&T_{a3,a}^{(0)} + F_3^{(0)} = 2h\rho\ddot{u}_3^{(0)}, \\
&D_{a,a}^{(0)} + D^{(0)} = 0, \\
&D_{a,a}^{(1)} - D_3^{(0)} + D^{(1)} = 0,
\end{aligned}
\tag{2.4.38}
$$

$$
\begin{aligned}
&T_r^{(0)} = 2h(\bar{c}_{rs}' S_s^{(0)} - \bar{e}_{kr}' E_k^{(0)}), \quad r,s = 1,2,6, \\
&D_i^{(0)} = 2h(\bar{e}_{is}' S_s^{(0)} + \bar{\varepsilon}_{ij} E_j^{(0)}), \\
&T_{3b}^{(0)} = T_{ab,a}^{(1)} + F_b^{(1)}, \\
&T_r^{(1)} = \frac{2h^3}{3}(\gamma_{rs} S_s^{(1)} - \psi_{kr} E_k^{(1)}), \\
&D_i^{(1)} = \frac{2h^3}{3}(\psi_{is} S_s^{(1)} + \zeta_{ij} E_j^{(1)}),
\end{aligned}
\tag{2.4.39}
$$

$$
\begin{aligned}
&S_1^{(0)} = u_{1,1}^{(0)}, \quad S_2^{(0)} = u_{2,2}^{(0)}, \quad S_6^{(0)} = u_{1,2}^{(0)} + u_{2,1}^{(0)}, \\
&S_{ab}^{(1)} = -u_{3,ab}^{(0)}, \\
&E_1^{(0)} = -\phi_{,1}^{(0)}, \quad E_2^{(0)} = -\phi_{,2}^{(0)}, \quad E_3^{(0)} = -\phi^{(1)}, \\
&E_1^{(1)} = -\phi_{,1}^{(1)}, \quad E_2^{(1)} = -\phi_{,2}^{(1)}, \quad E_3^{(1)} = 0.
\end{aligned}
\tag{2.4.40}
$$

With successive substitutions from Equations (2.4.39) and (2.4.40), Equations (2.4.38) can be written as five equations for $u_i^{(0)}$, $\phi^{(0)}$ and $\phi^{(1)}$. At the boundary of a plate with an in-plane unit exterior normal $\mathbf{n}$ and an in-plane unit tangent $\mathbf{s}$ (see Figure 2.2.1), we may prescribe

$$T_{nn}^{(0)} \quad \text{or} \quad u_n^{(0)}, \quad T_{ns}^{(0)} \quad \text{or} \quad u_s^{(0)},$$
$$T_{n3}^{(0)} + T_{ns,s}^{(1)} \quad \text{or} \quad u_3^{(0)}, \quad T_{nn}^{(1)} \quad \text{or} \quad u_{3,n}^{(0)}. \tag{2.4.41}$$

2.4.3.2 Dispersion curves of flexural waves

The above procedure for reduction to classical flexure is more readily seen in the special case of isotropic materials. We begin with the following equations from Equation (2.4.28) for coupled flexure and thickness-shear:

$$2h\mu\kappa^2(u_{3,11}^{(0)} + u_{1,1}^{(1)}) = 2h\rho\ddot{u}_3^{(0)},$$
$$Du_{1,11}^{(1)} - 2h\mu\kappa^2(u_{3,1}^{(0)} + u_1^{(1)}) = \frac{2h^3}{3}\rho\ddot{u}_1^{(1)}. \tag{2.4.42}$$

First we set the rotatory inertia to zero in $(2.4.42)_2$. This results in

$$Du_{1,11}^{(1)} - 2h\mu\kappa^2(u_{3,1}^{(0)} + u_1^{(1)}) = 0. \tag{2.4.43}$$

The elimination of $2h(\mu u_{3,1}^{(0)} + u_1^{(1)})$ from Equations (2.4.43) and $(2.4.42)_1$ yields

$$Du_{1,111}^{(1)} = 2h\rho\ddot{u}_3^{(0)}. \tag{2.4.44}$$

Next we set the plate shear strain to zero

$$u_{3,1}^{(0)} + u_1^{(1)} = 0, \tag{2.4.45}$$

with which we can express $u_1^{(1)}$ in terms of $u_3^{(0)}$ in Equation (2.4.44)

$$-Du_{3,1111}^{(0)} = 2h\rho\ddot{u}_3^{(0)}, \tag{2.4.46}$$

which is the well-known equation for classical flexure. In Equation (2.4.46), letting

$$u_3^{(0)} = \exp i(\xi x_1 - \omega t), \tag{2.4.47}$$

we obtain the dispersion relation for flexural waves as

$$\omega^2 = \frac{D}{2h\rho}\xi^4,$$ (2.4.48)

or

$$\Omega^2 = \frac{\hat{\gamma}_{11}}{3}X^4,$$ (2.4.49)

which is exactly the asymptotic expression in Equation (2.4.33). The dispersion curve determined by Equation (2.4.49) is plotted in Figure 2.4.2 in a dotted line, with a qualitative comparison to the result of Equation (2.4.31). It can be seen that the dispersion curve from the classical theory for flexure, and the flexural branch of the dispersion curves from the theory for coupled flexure and thickness-shear, agree for short waves at low frequencies.

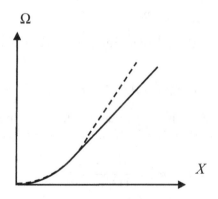

Figure 2.4.2. Dispersion curves for flexure. Dotted line: Classical flexure. Solid line: Coupled flexure and thickness-shear.

2.4.4 Thickness-shear approximation

In vibrations of a finite plate, flexure and thickness-shear deformations are usually coupled, whether the plate is isotropic or anisotropic. In piezoelectric devices, thickness-shear vibrations are often used as the operating modes. These modes are dominated by thickness-shear deformations, with a small coupling to flexure. There is a procedure for eliminating the weak coupling to flexure from the equations for coupled flexure and thickness-shear to obtain equations for the dominating thickness-shear waves only. This procedure is called the thickness-shear approximation [27], and is particularly useful in analyzing piezoelectric devices. Thickness-shear approximation was introduced in the analysis of quartz thickness-shear piezoelectric devices [27]. The procedure can be well explained in the case of an isotropic plate, which will be presented below. We begin with the equations for coupled flexure and thickness-shear (Equation (2.4.28))

$$2h\mu\kappa^2(u_{3,11}^{(0)} + u_{1,1}^{(1)}) = 2h\rho\ddot{u}_3^{(0)},$$

$$Du_{1,11}^{(1)} - 2h\mu\kappa^2(u_{3,1}^{(0)} + u_1^{(1)}) = \frac{2h^3}{3}\rho\ddot{u}_1^{(1)}.$$

(2.4.50)

Let

$$u_1^{(1)} = A_1 \exp i(\xi x_1 - \omega t),$$

$$u_3^{(0)} = A_3 \exp i(\xi x_1 - \omega t).$$

(2.4.51)

Substituting Equation (2.4.51) into Equation (2.4.50)$_1$, we obtain

$$\mu\kappa^2(-A_3\xi^2 + A_1 i\xi) = -\rho\omega^2 A_3.$$

(2.4.52)

For long waves we drop the term quadratic in ξ. For frequencies close to the lowest thickness-shear frequency ω_∞ we set

$$\omega^2 \approx \omega_\infty^2 = \frac{\pi^2}{4h^2}\frac{\mu}{\rho}$$

(2.4.53)

in Equation (2.4.52). Then Equation (2.4.52) becomes

$$A_3 = -\frac{h^2}{3}A_1 i\xi,$$

(2.4.54)

which is equivalent to the differential relation

$$u_3^{(0)} = -\frac{h^2}{3}u_{1,1}^{(1)}.$$

(2.4.55)

Substituting Equation (2.4.55) into Equation (2.4.50)$_2$ we arrive at

$$(D + \frac{2h^3}{3}\mu\kappa^2)u_{1,11}^{(1)} - 2h\mu\kappa^2 u_1^{(1)} = \frac{2h^3}{3}\rho\ddot{u}_1^{(1)}, \qquad (2.4.56)$$

which is an equation for the thickness-shear displacement $u_1^{(1)}$. In Equation (2.4.56), letting

$$u_1^{(1)} = \exp i(\xi x_1 - \omega t), \qquad (2.4.57)$$

we obtain

$$\Omega^2 = 1 + (\hat{\gamma}_{11} + \frac{1}{3})X^2, \qquad (2.4.58)$$

which is exactly the asymptotic expression in Equation (2.4.34). The dispersion curve determined by Equation (2.4.58) is plotted in Figure 2.4.3 in a dotted line, with a qualitative comparison to the result of Equation (2.4.34). It can be seen that the dispersion relation of the equation with the thickness-shear approximation and the thickness-shear branch from the theory for coupled flexure and thickness-shear agree for short waves with frequencies near $\Omega = 1$. Note that in Figure 2.4.3 we have also included the case of imaginary wave numbers. The complex branch shows that the thickness-shear approximation is good near $\Omega = 1$. Its low frequency behavior is not accurate.

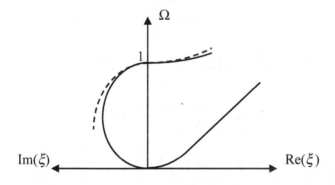

Figure 2.4.3. Dispersion curves for thickness-shear and flexure. The dotted line represents the thickness-shear approximation. The solid line is the coupled flexure and thickness-shear.

2.4.5 Equations for ceramic plates

We consider two cases of ceramic plates with thickness and in-plane poling, respectively.

2.4.5.1 Thickness poling

First consider a ceramic plate poled in the thickness direction (see Figure 2.4.4).

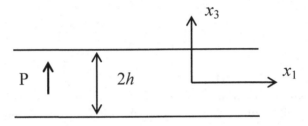

Figure 2.4.4. A ceramic plate with thickness poling.

From the material matrices in Equations (2.3.35) and (2.3.42) through (2.3.44), the constitutive relations take the following form:

$$T_{11}^{(0)} = 2h(\bar{c}_{11}u_{1,1}^{(0)} + \bar{c}_{12}u_{2,2}^{(0)} - \bar{e}_{31}E_3^{(0)}),$$

$$T_{22}^{(0)} = 2h(\bar{c}_{12}u_{1,1}^{(0)} + \bar{c}_{11}u_{2,2}^{(0)} - \bar{e}_{31}E_3^{(0)}), \qquad (2.4.59)$$

$$T_{12}^{(0)} = 2hc_{66}(u_{1,2}^{(0)} + u_{2,1}^{(0)}),$$

$$T_{3a}^{(0)} = 2h[\kappa^2 c_{44}(u_{3,a}^{(0)} + u_a^{(1)}) - \kappa e_{15}E_a^{(0)}], \qquad (2.4.60)$$

$$D_a^{(0)} = 2h[\kappa e_{15}(u_{3,a}^{(0)} + u_a^{(1)}) + \varepsilon_{11}E_a^{(0)}], \qquad (2.4.61)$$

$$D_3^{(0)} = 2h(\bar{e}_{31}u_{a,a}^{(0)} + \varepsilon_{33}^P E_3^{(0)}), \qquad (2.4.62)$$

$$T_{11}^{(1)} = \frac{2}{3} h^3 (\overline{c}_{11} u_{1,1}^{(1)} + \overline{c}_{12} u_{2,2}^{(1)}),$$

$$T_{22}^{(1)} = \frac{2}{3} h^3 (\overline{c}_{12} u_{1,1}^{(1)} + \overline{c}_{11} u_{2,2}^{(1)}), \tag{2.4.63}$$

$$T_{12}^{(1)} = \frac{2}{3} h^3 c_{66} (u_{1,2}^{(1)} + u_{2,1}^{(1)}),$$

$$D_a^{(1)} = -\frac{2}{3} h^3 \varepsilon_{11}^p \phi_{,a}^{(1)}. \tag{2.4.64}$$

Substitution of the above into the equations of motion and charge gives the equations for extension:

$$\overline{c}_{11} u_{1,11}^{(0)} + c_{66} u_{1,22}^{(0)} + (\overline{c}_{12} + c_{66}) u_{2,21}^{(0)} + \overline{e}_{31} \phi_{,1}^{(1)}$$

$$+ \frac{1}{2h} F_1^{(0)} = \rho \ddot{u}_1^{(0)},$$

$$c_{66} u_{2,11}^{(0)} + \overline{c}_{11} u_{2,22}^{(0)} + (\overline{c}_{12} + c_{66}) u_{1,12}^{(0)} + \overline{e}_{31} \phi_{,2}^{(1)} \tag{2.4.65}$$

$$+ \frac{1}{2h} F_2^{(0)} = \rho \ddot{u}_2^{(0)},$$

flexure:

$$\kappa^2 c_{44} (u_{3,aa}^{(0)} + u_{a,a}^{(1)}) + \kappa e_{15} \phi_{,aa}^{(0)} + \frac{1}{2h} F_3^{(0)} = \rho \ddot{u}_3^{(0)}, \tag{2.4.66}$$

thickness-shear:

$$\overline{c}_{11} u_{1,11}^{(1)} + c_{66} u_{1,22}^{(1)} + (\overline{c}_{12} + c_{66}) u_{2,21}^{(1)}$$

$$- 3h^{-2} \kappa^2 c_{44} (u_1^{(1)} + u_{3,1}^{(0)}) - 3h^{-2} \kappa e_{15} \phi_{,1}^{(0)}$$

$$+ \frac{3}{2h^3} F_1^{(1)} = \rho \ddot{u}_1^{(1)},$$

$$c_{66} u_{2,11}^{(1)} + \overline{c}_{11} u_{2,22}^{(1)} + (\overline{c}_{12} + c_{66}) u_{1,12}^{(1)} \tag{2.4.67}$$

$$- 3h^{-2} \kappa^2 c_{44} (u_2^{(1)} + u_{3,2}^{(0)}) - 3h^{-2} \kappa e_{15} \phi_{,2}^{(0)}$$

$$+ \frac{3}{2h^3} F_2^{(1)} = \rho \ddot{u}_2^{(1)},$$

and electrostatics:

$$- \varepsilon_{11} \phi_{,aa}^{(0)} + \kappa e_{15} (u_{3,aa}^{(0)} + u_{a,a}^{(1)}) + \frac{1}{2h} D^{(0)} = 0,$$

$$- \varepsilon_{11}^p \phi_{,aa}^{(1)} + 3h^{-2} \overline{\varepsilon}_{33} \phi^{(1)} - 3h^{-2} \overline{e}_{31} u_{a,a}^{(0)} + \frac{3}{2h^3} D^{(1)} = 0. \tag{2.4.68}$$

The thickness-shear approximation of a ceramic plate is often used for device modeling [28]. The weak flexural deformation accompanying thickness-shear is eliminated as follows [28], which simplifies the equations. We consider the case when there are no surface loads. This means that the major surfaces of the plate are traction-free, unelectroded and are without surface free charge. From Equations (2.4.66) and (2.4.68)$_1$, we obtain, by eliminating $\phi^{(0)}$,

$$\kappa^2 (c_{44} + \frac{e_{15}^2}{\varepsilon_{11}})(u_{3,aa}^{(0)} + u_{a,a}^{(1)}) = \rho \ddot{u}_3^{(0)} . \tag{2.4.69}$$

Consider the following wave solution:

$$u_3^{(0)} = A_3 \exp i(\xi_a x_a + \omega t),$$

$$u_b^{(1)} = A_b \exp i(\xi_a x_a + \omega t), \tag{2.4.70}$$

where A_i are constants. Substitution of Equation (2.4.70) into Equation (2.4.69) results in:

$$\kappa^2 (c_{44} + \frac{e_{15}^2}{\varepsilon_{11}})(-A_3 \xi_a \xi_a + A_a i \xi_a) = -\rho \omega^2 A_3 . \tag{2.4.71}$$

We are interested in long waves with small wave numbers ξ_a. The term quadratic in ξ_a in the above equation can be dropped. Also, since for long thickness-shear waves the frequency ω is very close to the exact pure thickness-shear frequency ω_∞ of an infinite plate, we make the following substitution in Equation (2.4.71):

$$\omega^2 \approx \omega_\infty^2 = \frac{3\kappa^2 c_{44}}{\rho h^2}, \tag{2.4.72}$$

where ω_∞ is determined from Equation (2.4.67)$_1$ by dropping all x_1 and x_2 dependence as well as the surface load. The approximate version of Equation (2.4.71) is then

$$A_3 = -\frac{h^2}{3}(1 + \frac{e_{15}^2}{c_{44}\varepsilon_{11}})i\xi_a A_a,$$ (2.4.73)

which is equivalent to the differential relation

$$u_3^{(0)} = -\frac{h^2}{3}(1 + k_{15}^2)u_{a,a}^{(1)},$$ (2.4.74)

where we have denoted

$$k_{15}^2 = \frac{e_{15}^2}{c_{44}\varepsilon_{11}}.$$ (2.4.75)

Substituting Equation (2.4.74) into Equations (2.4.67) and (2.4.68)$_1$, neglecting the third derivatives of $u_a^{(1)}$ under the long wave approximation, we obtain the following equations under the thickness-shear approximation:

$$c_{11}^* u_{1,11}^{(1)} + c_{66} u_{1,22}^{(1)} + c_{12}^* u_{2,12}^{(1)} - \rho\omega_\infty^2 u_1^{(1)} - 3h^{-2}\kappa e_{15}\phi_{,1}^{(0)} = \rho\ddot{u}_1^{(1)},$$

$$c_{66} u_{2,11}^{(1)} + c_{11}^* u_{2,22}^{(1)} + c_{12}^* u_{1,12}^{(1)} - \rho\omega_\infty^2 u_2^{(1)} - 3h^{-2}\kappa e_{15}\phi_{,2}^{(0)} = \rho\ddot{u}_2^{(1)},$$ (2.4.76)

$$-\varepsilon_{11}\phi_{,aa}^{(0)} + \kappa e_{15}u_{a,a}^{(1)} = 0,$$

where

$$c_{11}^* = \bar{c}_{11} + \kappa^2 c_{44}(1 + k_{15}^2),$$

$$c_{12}^* = \bar{c}_{12} + c_{66} + \kappa^2 c_{44}(1 + k_{15}^2).$$ (2.4.77)

Under the thickness-shear approximation, we approximately have

$$T_{3a}^{(0)} = 2h(\kappa^2 c_{44}u_a^{(1)} - \kappa e_{15}E_a^{(0)}),$$

$$D_a^{(0)} = 2h(\kappa e_{15}u_a^{(1)} + \varepsilon_{11}E_a^{(0)}).$$ (2.4.78)

2.4.5.2 In-plane poling

Next consider the case of a ceramic plate with in-plane poling in the x_1 direction (see Figure 2.4.5).

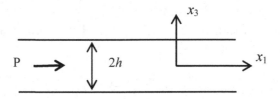

Figure 2.4.5. A ceramic plate with in-plane poling.

From the material matrices in Equations (2.3.53) through (2.3.55) and (2.3.46), the constitutive relations take the following form:

$$T_{11}^{(0)} = 2h(\bar{c}_{11}u_{1,1}^{(0)} + \bar{c}_{12}u_{2,2}^{(0)} - \bar{e}_{11}E_1^{(0)}),$$

$$T_{22}^{(0)} = 2h(\bar{c}_{12}u_{1,1}^{(0)} + \bar{c}_{22}u_{2,2}^{(0)} - \bar{e}_{12}E_1^{(0)}),$$

$$T_{23}^{(0)} = 2h\kappa_2^2 c_{66}(u_{3,2}^{(0)} + u_2^{(1)}), \qquad\qquad (2.4.79)$$

$$T_{31}^{(0)} = 2h[\kappa_1^2 c_{44}(u_{3,1}^{(0)} + u_1^{(1)}) - \kappa_1 e_{15}E_3^{(0)}],$$

$$T_{12}^{(0)} = 2h[c_{44}(u_{1,2}^{(0)} + u_{2,1}^{(0)}) - e_{15}E_2^{(0)}],$$

$$D_1^{(0)} = 2h(\bar{e}_{11}u_{1,1}^{(0)} + \bar{e}_{12}u_{2,2}^{(0)} + \bar{\varepsilon}_{11}E_1^{(0)}),$$

$$D_2^{(0)} = 2h[e_{15}(u_{1,2}^{(0)} + u_{2,1}^{(0)}) + \varepsilon_{11}E_2^{(0)}], \qquad\qquad (2.4.80)$$

$$D_3^{(0)} = 2h[\kappa_1 e_{15}(u_{3,1}^{(0)} + u_1^{(1)}) + \varepsilon_{11}E_3^{(0)}],$$

$$T_{11}^{(1)} = \frac{2h^3}{3}(\gamma_{11}S_{11}^{(1)} + \gamma_{12}S_{22}^{(1)} - \psi_{11}E_1^{(1)}),$$

$$T_{22}^{(1)} = \frac{2h^3}{3}(\gamma_{12}S_{11}^{(1)} + \gamma_{22}S_{22}^{(1)} - \psi_{12}E_1^{(1)}), \qquad\qquad (2.4.81)$$

$$T_{12}^{(1)} = \frac{2h^3}{3}(c_{44}S_6^{(1)} - e_{15}E_2^{(1)}),$$

$$D_1^{(1)} = \frac{2h^3}{3}(\psi_{11}S_{11}^{(1)} + \psi_{12}S_{22}^{(1)} + \zeta_{11}E_1^{(1)}),$$

$$D_2^{(1)} = \frac{2h^3}{3}(e_{15}S_6^{(1)} + \varepsilon_{11}E_2^{(1)}), \qquad (2.4.82)$$

$$D_3^{(1)} = \frac{2h^3}{3}\zeta_{33}E_3^{(1)}.$$

Substitution of the above into the equations of motion and charge gives the equations of extension:

$$\bar{c}_{11}u_{1,11}^{(0)} + c_{44}u_{1,22}^{(0)} + (\bar{c}_{12} + c_{44})u_{2,21}^{(0)}$$
$$+ \bar{e}_{11}\phi_{,11}^{(0)} + e_{15}\phi_{,22}^{(0)} + \frac{1}{2h}F_1^{(0)} = \rho\ddot{u}_1^{(0)},$$
$$(\bar{c}_{12} + c_{44})u_{1,12}^{(0)} + c_{44}u_{2,11}^{(0)} + \bar{c}_{22}u_{2,22}^{(0)} \qquad (2.4.83)$$
$$+ (e_{15} + \bar{e}_{12})\phi_{,12}^{(0)} + \frac{1}{2h}F_2^{(0)} = \rho\ddot{u}_2^{(0)},$$

flexure:

$$\kappa_1^2 c_{44}(u_{3,11}^{(0)} + u_{1,1}^{(1)}) + \kappa_2^2 c_{66}(u_{3,22}^{(0)} + u_{2,2}^{(1)})$$
$$+ \kappa_1 e_{15}\phi_{,1}^{(1)} + \frac{1}{2h}F_3^{(0)} = \rho\ddot{u}_3^{(0)}, \qquad (2.4.84)$$

thickness-shear:

$$\gamma_{11}u_{1,11}^{(1)} + c_{44}u_{1,22}^{(1)} + (\gamma_{12} + c_{44})u_{2,21}^{(1)} + \psi_{11}\phi_{,11}^{(1)}$$
$$+ e_{15}\phi_{,22}^{(1)} - 3h^{-2}[\kappa_1^2 c_{44}(u_{3,1}^{(0)} + u_1^{(1)}) + \kappa_1 e_{15}\phi^{(1)}]$$
$$+ \frac{3}{2h^3}F_1^{(1)} = \rho\ddot{u}_1^{(1)}, \qquad (2.4.85)$$
$$(\gamma_{12} + c_{44})u_{1,12}^{(1)} + c_{44}u_{2,11}^{(1)} + \gamma_{22}u_{2,22}^{(1)} + (e_{15} + \psi_{12})\phi_{,12}^{(1)}$$
$$- 3h^{-2}\kappa_2^2 c_{66}(u_{3,2}^{(0)} + u_2^{(1)}) + \frac{3}{2h^3}F_2^{(1)} = \rho\ddot{u}_2^{(1)},$$

and electrostatics:

$$\bar{e}_{11}u_{1,11}^{(0)} + e_{15}u_{1,22}^{(0)} + (\bar{e}_{12} + e_{15})u_{2,12}^{(0)}$$

$$- \bar{\varepsilon}_{11}\phi_{,11}^{(0)} - \varepsilon_{11}\phi_{,22}^{(0)} + \frac{1}{2h}D^{(0)} = 0,$$

$$\psi_{11}u_{1,11}^{(1)} + e_{15}u_{1,22}^{(1)} + (\psi_{12} + e_{15})u_{2,12}^{(1)} - \zeta_{11}\phi_{,11}^{(1)} - \varepsilon_{11}\phi_{,22}^{(1)}$$

$$- 3h^{-2}[\kappa_1 e_{15}(u_{3,1}^{(0)} + u_1^{(1)}) - \varepsilon_{11}\phi^{(1)}] + \frac{3}{2h^3}D^{(1)} = 0.$$

(2.4.86)

2.4.6 Shear correction factors for ceramics plates

In order to determine the shear correction factors, we solve the same problem of thickness-shear vibrations of a ceramic plate by two procedures using the three-dimensional equations and the two-dimensional equations, respectively, and match the results. We consider two cases of ceramic plates with thickness and in-plane poling, respectively.

2.4.6.1 Thickness poling

First consider the case of an infinite, unelectroded plate with the electric polarization in the thickness direction x_3 (see Figure 2.4.6).

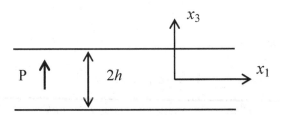

Figure 2.4.6. A ceramic plate with thickness poling.

The plate has traction-free and vanishing normal electric displacement boundary conditions at $x_3 = \pm h$. The two edges at $x_1 = \pm\infty$ are electrically shorted. As an exercise, we use the constitutive relations in terms of the following matrices:

$$\begin{pmatrix} s_{11} & s_{12} & s_{13} & 0 & 0 & 0 \\ s_{12} & s_{11} & s_{13} & 0 & 0 & 0 \\ s_{13} & s_{13} & s_{33} & 0 & 0 & 0 \\ 0 & 0 & 0 & s_{44} & 0 & 0 \\ 0 & 0 & 0 & 0 & s_{44} & 0 \\ 0 & 0 & 0 & 0 & 0 & s_{66} \end{pmatrix},$$

$$\begin{pmatrix} 0 & 0 & 0 & 0 & d_{15} & 0 \\ 0 & 0 & 0 & d_{15} & 0 & 0 \\ d_{31} & d_{31} & d_{33} & 0 & 0 & 0 \end{pmatrix}, \begin{pmatrix} \varepsilon_{11}^T & 0 & 0 \\ 0 & \varepsilon_{11}^T & 0 \\ 0 & 0 & \varepsilon_{33}^T \end{pmatrix}, \qquad (2.4.87)$$

where $s_{66} = 2(s_{11}-s_{12})$. We look for solutions in the following form:

$$u_1 = u_1(x_3,t), \quad u_2 = u_3 = 0, \quad \phi = 0, \qquad (2.4.88)$$

which lead to

$$S_1 = S_2 = S_3 = S_4 = S_6 = 0, \quad S_5 = u_{1,3},$$
$$E_1 = E_2 = E_3 = 0,$$
$$T_1 = T_2 = T_3 = T_4 = T_6 = 0, \quad T_5 = \frac{1}{s_{44}}u_{1,3}, \qquad (2.4.89)$$
$$D_1 = \frac{d_{15}}{s_{44}}u_{1,3}, \quad D_2 = D_3 = 0.$$

The equation and boundary conditions to be satisfied are

$$\frac{1}{s_{44}}u_{1,33} = \rho\ddot{u}_1, \quad -h < x_3 < h,$$

$$T_{31} = \frac{1}{s_{44}}u_{1,3} = 0, \quad x_3 = \pm h, \qquad (2.4.90)$$

which allow the following simple solution for the lowest thickness-shear mode

$$u_1 = \sin\frac{\pi x_3}{2c}\exp(i\omega t), \quad \omega^2 = \omega_\infty^2 = \frac{\pi^2}{4h^2\rho s_{44}}. \qquad (2.4.91)$$

Note that the solution is not pure elastic because of the following nontrivial electric displacement component:

$$D_1 = \frac{d_{15}}{s_{44}} S_5 = \frac{d_{15}}{s_{44}} \frac{\pi}{2c} \cos \frac{\pi x_3}{2c} \exp(i\omega t), \qquad (2.4.92)$$

which is responsible for the current flowing through the shorted edges at $x_1 = \pm\infty$. To determine the correction factor κ, we also consider the above thickness-shear mode from the plate theory. From Equation $(2.4.67)_1$ we have, with $\partial_1 = \partial_2 = 0$, $\phi^{(0)} = 0$, and $\phi^{(1)} = 0$,

$$-\frac{3\kappa^2}{h^2 s_{44}} u_1^{(1)} = \rho \ddot{u}_1^{(1)}, \qquad (2.4.93)$$

which implies that

$$\omega^2 = \frac{3\kappa^2}{\rho h^2 s_{44}}. \qquad (2.4.94)$$

Comparing the frequencies in Equations (2.4.91) and (2.4.94), we determine the correction factor as

$$\kappa^2 = \frac{\pi^2}{12}. \qquad (2.4.95)$$

The above solution is for the case when the two edges of the plate at $x_1 = \pm\infty$ are shorted. If the two ends of the plate are open (which implies that $D_1^{(0)}$ vanishes but $E_1^{(0)}$ survives in the plate theory), then from the plate equations we have

$$\omega_\infty^2 \approx \frac{\pi^2}{4c^2 \rho s_{44}(1 - k_{15}^2)}, \quad k_{15}^2 = \frac{d_{15}^2}{\varepsilon_{11} s_{44}}. \qquad (2.4.96)$$

2.4.6.2 In-plane poling

Next consider the case of a ceramic plate with in-plane poling in the x_1 direction (see Figure 2.4.7). The plate is unelectroded at its major surfaces.

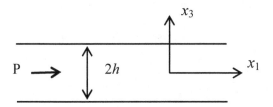

Figure 2.4.7. A ceramic plate with in-plane poling.

For ceramics poled in the x_1 direction, we have [9]

$$\begin{pmatrix} s_{33} & s_{13} & s_{13} & 0 & 0 & 0 \\ s_{13} & s_{11} & s_{12} & 0 & 0 & 0 \\ s_{13} & s_{12} & s_{11} & 0 & 0 & 0 \\ 0 & 0 & 0 & s_{66} & 0 & 0 \\ 0 & 0 & 0 & 0 & s_{44} & 0 \\ 0 & 0 & 0 & 0 & 0 & s_{44} \end{pmatrix},$$

$$\begin{pmatrix} d_{33} & d_{31} & d_{31} & 0 & 0 & 0 \\ 0 & 0 & 0 & 0 & 0 & d_{15} \\ 0 & 0 & 0 & 0 & d_{15} & 0 \end{pmatrix}, \begin{pmatrix} \varepsilon_{33}^T & 0 & 0 \\ 0 & \varepsilon_{11}^T & 0 \\ 0 & 0 & \varepsilon_{11}^T \end{pmatrix}. \tag{2.4.97}$$

We look for

$$u_1 = u_1(x_3,t), \quad u_2 = u_3 = 0, \quad \phi = \phi(x_3,t), \tag{2.4.98}$$

which leads to

$$S_1 = S_2 = S_3 = S_4 = S_6 = 0, \quad S_5 = u_{1,3}, \\ E_1 = E_2 = 0, \quad E_3 = -\phi_{,3}, \tag{2.4.99}$$

$$T_1 = T_2 = T_3 = T_4 = T_6 = 0, \quad T_5 = \frac{1}{s_{44}}(u_{1,3} - d_{15}E_3),$$

$$D_1 = D_2 = 0, \quad D_3 = \frac{d_{15}}{s_{44}}u_{1,3} + \varepsilon_{33}^T(1 - k_{15}^2)E_3. \tag{2.4.100}$$

The equation and boundary conditions to be satisfied are

$$\frac{1}{s_{44}(1-k_{15}^2)}u_{1,33} = \rho\ddot{u}_1, \quad -h < x_3 < h,$$

$$T_{31} = \frac{1}{s_{44}(1-k_{15}^2)}u_{1,3} = 0, \quad x_3 = \pm h, \tag{2.4.101}$$

which allow the following simple solution for the lowest thickness-shear mode

$$u_1 = \sin\frac{\pi x_3}{2h}\exp(i\omega t),$$

$$\omega^2 = \omega_\infty^2 = \frac{\pi^2}{4h^2\rho s_{44}(1-k_{15}^2)}. \tag{2.4.102}$$

The corresponding plate solution can be found from Equations $(2.4.85)_1$ and $(2.4.86)_2$ as

$$\omega^2 = \frac{3\kappa_1^2}{\rho h^2 s_{44}(1-k_{15}^2)}. \tag{2.4.103}$$

Comparing Equations (2.4.102) and (2.4.103) we obtain

$$\kappa_1^2 = \frac{\pi^2}{12}. \tag{2.4.104}$$

If the plate is electroded at $x_3 = \pm h$ and the electrodes are shorted, from the plate equations we have

$$\omega_\infty^2 \approx \frac{\pi^2}{4\rho h^2 s_{44}}. \tag{2.4.105}$$

2.4.7 Thickness-shear vibration of an inhomogeneous ceramic plate

As an application of the equations obtained, we analyze thickness-shear vibrations of an inhomogeneous ceramic plate with thickness poling (see Figure 2.4.8) [25]. The thick lines represent electrodes. The analysis is also valid for a plate of 6mm crystals with the six-fold axis along the plate thickness. This structure has various applications in piezoelectric devices operating with thickness-shear modes. When $a \gg$

$h = \max\{h_1, h_2\}$, i.e., thin plates, the flexural deformation accompanying thickness-shear can be essentially avoided by properly choosing the aspect ratios of a/h and b/h. Therefore, in the following we neglect the coupling to flexure and consider thickness-shear vibrations.

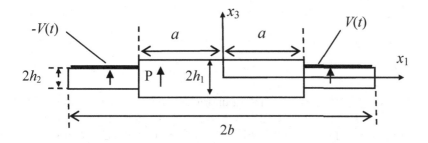

Figure 2.4.8. An inhomogeneous ceramic plate with thickness poling.

For a plate with the material orientation and electrode configuration as shown in Figure 2.4.8, the thickness-shear deformation is coupled to the electric field E_1. Hence the displacement and potential fields can be approximated by

$$u_1 \cong x_3 u_1^{(1)}(x_1, t), \quad u_2 = u_3 \cong 0,$$
$$\phi \cong \phi^{(0)}(x_1, t). \tag{2.4.106}$$

Under Equation (2.4.106), in an electroded region, we have $\phi^{(0)} =$ constant. We are interested in free vibration modes when the two electrodes in the figure are either shorted or open, i.e., $V = 0$ or $D_1^{(0)} = 0$. We discuss these two cases separately below.

2.4.7.1 Shorted electrodes

In the case of shorted electrodes, from Equations $(2.4.67)_1$ and $(2.4.68)_1$, we have the following equations for free vibrations:

$$\bar{c}_{11}u_{1,11}^{(1)} - 3h_1^{-2}\kappa^2 c_{44}u_1^{(1)} - 3h_1^{-2}\kappa e_{15}\phi_{,1}^{(0)} = \rho\ddot{u}_1^{(1)}, \quad |x_1| < a,$$

$$-\varepsilon_{11}\phi_{,11}^{(0)} + \kappa e_{15}u_{1,1}^{(1)} = 0, \quad |x_1| < a,$$

$$\bar{c}_{11}u_{1,11}^{(1)} - 3h_2^{-2}\kappa^2 c_{44}u_1^{(1)} = \rho\ddot{u}_1^{(1)}, \quad a < |x_1| < b,$$

$$u_1^{(1)}(\pm a^-) = u_1^{(1)}(\pm a^+),$$ (2.4.107)

$$u_{1,1}^{(1)}(\pm a^-) = u_{1,1}^{(1)}(\pm a^+),$$

$$\phi^{(0)}(\pm a^-) = 0,$$

$$u_1^{(1)}(\pm b) = 0,$$

where $\kappa^2 = \pi^2/12$. Note that Equation (2.4.107)$_5$ is of an approximate nature. In Equation (2.4.107) we directly neglected the coupling to flexure and did not use the thickness-shear approximation to eliminate flexure. Consider time-harmonic motions with a frequency ω. Then Equations (2.4.107)$_{1,3}$ can be written as

$$\bar{c}_{11}u_{1,11}^{(1)} + \rho(\omega^2 - \omega_1^2)u_1^{(1)} - 3h_1^{-2}\kappa e_{15}\phi_{,1}^{(0)} = 0, \quad |x_1| < a,$$

$$\bar{c}_{11}u_{1,11}^{(1)} + \rho(\omega^2 - \omega_2^2)u_1^{(1)} = 0, \quad a < |x_1| < b,$$ (2.4.108)

where

$$\omega_1^2 = \frac{3\kappa^2 c_{44}}{h_1^2\rho} = \frac{\pi^2 c_{44}}{4h_1^2\rho}, \quad \omega_2^2 = \frac{3\kappa^2 c_{44}}{h_2^2\rho} = \frac{\pi^2 c_{44}}{4h_2^2\rho}$$ (2.4.109)

are the pure thickness-shear frequencies of a plate of thickness $2h_1$ or $2h_2$ when there is no piezoelectric stiffening due to E_1. Equation (2.4.107)$_2$ can be integrated to give

$$\phi_{,1}^{(0)} = \frac{\kappa e_{15}}{\varepsilon_{11}}u_1^{(1)} + C_1, \quad |x_1| < a,$$ (2.4.110)

where C_1 is an integration constant. Substitution of Equation (2.4.110) into Equation (2.4.108)$_1$ yields

$$\bar{c}_{11}u_{1,11}^{(1)} + \rho(\omega^2 - \bar{\omega}_1^2)u_1^{(1)} - \frac{3\kappa e_{15}}{h_1^2}C_1 = 0, \quad |x_1| < a,$$ (2.4.111)

where

$$\overline{\omega}_1^2 = \omega_1^2(1 + k_{15}^2), \quad k_{15}^2 = \frac{e_{15}^2}{\varepsilon_{11}c_{44}}. \tag{2.4.112}$$

$\overline{\omega}_1$ is the pure thickness-shear frequency of a plate of thickness $2h_1$ when there is piezoelectric stiffening due to E_1. The general solutions to Equation (2.4.111) and Equation (2.4.108)$_2$ are

$$u_1^{(1)} = C_3 \cosh \alpha x_1 + C_4 \sinh \alpha x_1 + \gamma C_1, \quad |x_1| < a,$$

$$u_1^{(1)} = C_5 \cos \beta x_1 + C_6 \sin \beta x_1, \quad a < x_1 < b, \tag{2.4.113}$$

$$u_1^{(1)} = C_7 \cos \beta x_1 + C_8 \sin \beta x_1, \quad -b < x_1 < -a,$$

where C_3 through C_6 are integration constants, and

$$\alpha^2 = \frac{\rho(\overline{\omega}_1^2 - \omega^2)}{\overline{c}_{11}},$$

$$\beta^2 = \frac{\rho(\omega^2 - \omega_2^2)}{\overline{c}_{11}}, \tag{2.4.114}$$

$$\gamma = \frac{3\kappa e_{15}}{\rho(\omega^2 - \overline{\omega}_1^2)h_1^2}.$$

With Equation (2.4.113)$_1$, Equation (2.4.110) can be integrated once more to yield

$$\phi^{(0)} = \frac{\kappa e_{15}}{\varepsilon_{11}} \left(C_3 \frac{1}{\alpha} \sinh \alpha x_1 + C_4 \frac{1}{\alpha} \cosh \alpha x_1 + \gamma C_1 x_1 \right)$$
$$+ C_1 x_1 + C_2, \quad |x_1| < a, \tag{2.4.115}$$

where C_2 is another integration constant. Substitution of Equations (2.4.113) and (2.4.115) into the boundary and continuity conditions in Equation (2.4.107) results in a system of homogeneous equations for C_1 through C_6. For nontrivial solutions the determinant of the coefficient matrix has to vanish, which determines the eigenvalues or resonant frequencies. Mode shapes are determined by the eigenvectors. Because of the symmetry in Figure 2.4.8, the modes can be separated into two groups. One is called symmetric and the other anti-symmetric in x_1. In applications the symmetric modes when $u_1^{(1)}$ is an even function of x_1 are useful, especially the lowest one. For these modes we let

$$C_2 = C_4 = 0, \quad C_7 = C_5, \quad C_8 = -C_6. \tag{2.4.116}$$

Then from the continuity and boundary conditions at a and b, we obtain the following equations for C_1, C_3, C_5 and C_6:

$$
\begin{aligned}
&C_3 \cosh \alpha a + \gamma C_1 = C_5 \cos \beta a + C_6 \sin \beta a, \\
&C_3 \alpha \sinh \alpha a = -C_5 \beta \sin \beta a + C_6 \beta \cos \beta a, \\
&\frac{\kappa e_{15}}{\varepsilon_{11}} \left(C_3 \frac{1}{\alpha} \sinh \alpha a + \gamma C_1 a \right) + C_1 a = 0, \\
&C_5 \cos \beta b + C_6 \sin \beta b = 0.
\end{aligned}
\tag{2.4.117}
$$

The frequency equation is

$$
\beta a \left(1 + \frac{\gamma \kappa e_{15}}{\varepsilon_{11}} - \frac{\gamma \kappa e_{15} \tanh \alpha a}{\alpha a \varepsilon_{11}} \right)
$$
$$
+ \alpha a \tan \beta (b - a) \tanh \alpha a \left(1 + \frac{\gamma \kappa e_{15}}{\varepsilon_{11}} \right) = 0.
\tag{2.4.118}
$$

First consider a plate of uniform thickness ($h_1 = h_2$). In this case $\bar{\omega}_1 > \omega_2$. As a numerical example, we consider the case of $h_1 = h_2 = 1$ mm, $a = 22$ mm, $b = 30$ mm, and the electrodes are shorted. Cadmium Sulfide (CdS) is used in the calculation. The material constants are given in Appendix 2. For $\omega_2 < \omega < \bar{\omega}_1$, the modes are hyperbolic (or exponential) in the unelectroded region and sinusoidal in the electroded region. The first mode of $u_1^{(1)}$ is plotted in Figure 2.4.9. The vibration is essentially in the electroded region; once it enters the unelectroded region it decays rapidly. The cause of this type of behavior is that the piezoelectric stiffening effect in the unelectroded central region makes $\bar{\omega}_1 > \omega_2$. The corresponding $\phi^{(0)}$ is shown in Figure 2.4.10.

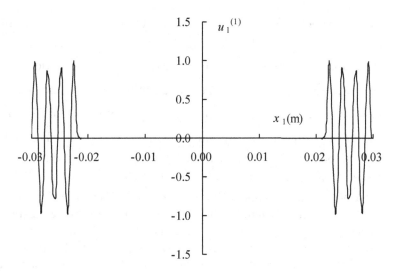

Figure 2.4.9. $u_1^{(1)}$ of the first thickness-shear mode in $(\omega_2, \overline{\omega}_1)$ when $h_1 = h_2$ (shorted electrodes).

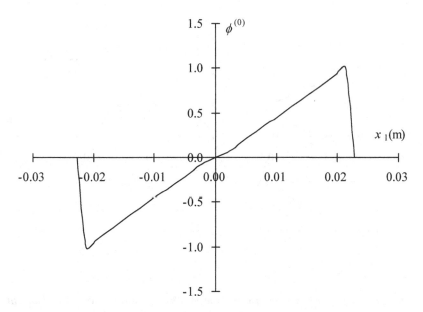

Figure 2.4.10. $\phi^{(0)}$ of the first thickness-shear mode in $(\omega_2, \overline{\omega}_1)$ when $h_1 = h_2$ (shorted electrodes).

Next consider a plate of non-uniform thickness ($h_1 > h_2$), still with shorted electrodes. When h_1 is different from h_2, the case of $h_1 > h_2$ is more useful in applications. In this case, when h_1 is sufficiently larger than h_2, we have $\overline{\omega}_1 < \omega_2$. As an example, we consider the case of $h_1 = 2h_2 = 2$ mm, $a = 22$ mm, and $b = 30$ mm so that $\overline{\omega}_1 < \omega_2$. For $\overline{\omega}_1 < \omega < \omega_2$, the modes are sinusoidal in the unelectroded central region and exponential in the electroded region. The first three symmetric modes of $u_1^{(1)}$ are plotted in Figure 2.4.11. They are the sum of a cosine function and a constant in the central region and decay rapidly in the electroded region. Therefore the vibration is trapped in the central region. Near the edge of the plate there is essentially no vibration. This important phenomenon is called energy trapping [27,29,30] and is very useful in device applications. When the vibration is trapped to the center, mounting the device at the edge will not affect the vibration. Energy

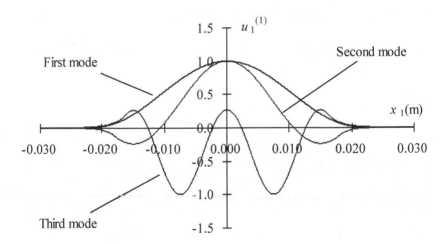

Figure 2.4.11. $u_1^{(1)}$ of the first three thickness-shear modes in $(\overline{\omega}_1, \omega_2)$ when $h_1 = 2h_2$ (shorted electrodes).

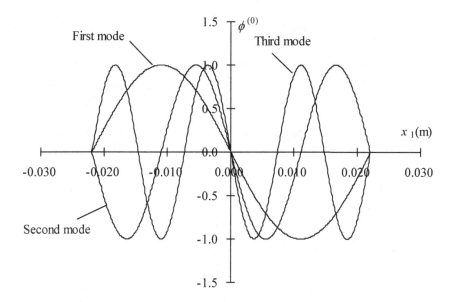

Figure 2.4.12. $\phi^{(0)}$ of the first three thickness-shear modes in $(\overline{\omega}_1, \omega_2)$ when $h_1 = 2h_2$ (shorted electrodes).

trapping can be due to either the mass effect in a plate with nonuniform thickness or piezoelectric stiffening. In the structure discussed here, the piezoelectric stiffening in the central region is against energy trapping in the central region. A thicker central region is needed to overcome the piezoelectric stiffening effect and produce energy trapping in the central region. The corresponding $\phi^{(0)}$ is shown in Figure 2.4.12.

2.4.7.2 Open electrodes

Still consider the case of $h_1 = 2h_2 = 2$ mm, $a = 22$ mm, and $b = 30$ mm so that $\overline{\omega}_1 < \omega_2$. If the two electrodes are open, then $D_1^{(0)} = 0$ which implies that $C_1 = 0$. In this case the boundary conditions on $\phi^{(0)}$,

namely $(2.4.107)_6$ and $(2.4.117)_3$, should be dropped. The frequency
equation takes the following form:

$$\beta(1 - \tanh \beta a \tanh \beta b) + \alpha \tan \alpha a(\tanh \beta a - \tanh \beta b) = 0. \quad (2.4.119)$$

The first three symmetric modes of $u_1^{(1)}$ are plotted in Figure 2.4.13.
They are a cosine function only without C_1 (which is zero) in the central
region and decay rapidly in the electroded region. Compared to the
modes in Figure 2.4.11, the modes in Figure 2.4.13 are not trapped as
strongly. This is because for the case in Figure 2.4.13 the open electrodes
allow stronger electric fields in the central region, which cause more
piezoelectric stiffening there against energy trapping. The corresponding
$\phi^{(0)}$ is shown in Figure 2.4.14. The two electrodes now have a difference in their electric potentials.

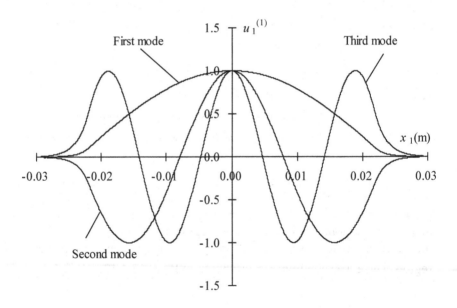

Figure 2.4.13. $u_1^{(1)}$ of the first three thickness-shear modes in $(\overline{\omega}_1, \omega_2)$
when $h_1 = 2h_2$ (open electrodes).

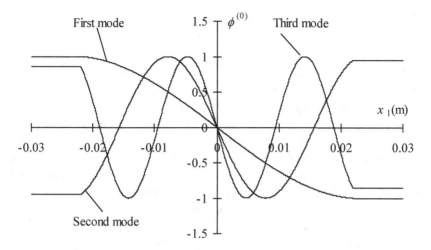

Figure 2.4.14. $\phi^{(0)}$ of the first three thickness-shear modes in $(\overline{\omega}_1, \omega_2)$
when $h_1 = 2h_2$ (open electrodes).

2.4.8 A ceramic plate piezoelectric gyroscope

Consider a rectangular ceramic plate poled in the thickness direction
as shown in Figure 2.4.15 [28].

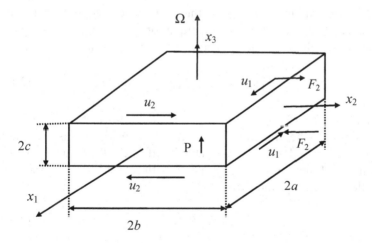

Figure 2.4.15. A ceramic plate piezoelectric gyroscope.

The plate can vibrate at the fundamental thickness-shear modes in both the x_1 and the x_2 directions. For a ceramic plate poled in the thickness direction x_3, these thickness-shear modes can be excited or detected electrically by lateral electrodes on the sides at $x_1 = \pm a$ or $x_2 = \pm b$. Suppose a thickness-shear vibration u_1 in the x_1 direction is excited by a time-harmonic voltage $2V_1 \exp(i\omega t)$ applied across a pair of electrodes at $x_1 = \pm a$. If the plate is rotating about its normal with an angular rate Ω, the Coriolis force F_2 will excite a thickness-shear motion u_2 in the x_2 direction. This secondary thickness-shear will produce a voltage $2V_2 \exp(i\omega t)$ across a pair of electrodes at $x_2 = \pm b$. V_2 can be shown to be proportional Ω and therefore can be used to detect Ω.

We are going to use an approximate procedure to analyze the gyroscope, which can be considered as a perturbation procedure [28]. For the thickness-shear motion excited by V_1, the main motion is $u_1^{(1)}$. We approximately have the following one-dimensional problem from Equation (2.4.76):

$$c_{11}^* u_{1,11}^{(1)} - 3c^{-2}\kappa^2 c_{44} u_1^{(1)} - 3c^{-2}\kappa e_{15}\phi_{,1}^{(0)} = \rho\ddot{u}_1^{(1)}, \quad |x_1| < a,$$

$$\kappa e_{15} u_{1,1}^{(1)} - \varepsilon_{11}\phi_{,11}^{(0)} = 0, \quad |x_1| < a,$$

$$T_{13}^{(0)} = 2c(\kappa^2 c_{44} u_1^{(1)} + \kappa e_{15}\varphi_{,1}^{(0)}) = 0, \quad x_1 = \pm a, \qquad (2.4.120)$$

$$\phi^{(0)} = \pm V_1 \sin \omega t, \quad x_1 = \pm a.$$

As an approximation, we neglect the piezoelectric coupling term to $u_1^{(1)}$ in the electrostatic equation in Equation (2.4.120)$_2$. This gives an approximate solution of the driving electric potential as

$$\phi^{(0)} = \frac{x_1}{a} V_1 \sin \omega t \cong \frac{8}{\pi^2} V_1 \sin \frac{\pi x_1}{2a} \sin \omega t. \qquad (2.4.121)$$

In Equation (2.4.121) we have approximated a linear function of x_1 over $[-a,a]$ by a sine function. It can be considered as a one-term approximation by Fourier series. This is sufficient for a qualitative study. Substituting Equation (2.4.121) into Equation (2.4.120)$_1$, we then obtain the following expression for $u_1^{(1)}$:

$$u_1^{(1)} = B_1 \cos \frac{\pi x_1}{2a} \sin \omega t, \qquad (2.4.122)$$

where

$$B_1 = \frac{2\sqrt{3}}{\omega^2 - \omega_\infty^2 (1 + \dfrac{c_{11}^* c^2}{c_{44} a^2})} \frac{e_{15} V_1 / a}{\rho c^2}, \quad \omega_\infty^2 = \frac{\pi^2 c_{44}}{4\rho c^2}. \tag{2.4.123}$$

With $u_1^{(1)}$ known, we have the displacement field

$$u_1 = x_3 u_1^{(1)} = B_1 x_3 \cos\frac{\pi x_1}{2a} \sin \omega t, \tag{2.4.124}$$

which leads to a Coriolis force field

$$\vec{f} = -2\Omega \vec{e}_3 \times \dot{u}_1 \vec{e}_1 = -2\Omega \dot{u}_1 \vec{e}_2 \tag{2.4.125}$$

or

$$f_2 = -2\Omega\omega B_1 x_3 \cos\frac{\pi x_1}{2a} \cos \omega t$$

$$\cong -\frac{4}{\pi}\Omega\omega B_1 x_3 \cos \omega t \cong -2\Omega\omega B_1 x_3 \cos\frac{\pi x_2}{2b} \cos \omega t, \tag{2.4.126}$$

where we have approximated a cosine function over $[-a, a]$ by a constant, and similarly a constant over $[-b, b]$ by a cosine function. With F_2 we calculate the plate resultant

$$F_2^{(1)} = \int_{-c}^{c} x_3 \rho f_2 dx_3 = -\frac{2c^3}{3}\rho 2\Omega\omega B_1 \cos\frac{\pi x_2}{2b} \cos \omega t. \tag{2.4.127}$$

Then the boundary value problem for the thickness-shear in the x_2 direction can be written as

$$c_{11}^* u_{2,22}^{(1)} - 3c^{-2}\kappa^2 c_{44} u_2^{(1)}$$

$$- 3c^{-2}\kappa e_{15} \phi_{,2}^{(0)} + \frac{3}{2c^3} F_2^{(1)} = \rho \ddot{u}_2^{(1)}, \quad |x_2| < b,$$

$$\kappa e_{15} u_{2,2}^{(1)} - \varepsilon_{11} \phi_{,22}^{(0)} = 0, \quad |x_2| < b, \tag{2.4.128}$$

$$T_{23}^{(0)} = 2c(\kappa^2 c_{44} u_2^{(1)} + \kappa e_{15} \phi_{,2}^{(0)}) = 0, \quad x_2 = \pm b$$

$$\phi^{(0)} = \pm V_2 \sin \omega t, \quad x_2 = \pm b,$$

$$D_2^{(0)} = 2c(\kappa e_{15} u_2^{(1)} - \varepsilon_{11} \phi_{,2}^{(0)}) = 0, \quad x_2 = \pm b.$$

In Equation (2.4.128) where V_2 is unknown, we need some circuit condition joining the electrodes at $x_2 = \pm b$ to determine V_2. We consider the simple case of open circuit in which $D_2^{(0)}$ or the charge, and hence the current on the electrodes, vanishes. As an approximation, we neglect the piezoelectric coupling term to $\phi^{(0)}$ in Equation (2.4.128)$_1$. Then, under the Coriolis force $F_2^{(1)}$, we obtain

$$u_2^{(1)} = B_2 \cos\frac{\pi x_2}{2b}\cos\omega t, \qquad (2.4.129)$$

where

$$B_2 = \frac{2\Omega\omega}{\omega^2 - \omega_\infty^2(1 + \dfrac{c_{11}^* c^2}{c_{44}b^2})} B_1. \qquad (2.4.130)$$

With $u_2^{(1)}$ known, from Equation (2.4.128)$_2$, we obtain

$$\phi^{(0)} = A_2 \sin\frac{\pi x_2}{2b}\cos\omega t \cong \frac{\pi^2}{8} A_2 \frac{x_2}{b}\cos\omega t, \qquad (2.4.131)$$

where

$$A_2 = \frac{be_{15}}{\sqrt{3}\varepsilon_{11}} B_2. \qquad (2.4.132)$$

The mechanical boundary conditions and the open circuit condition in Equation (2.3.128) are satisfied. The electric boundary condition (2.4.128)$_4$ gives the output voltage

$$V_2 = \phi^{(0)}(x_2 = b)$$

$$= \left[\frac{\pi^2 be_{15}}{8\sqrt{3}\varepsilon_{11}}\right]\left[\frac{2\Omega\omega}{\omega^2 - \omega_\infty^2(1 + \dfrac{c_{11}^* c^2}{c_{44}b^2})}\right] \qquad (2.4.133)$$

$$\times \left[\frac{2\sqrt{3}}{\omega^2 - \omega_\infty^2(1 + \dfrac{c_{11}^* c^2}{c_{44}a^2})}\frac{e_{15}V_1/a}{\rho c^2}\right]\cos\omega t,$$

or, for voltage sensitivity

$$
\frac{V_2}{V_1\Omega} = \left[2k_{15}^2\,\frac{b}{a}\right] \left[\frac{\omega}{\omega^2 - \omega_\infty^2\left(1 + \dfrac{c_{11}^{*}c^2}{c_{44}b^2}\right)}\right]
$$

$$
\times \left[\frac{\omega_\infty^2}{\omega^2 - \omega_\infty^2\left(1 + \dfrac{c_{11}^{*}c^2}{c_{44}a^2}\right)}\right]\cos\omega t.
$$

(2.4.134)

In Equation (2.4.133), the three pairs of brackets on the right hand side represent, from right to left, respectively, the driving of the thickness-shear motion in the x_1 direction by the applied voltage V_1, the driving of the thickness-shear motion in the x_2 direction by the Coriolis force, and the sensing of the thickness-shear motion in the x_2 direction. The roles of various material constants and geometric parameters are clearly exhibited. At resonant frequencies, the above expressions become singular. From Equation (2.4.134) it can be seen that for large voltage sensitivity, the driving frequency and the two resonant frequencies of the thickness shear motions in two directions must all be very close (double resonance). This implies that a must be very close to b and the plate is almost square. When that is the case, we can write

$$
\omega_\infty^2\left(1 + \frac{c_{11}^{*}c^2}{c_{44}a^2}\right) \cong (\omega + \Delta\omega_1)^2,
$$

$$
\omega_\infty^2\left(1 + \frac{c_{11}^{*}c^2}{c_{44}b^2}\right) \cong (\omega + \Delta\omega_2)^2,
$$

(2.4.135)

which implies that

$$
\frac{V_2}{V_1\Omega} \cong k_{15}^2\,\frac{b}{2a}\,\frac{\omega}{(\Delta\omega_1)(\Delta\omega_2)}.
$$

(2.4.136)

2.4.9 Equations for a quartz plate

Rotated Y-cut quartz exhibits monoclinic symmetry in a coordinate system (x_1, x_2) in and normal to the plane of the plate (see Figure 2.4.16).

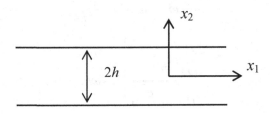

Figure 2.4.16. A quartz plate and coordinate system.

2.4.9.1 Equations for a plate with the x_2 axis along its normal

When the normal of the plate is along the x_2 axis, the stress relaxation $T_{22}^{(0)} = 0$ leads to the following relaxed material constants:

$$\begin{aligned}
\bar{c}_{ijkl} &= c_{ijkl} - c_{ij22}c_{22kl} / c_{2222}, \\
\bar{e}_{kij} &= e_{kij} - e_{k22}c_{22ij} / c_{2222}, \\
\bar{\varepsilon}_{ij} &= \varepsilon_{ij} + e_{i22}e_{j22} / c_{2222}.
\end{aligned} \tag{2.4.137}$$

Introduction of the shear corrections factors

$$S_{21}^{(0)} \to \kappa_1 S_{21}^{(0)}, \quad S_{23}^{(0)} \to \kappa_3 S_{23}^{(0)} \tag{2.4.138}$$

further modifies the relaxed constants into [17]

$$\begin{aligned}
\bar{c}'_{ijkl} &= \kappa_{i+j-2}^{\mu} \kappa_{k+l-2}^{\nu} \bar{c}_{ijkl}, \quad \text{(not summed)}, \\
\bar{e}'_{kij} &= \kappa_{i+j-2}^{\mu} \bar{e}_{kij}, \quad \text{(not summed)}, \\
\mu &= \cos^2(ij\pi/2), \quad \nu = \cos^2(kl\pi/2).
\end{aligned} \tag{4.2.139}$$

κ_{i+j-2}^{μ} (or κ_{k+l-2}^{ν}) is equal to κ_1, κ_3 or unity according as $i+j$ (or $k+l$) is 3, 5, or neither, respectively. Equation (4.2.139) can be represented by

$$[\bar{c}'_{pq}] = \begin{pmatrix} \bar{c}_{11} & 0 & \bar{c}_{13} & \kappa_3\bar{c}_{14} & \bar{c}_{15} & \kappa_1\bar{c}_{16} \\ 0 & 0 & 0 & 0 & 0 & 0 \\ \bar{c}_{13} & 0 & \bar{c}_{33} & \kappa_3\bar{c}_{34} & \bar{c}_{35} & \kappa_1\bar{c}_{36} \\ \kappa_3\bar{c}_{14} & 0 & \kappa_3\bar{c}_{34} & \kappa_3\kappa_3\bar{c}_{44} & \kappa_3\bar{c}_{45} & \kappa_3\kappa_1\bar{c}_{46} \\ \bar{c}_{15} & 0 & \bar{c}_{35} & \kappa_3\bar{c}_{45} & \bar{c}_{55} & \kappa_1\bar{c}_{56} \\ \kappa_1\bar{c}_{16} & 0 & \kappa_1\bar{c}_{36} & \kappa_1\kappa_3\bar{c}_{46} & \kappa_1\bar{c}_{56} & \kappa_1\kappa_1\bar{c}_{66} \end{pmatrix}, \tag{2.4.140}$$

$$[\bar{e}'_{iq}] = \begin{bmatrix} \bar{e}_{11} & 0 & \bar{e}_{13} & \kappa_3\bar{e}_{14} & \bar{e}_{15} & \kappa_1\bar{e}_{16} \\ \bar{e}_{21} & 0 & \bar{e}_{23} & \kappa_3\bar{e}_{24} & \bar{e}_{25} & \kappa_1\bar{e}_{26} \\ \bar{e}_{31} & 0 & \bar{e}_{33} & \kappa_3\bar{e}_{34} & \bar{e}_{35} & \kappa_1\bar{e}_{36} \end{bmatrix}. \tag{2.4.141}$$

The first-order plate material constants are defined by

$$\begin{aligned} \gamma_{rs} &= c_{rs} - c_{rv}c_{vw}^{-1}c_{ws}, \quad r,s = 1,3,5, \\ \psi_{ks} &= e_{ks} - e_{kw}c_{wv}^{-1}c_{vs}, \quad v,w = 2,4,6, \\ \varsigma_{kj} &= \varepsilon_{kj} + e_{kv}c_{vw}^{-1}e_{jw}. \quad j,k = 1,2,3. \end{aligned} \tag{2.4.142}$$

The equations of the first-order theory take the following form:

$$\begin{aligned} T_{ab,a}^{(0)} + F_b^{(0)} &= 2h\rho\ddot{u}_b^{(0)}, \quad a,b = 1,3, \\ T_{a2,a}^{(0)} + F_2^{(0)} &= 2h\rho\ddot{u}_2^{(0)}, \\ T_{ab,a}^{(1)} - T_{2b}^{(0)} + F_b^{(1)} &= \frac{2h^3}{3}\rho\ddot{u}_b^{(1)}, \\ D_{a,a}^{(0)} + D^{(0)} &= 0, \\ D_{a,a}^{(1)} - D_2^{(0)} + D^{(1)} &= 0, \end{aligned} \tag{2.4.143}$$

$$\begin{aligned} T_{ij}^{(0)} &= 2h(\bar{c}'_{ijkl}S_{kl}^{(0)} - \bar{e}'_{kij}E_k^{(0)}), \\ D_i^{(0)} &= 2h(\bar{e}'_{ikl}S_{kl}^{(0)} + \bar{\varepsilon}_{ij}E_j^{(0)}), \\ T_r^{(1)} &= \frac{2h^3}{3}(\gamma_{rs}S_s^{(1)} - \psi_{kr}E_k^{(1)}), \quad r,s = 1,3,5, \\ D_i^{(1)} &= \frac{2h^3}{3}(\psi_{is}S_s^{(1)} + \varsigma_{ij}E_j^{(1)}), \end{aligned} \tag{2.4.144}$$

$$S_1^{(0)} = u_{1,1}^{(0)}, \quad S_3^{(0)} = u_{3,3}^{(0)},$$

$$S_4^{(0)} = u_{2,3}^{(0)} + u_3^{(1)}, \quad S_5^{(0)} = u_{1,3}^{(0)} + u_{3,1}^{(0)}, \quad S_6^{(0)} = u_{2,1}^{(0)} + u_1^{(1)},$$

$$S_1^{(1)} = u_{1,1}^{(1)}, \quad S_3^{(1)} = u_{3,3}^{(1)}, \quad S_5^{(1)} = u_{1,3}^{(1)} + u_{3,1}^{(1)}, \qquad (2.4.145)$$

$$E_3^{(0)} = -\phi_{,3}^{(0)}, \quad E_1^{(0)} = -\phi_{,1}^{(0)}, \quad E_2^{(0)} = -\phi^{(1)},$$

$$E_3^{(1)} = -\phi_{,3}^{(1)}, \quad E_1^{(1)} = -\phi_{,1}^{(1)}, \quad E_2^{(1)} = 0.$$

With successive substitutions, Equations (2.4.143) can be written as seven equations for $u_i^{(0)}$, $u_a^{(1)}$, $\phi^{(0)}$ and $\phi^{(1)}$. At the boundary of a plate with an in-plane unit exterior normal **n** and an in-plane unit tangent **s**, we may prescribe

$$T_{nn}^{(0)} \quad \text{or} \quad u_n^{(0)}, \quad T_{ns}^{(0)} \quad \text{or} \quad u_s^{(0)}, \quad T_{n2}^{(0)} \quad \text{or} \quad u_2^{(0)},$$

$$T_{nn}^{(1)} \quad \text{or} \quad u_n^{(1)}, \quad T_{ns}^{(1)} \quad \text{or} \quad u_s^{(1)}, \qquad (2.4.146)$$

$$D_n^{(0)} \quad \text{or} \quad \phi^{(0)}, \quad D_n^{(1)} \quad \text{or} \quad \phi^{(1)}.$$

2.4.9.2 Monoclinic crystals

For monoclinic crystals, the material tensors c_{ijkl}^E, e_{ijk} and ε_{ij}^S are given by Equation (2.1.2). Then the zero-order plate material constants are found to be

$$[\bar{c}_{pq}'] = \begin{pmatrix} \bar{c}_{11} & 0 & \bar{c}_{13} & \kappa_3\bar{c}_{14} & 0 & 0 \\ 0 & 0 & 0 & 0 & 0 & 0 \\ \bar{c}_{13} & 0 & \bar{c}_{33} & \kappa_3\bar{c}_{34} & 0 & 0 \\ \kappa_3\bar{c}_{14} & 0 & \kappa_3\bar{c}_{34} & \kappa_3\kappa_3\bar{c}_{44} & 0 & 0 \\ 0 & 0 & 0 & 0 & c_{55} & \kappa_1 c_{56} \\ 0 & 0 & 0 & 0 & \kappa_1 c_{56} & \kappa_1\kappa_1 c_{66} \end{pmatrix}, \qquad (2.4.147)$$

$$[\bar{e}_{iq}'] = \begin{bmatrix} \bar{e}_{11} & 0 & \bar{e}_{13} & \kappa_3\bar{e}_{14} & 0 & 0 \\ 0 & 0 & 0 & 0 & e_{25} & \kappa_1 e_{26} \\ 0 & 0 & 0 & 0 & e_{35} & \kappa_1 e_{36} \end{bmatrix}, \qquad (2.4.148)$$

$$\begin{pmatrix} \bar{\varepsilon}_{11} & 0 & 0 \\ 0 & \varepsilon_{11} & \varepsilon_{23} \\ 0 & \varepsilon_{32} & \varepsilon_{33} \end{pmatrix}, \tag{2.4.149}$$

where

$$\bar{c}_{11} = c_{11} - c_{12}^2 / c_{22}, \quad \bar{c}_{13} = c_{13} - c_{12}c_{23} / c_{22},$$
$$\bar{c}_{33} = c_{33} - c_{23}^2 / c_{22}, \quad \bar{c}_{14} = c_{14} - c_{12}c_{24} / c_{22}, \tag{2.4.150}$$
$$\bar{c}_{44} = c_{44} - c_{24}^2 / c_{22}, \quad \bar{c}_{34} = c_{34} - c_{32}c_{24} / c_{22},$$

$$\bar{e}_{11} = e_{11} - e_{12}c_{12} / c_{22}, \quad \bar{e}_{13} = e_{13} - e_{12}c_{23} / c_{22},$$
$$\bar{e}_{14} = e_{14} - e_{12}c_{24} / c_{22}, \quad \bar{\varepsilon}_{11} = \varepsilon_{11} + e_{12}^2 / c_{22}. \tag{2.4.151}$$

The first-order plate material constants are

$$[\gamma_{rs}] = \begin{bmatrix} \gamma_{11} & \gamma_{13} & 0 \\ \gamma_{13} & \gamma_{33} & 0 \\ 0 & 0 & \gamma_{55} \end{bmatrix}, \quad [\psi_{ks}] = \begin{bmatrix} \psi_{11} & \psi_{13} & 0 \\ 0 & 0 & \psi_{25} \\ 0 & 0 & \psi_{35} \end{bmatrix},$$

$$[\varsigma_{kj}] = \begin{bmatrix} \varsigma_{11} & 0 & 0 \\ 0 & \zeta_{22} & \zeta_{23} \\ 0 & \varsigma_{23} & \zeta_{33} \end{bmatrix}, \quad r,s = 1,3,5, \quad k,j = 1,2,3, \tag{2.4.152}$$

where

$$\gamma_{11} = c_{11} - \frac{c_{12}^2 c_{44} + c_{14}^2 c_{22} - 2c_{12}c_{24}c_{14}}{c_{22}c_{44} - c_{24}^2} = s_{33} / (s_{11}s_{33} - s_{13}^2),$$

$$\gamma_{13} = c_{13} - \frac{c_{23}c_{12}c_{44} + c_{34}c_{14}c_{22} - c_{23}c_{14}c_{24} - c_{34}c_{12}c_{24}}{c_{22}c_{44} - c_{24}^2},$$

$$= -s_{13} / (s_{11}s_{33} - s_{13}^2), \tag{2.4.153}$$

$$\gamma_{33} = c_{33} - \frac{c_{23}^2 c_{44} + c_{34}^2 c_{22} - 2c_{23}c_{34}c_{24}}{c_{22}c_{44} - c_{24}^2} = s_{11} / (s_{11}s_{33} - s_{13}^2),$$

$$\gamma_{55} = c_{55} - c_{56}^2 / c_{66} = 1/ s_{55},$$

$$\psi_{11} = e_{11} - \frac{e_{12}(c_{12}c_{44} - c_{14}c_{24}) + e_{14}(c_{14}c_{22} - c_{12}c_{24})}{c_{22}c_{44} - c_{24}^2}$$

$$= d_{11}\gamma_{11} + d_{13}\gamma_{13},$$

$$\psi_{13} = e_{13} - \frac{e_{12}(c_{32}c_{44} - c_{34}c_{24}) + e_{14}(c_{34}c_{22} - c_{32}c_{24})}{c_{22}c_{44} - c_{24}^2} \qquad (2.4.154)$$

$$= d_{11}\gamma_{13} + d_{13}\gamma_{33},$$

$$\psi_{25} = e_{25} - e_{26}c_{56} / c_{66},$$

$$\psi_{35} = e_{36} - e_{36}c_{56} / c_{66} = d_{35}\gamma_{55},$$

$$\zeta_{11} = \varepsilon_{11} + \frac{e_{12}^2 c_{44} + e_{14}^2 c_{22} - 2e_{12}e_{14}c_{24}}{c_{22}c_{44} - c_{24}^2},$$

$$\zeta_{22} = \varepsilon_{22} + e_{26}^2 / c_{66}, \qquad (2.4.155)$$

$$\zeta_{23} = \varepsilon_{23} + e_{26}e_{36} / c_{66},$$

$$\zeta_{33} = \varepsilon_{33} + e_{36}^2 / c_{66}.$$

The constitutive relations are

$$T_1^{(0)} = 2h(\bar{c}_{11}S_1^{(0)} + \bar{c}_{13}S_3^{(0)} + \kappa_3\bar{c}_{14}S_4^{(0)} - \bar{e}_{11}E_1^{(0)}),$$

$$T_3^{(0)} = 2h(\bar{c}_{13}S_1^{(0)} + \bar{c}_{33}S_3^{(0)} + \kappa_3\bar{c}_{34}S_4^{(0)} - \bar{e}_{13}E_1^{(0)}),$$

$$T_4^{(0)} = 2h(\kappa_3\bar{c}_{14}S_1^{(0)} + \kappa_3\bar{c}_{34}S_3^{(0)} + \kappa_3^2\bar{c}_{44}S_4^{(0)} - \kappa_3\bar{e}_{14}E_1^{(0)}), \qquad (2.4.156)$$

$$T_5^{(0)} = 2h(c_{55}S_5^{(0)} + \kappa_1 c_{56}S_6^{(0)} - e_{25}E_2^{(0)} - e_{35}E_3^{(0)}),$$

$$T_6^{(0)} = 2h(\kappa_1 c_{56}S_5^{(0)} + \kappa_1^2 c_{66}S_6^{(0)} - \kappa_1 e_{26}E_2^{(0)} - \kappa_1 e_{36}E_3^{(0)}),$$

$$D_1^{(0)} = 2h(\bar{e}_{11}S_1^{(0)} + \bar{e}_{13}S_3^{(0)} + \kappa_3\bar{e}_{14}S_4^{(0)} + \bar{\varepsilon}_{11}E_1^{(0)}),$$

$$D_2^{(0)} = 2h(e_{25}S_5^{(0)} + \kappa_1 e_{26}S_6^{(0)} + \varepsilon_{22}E_2^{(0)} + \varepsilon_{23}E_3^{(0)}), \qquad (2.4.157)$$

$$D_3^{(0)} = 2h(e_{35}S_5^{(0)} + \kappa_1 e_{36}S_6^{(0)} + \varepsilon_{23}E_2^{(0)} + \varepsilon_{33}E_3^{(0)}),$$

$$T_1^{(1)} = \frac{2h^3}{3}(\gamma_{11}S_1^{(1)} + \gamma_{13}S_3^{(1)} - \psi_{11}E_1^{(1)}),$$

$$T_3^{(1)} = \frac{2h^3}{3}(\gamma_{13}S_1^{(1)} + \gamma_{33}S_3^{(1)} - \psi_{13}E_1^{(1)}), \qquad (2.4.158)$$

$$T_5^{(1)} = \frac{2h^3}{3}(\gamma_{55}S_5^{(1)} - \psi_{25}E_2^{(1)} - \psi_{35}E_3^{(1)}),$$

$$D_1^{(1)} = \frac{2h^3}{3}(\psi_{11}S_1^{(0)} + \psi_{13}S_3^{(0)} + \zeta_{11}E_1^{(1)}),$$

$$D_3^{(1)} = \frac{2h^3}{3}(\psi_{35}S_5^{(0)} + \zeta_{32}E_2^{(1)} + \zeta_{33}E_3^{(1)}). \qquad (2.4.159)$$

The equations of motion and charge take the following form:

$$\bar{c}_{11}u_{1,11}^{(0)} + c_{55}u_{1,33}^{(0)} + (\kappa_1 c_{56} + \kappa_3\bar{c}_{14})u_{2,13}^{(0)} + (\bar{c}_{13} + c_{55})u_{3,31}^{(0)}$$
$$+ \bar{e}_{11}\phi_{,11}^{(0)} + e_{35}\phi_{,33}^{(0)} + \kappa_3\bar{c}_{14}u_{3,1}^{(1)} + \kappa_1 c_{56}u_{1,3}^{(1)} \qquad (2.4.160)$$
$$+ e_{25}\phi_{,3}^{(1)} + \frac{1}{2h}F_1^{(0)} = \rho\ddot{u}_1^{(0)},$$

$$(\kappa_1 c_{56} + \kappa_3\bar{c}_{14})u_{1,13}^{(0)} + \kappa_1^2 c_{66}u_{2,11}^{(0)} + \kappa_3^2\bar{c}_{44}u_{2,33}^{(0)}$$
$$+ \kappa_1 c_{56}u_{3,11}^{(0)} + \kappa_3\bar{c}_{34}u_{3,33}^{(0)} + (\kappa_1 e_{36} + \kappa_3\bar{e}_{14})\phi_{,13}^{(0)} \qquad (2.4.161)$$
$$+ \kappa_1^2 c_{66}u_{1,1}^{(1)} + \kappa_3^2\bar{c}_{44}u_{3,3}^{(1)} + \kappa_1 e_{26}\phi_{,1}^{(1)} + \frac{1}{2h}F_2^{(0)} = \rho\ddot{u}_2^{(0)},$$

$$(c_{55} + \bar{c}_{13})u_{1,13}^{(0)} + \kappa_1 c_{56}u_{2,11}^{(0)} + \kappa_3\bar{c}_{34}u_{2,33}^{(0)} + c_{55}u_{3,11}^{(0)} + \bar{c}_{33}u_{3,33}^{(0)}$$
$$+ (e_{35} + \bar{e}_{13})\phi_{,13}^{(0)} + \kappa_1 c_{56}u_{1,1}^{(1)} + \kappa_3\bar{c}_{34}u_{3,3}^{(1)} \qquad (2.4.162)$$
$$+ e_{25}\phi_{,1}^{(1)} + \frac{1}{2h}F_3^{(0)} = \rho\ddot{u}_3^{(0)},$$

$$\bar{e}_{11}u_{1,11}^{(0)} + e_{15}u_{1,33}^{(0)} + (\kappa_1 e_{36} + \kappa_3 \bar{e}_{14})u_{2,13}^{(0)} + (\bar{e}_{13} + e_{35})u_{3,31}^{(0)}$$
$$+ \kappa_1 e_{36}u_{1,3}^{(1)} + \kappa_3 \bar{e}_{14}u_{3,1}^{(1)} - \bar{\varepsilon}_{11}\phi_{,11}^{(0)} + \varepsilon_{33}\phi_{,33}^{(0)} \qquad (2.4.163)$$
$$- \varepsilon_{23}\phi_{,3}^{(1)} + \frac{1}{2h}D^{(0)} = 0,$$

$$\gamma_{11}u_{1,11}^{(1)} + \gamma_{55}u_{1,33}^{(1)} + (\gamma_{13} + \gamma_{55})u_{3,31}^{(1)} + \psi_{11}\phi_{,11}^{(1)} + \psi_{35}\phi_{,33}^{(1)}$$
$$- 3h^{-2}[\kappa_1 c_{56}(u_{1,3}^{(0)} + u_{3,1}^{(0)}) + \kappa_1^2 c_{66}(u_{2,1}^{(0)} + u_1^{(1)}) \qquad (2.4.164)$$
$$+ \kappa_1 e_{26}\phi^{(1)} + \kappa_1 e_{36}\phi_{,3}^{(0)}] + \frac{3}{2h^3}F_1^{(1)} = \rho\ddot{u}_1^{(1)},$$

$$(\gamma_{13} + \gamma_{55})u_{1,31}^{(1)} + \gamma_{55}u_{3,11}^{(1)} + \gamma_{33}u_{3,33}^{(1)} + (\psi_{11} + \psi_{35})\phi_{,13}^{(1)}$$
$$- 3h^{-2}[\kappa_3 \bar{c}_{14}u_{1,1}^{(0)} + \kappa_3 \bar{c}_{34}u_{3,3}^{(0)} + \kappa_3^2 \bar{c}_{44}(u_{2,3}^{(0)} + u_3^{(1)}) \qquad (2.4.165)$$
$$+ \kappa_4 \bar{e}_{14}\phi_{,1}^{(0)}] + \frac{3}{2h^3}F_3^{(1)} = \rho\ddot{u}_3^{(1)},$$

$$\psi_{11}u_{1,11}^{(1)} + \psi_{35}u_{1,33}^{(1)} + (\psi_{13} + \psi_{35})u_{3,31}^{(1)} - \varsigma_{11}\phi_{,11}^{(1)} - \varsigma_{33}\phi_{,33}^{(1)}$$
$$- 3h^{-2}[e_{25}(u_{1,3}^{(0)} + u_{3,1}^{(0)}) + \kappa_1 e_{26}(u_{2,1}^{(0)} + u_1^{(1)}) \qquad (2.4.166)$$
$$- \varepsilon_{22}\phi^{(1)} - \varepsilon_{23}\phi_{,3}^{(0)}] + \frac{3}{2h^3}D^{(1)} = 0.$$

2.4.10 A quartz piezoelectric resonator

We now revisit the thickness-shear vibration problem of a quartz plate analyzed in the first section of this chapter using the three-dimensional equations. Consider a rotated Y-cut quartz plate as shown in Figure 2.4.17.

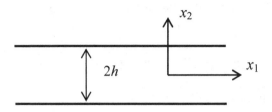

Figure 2.4.17. A rotated Y-cut quartz plate.

The plate is electroded, with a voltage $\pm 0.5V \exp(i\omega t)$ applied at $x_2 = \pm h$. We study pure thickness-shear motions described by $u_1^{(1)}$, without x_1 and x_3 dependence. The electric potentials are given by

$$\phi^{(0)} = 0, \quad \phi^{(1)} = \frac{V}{2h} \exp(i\omega t). \tag{2.4.167}$$

From Equation (2.4.164) we have the equation for $u_1^{(1)}$:

$$-3h^{-2}[\kappa_1^2 c_{66} u_1^{(1)} + \kappa_1 e_{26} \phi^{(1)}] = \rho \ddot{u}_1^{(1)}, \tag{2.4.168}$$

from which we obtain

$$u_1^{(1)} = \frac{3\kappa_1 e_{26}}{\rho h^2 (\omega^2 - \omega_\infty^2)} \frac{V}{2h} \exp(i\omega t),$$

$$\omega_\infty^2 = \frac{3\kappa_1^2 c_{66}}{\rho h^2}. \tag{2.4.169}$$

From Equation (2.4.157)$_2$

$$D_2^{(0)} = 2h(\kappa_1 e_{26} S_6^{(0)} + \varepsilon_{22} E_2^{(0)}) = 2h(\kappa_1 e_{26} u_1^{(1)} - \varepsilon_{22} \phi^{(1)})$$

$$= \varepsilon_{22} \left[\frac{k_{26}^2 \omega_\infty^2}{\omega^2 - \omega_\infty^2} - 1 \right] V \exp(i\omega t), \tag{2.4.170}$$

$$k_{26}^2 = \frac{e_{26}^2}{\varepsilon_{22} c_{66}}.$$

The surface free charge per unit area of the electrode at $x_2 = h$ is given by

$$\sigma_e = -\frac{D_2^{(0)}}{2h} = \frac{\varepsilon_{22}}{2h}\left[1 - \frac{k_{26}^2 \omega_\infty^2}{\omega^2 - \omega_\infty^2}\right] V \exp(i\omega t). \qquad (2.4.171)$$

Hence the capacitance per unit area is

$$C = \frac{\sigma_e}{V} = \frac{\varepsilon_{22}}{2h}\left[1 - \frac{k_{26}^2 \omega_\infty^2}{\omega^2 - \omega_\infty^2}\right] \cong \frac{\varepsilon_{22}}{2h}\left[1 - \frac{k_{26}^2 \omega_\infty}{2(\omega - \omega_\infty)}\right],$$

$$C = C_0 + C_m, \qquad (2.4.172)$$

$$C_0 = \frac{\varepsilon_{22}}{2h}, \quad C_m = \frac{\varepsilon_{22}}{2h}\frac{k_{26}^2 \omega_\infty^2}{\omega^2 - \omega_\infty^2},$$

where C_0 is the static capacitance and C_m is the motional capacitance.

2.4.11 Free vibration eigenvalue problem

One of the major applications of the two-dimensional equations developed is resonant piezoelectric devices. In these devices the eigenvalue problem for free vibration frequencies and modes is of fundamental importance. The free vibration eigenvalue problem of a piezoelectric body using the three-dimensional equations was studied in [9,31]. We discuss the two-dimensional version using the first-order plate equations below [31,32]. Consider a plate with the x_2 axis along its normal as shown in Figure 2.4.18. Let the two-dimensional area of

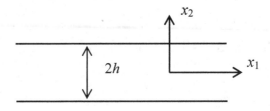

Figure 2.4.18. A piezoelectric plate and coordinate system.

the plate be A and the boundary curve of A be C, and C be partitioned as

$$C_u \cup C_T = C_\phi \cup C_D = C,$$
$$C_u \cap C_T = C_\phi \cap C_D = 0. \tag{2.4.173}$$

The eigenvalue problem for free vibrations of a piezoelectric plate by the first-order theory is

$$-T_{ji,j}^{(0)} = \omega^2 2h\rho u_i^{(0)} \quad \text{in} \quad A,$$

$$-T_{ba,b}^{(1)} + T_{2a}^{(0)} = \omega^2 \frac{2h^3}{3} \rho u_a^{(1)} \quad \text{in} \quad A,$$

$$-D_{i,i}^{(0)} = 0 \quad \text{in} \quad A, \tag{2.4.174}$$

$$-D_{a,a}^{(1)} - D_2^{(0)} = 0 \quad \text{in} \quad A,$$

$$-S_{ij}^{(0)} + \frac{1}{2}(u_{j,i}^{(0)} + u_{i,j}^{(0)} + \delta_{i2}u_j^{(1)} + \delta_{2j}u_i^{(1)}) = 0 \quad \text{in} \quad A,$$

$$-S_{ab}^{(1)} + \frac{1}{2}(u_{a,b}^{(0)} + u_{b,a}^{(0)}) = 0 \quad \text{in} \quad A, \tag{2.4.175}$$

$$E_i^{(0)} + \phi_{,i}^{(0)} + \delta_{2i}\phi^{(1)} = 0 \quad \text{in} \quad A,$$

$$E_a^{(1)} + \phi_{,a}^{(1)} = 0 \quad \text{in} \quad A,$$

$$-T_{ij}^{(0)} + 2h(\overline{c}_{ijkl}' S_{kl}^{(0)} - \overline{e}_{kij}' E_k^{(0)}) = 0 \quad \text{in} \quad A,$$

$$-T_r^{(1)} + \frac{2h^3}{3}(\gamma_{rs} S_s^{(1)} - \psi_{kr} E_k^{(1)}) = 0 \quad \text{in} \quad A, \tag{2.4.176}$$

$$D_i^{(0)} - 2h(\overline{e}_{ikl}' S_{kl}^{(0)} + \overline{\varepsilon}_{ij} E_j^{(0)}) = 0 \quad \text{in} \quad A,$$

$$D_a^{(1)} - \frac{2h^3}{3}(\psi_{as} S_s^{(1)} + \zeta_{ab} E_b^{(1)}) = 0 \quad \text{in} \quad A, \tag{2.4.177}$$

$$-u_i^{(0)} = 0, \quad -u_a^{(1)} = 0 \quad \text{on} \quad C_u,$$
$$n_j T_{ji}^{(0)} = 0, \quad n_b T_{ba}^{(0)} = 0 \quad \text{on} \quad C_T,$$
$$-\phi^{(0)} = 0, \quad -\phi^{(1)} = 0 \quad \text{on} \quad C_\phi, \tag{2.4.178}$$
$$n_i D_i^{(0)} = 0, \quad n_a D_a^{(1)} = 0 \quad \text{on} \quad C_D.$$

We look for values of ω^2 corresponding to which nontrivial solutions exist. For convenience we introduce the following vector $\mathbf{U}$:

$$\mathbf{U} = \{u_i^{(0)}, u_a^{(1)}, \phi^{(0)}, \phi^{(1)}, T_{ji}^{(0)}, T_{ba}^{(1)}, D_i^{(0)}, D_a^{(1)},$$
$$S_{ij}^{(0)}, S_{ba}^{(1)}, E_i^{(0)}, E_a^{(1)}\}, \tag{2.4.179}$$

and operators $\mathbf{A}$ and $\mathbf{B}$:

$$\mathbf{AU} = \{-T_{ji,j}^{(0)}, -T_{ba,b}^{(1)} + T_{2a}^{(0)}, -D_{i,i}^{(0)}, -D_{a,a}^{(1)} - D_2^{(0)},$$
$$-S_{ij}^{(0)} + \frac{1}{2}(u_{j,i}^{(0)} + u_{i,j}^{(0)} + \delta_{i2} u_j^{(1)} + \delta_{2j} u_i^{(1)}),$$
$$-S_{ab}^{(1)} + \frac{1}{2}(u_{a,b}^{(0)} + u_{b,a}^{(0)}), \, E_i^{(0)} + \phi_{,i}^{(0)} + \delta_{2i} \phi^{(1)}, \, E_a^{(1)} + \phi_{,a}^{(1)},$$
$$-T_{ij}^{(0)} + 2h(\overline{c}_{ijkl}' S_{kl}^{(0)} - \overline{e}_{kij}' E_k^{(0)}), \tag{2.4.180}$$
$$-T_r^{(1)} + \frac{2h^3}{3}(\gamma_{rs} S_s^{(1)} - \psi_{kr} E_k^{(1)}),$$
$$D_i^{(0)} - 2h(\overline{e}_{ikl}' S_{kl}^{(0)} + \overline{\varepsilon}_{ij} E_j^{(0)}),$$
$$D_a^{(1)} - \frac{2h^3}{3}(\psi_{as} S_s^{(1)} + \zeta_{ab} E_b^{(1)})\},$$

$$\mathbf{BU} = \{2h\rho u_i^{(0)}, \frac{2h^3}{3}\rho u_a^{(1)}, 0, 0, 0, 0, 0, 0, 0, 0, 0, 0\}. \tag{2.4.181}$$

We also define

$$\Sigma(A) = \{\mathbf{U} \mid \text{Equation (2.4.178) is satisfied}\}. \tag{2.4.182}$$

With the above definitions, the eigenvalue problem can be stated as: Find $\lambda = \omega^2$ for which there exists a nontrivial $\mathbf{U} \in \Sigma(A)$ such that

$$\mathbf{AU} = \lambda \mathbf{BU}. \tag{2.4.183}$$

For two vectors $\mathbf{U}, \mathbf{V} \in \Sigma(A)$, we introduce an inner product

$$
\begin{aligned}
< \mathbf{U}; \mathbf{V} >= \int_A \ & (u_i^{(0)} v_i^{(0)} + u_a^{(1)} v_a^{(1)} + \phi^{(0)} \psi^{(0)} + \phi^{(1)} \psi^{(1)} \\
& + T_{ji}^{(0)} G_{ji}^{(0)} + T_{ba}^{(1)} G_{ba}^{(1)} + D_i^{(0)} B_i^{(0)} + D_a^{(1)} B_a^{(1)} \\
& + S_{ji}^{(0)} F_{ji}^{(0)} + S_{ba}^{(1)} F_{ba}^{(1)} + E_i^{(0)} C_i^{(0)} + E_a^{(1)} C_a^{(1)}) dA,
\end{aligned}
\tag{2.4.184}
$$

where $\mathbf{U}$ is given by Equation (2.4.180) and

$$
\begin{aligned}
\mathbf{V} = \{ & v_i^{(0)}, v_a^{(1)}, \psi^{(0)}, \psi^{(1)}, G_{ji}^{(0)}, G_{ba}^{(1)}, B_i^{(0)}, B_a^{(1)}, \\
& F_{ij}^{(0)}, F_{ba}^{(1)}, C_i^{(0)}, C_a^{(1)} \}.
\end{aligned}
\tag{2.4.185}
$$

For two vectors $\mathbf{U}, \mathbf{V} \in \Sigma(A)$, it can be shown by integration by parts that

$$
\begin{aligned}
& < \mathbf{AU}; \mathbf{V} >=< \mathbf{U}; \mathbf{AV} >, \\
& < \mathbf{BU}; \mathbf{V} >=< \mathbf{U}; \mathbf{BV} >.
\end{aligned}
\tag{2.4.186}
$$

Therefore the operators $\mathbf{A}$ and $\mathbf{B}$ are self-adjoint. This is consistent with the self-adjointness of the three-dimensional operators [9] from which the plate equations are derived. With the above formulation in terms of the abstract vectors and operators, it can be shown in the manner of [9] that the eigenvalues are real and the eigenvectors associated with distinct eigenvalues $\lambda^{(m)}$ and $\lambda^{(n)}$ are orthogonal:

$$
< \mathbf{AU}^{(m)}; \mathbf{U}^{(n)} >= 0, \quad < \mathbf{BU}^{(m)}; \mathbf{U}^{(n)} >= 0, \quad m \neq n.
\tag{2.4.187}
$$

With the abstract formulation, in a way similar to [9], it can also be shown that the Rayleigh quotient of the variational formulation of the eigenvalue problem is

$$
\Pi(\mathbf{U}) = \frac{< \mathbf{AU}; \mathbf{U} >}{< \mathbf{BU}; \mathbf{U} >}.
\tag{2.4.188}
$$

2.5 Second-Order Theory

In this section we explore the effects of the second-order terms in the power series expansion. A second-order theory is outlined, and several special cases of the second-order theory are examined.

2.5.1 Equations for second-order theory

A second-order theory describes coupled extensional ($u_a^{(0)}$, $a = 1, 2$), flexural ($u_3^{(0)}$), fundamental thickness-shear ($u_a^{(1)}$), thickness-stretch ($u_3^{(1)}$), and symmetric thickness-shear ($u_a^{(2)}$) motions of a plate. For the electrical behavior of the plate, we consider $\phi^{(0)}$, $\phi^{(1)}$ and $\phi^{(2)}$. The approximate displacement and potential fields are

$$u_i \cong u_i^{(0)} + x_3 u_i^{(1)} + x_3^2 u_i^{(2)} + x_3^3 u_i^{(3)},$$
$$\phi \cong \phi^{(0)} + x_3 \phi^{(1)} + x_3^2 \phi^{(2)}. \tag{2.5.1}$$

We have included a few additional displacement components in Equation (2.5.1). $u_3^{(2)}$ and $u_3^{(3)}$ represent the thickness stretch or contraction due to Poisson's effect. From Equations (2.2.6) and (2.2.7) it can be seen that $u_3^{(2)}$ together with $u_a^{(3)}$ contribute to thickness-shear deformations $S_4^{(2)}$ and $S_5^{(2)}$ which may couple to the other second-order strains due to anisotropy. The two-dimensional plate equations we will obtain are for $u_i^{(0)}$, $u_i^{(1)}$ and $u_a^{(2)}$ only. Other displacements will be eliminated through stress relaxations. Within the approximation in Equation (2.5.1), the strains and electric fields in Equations (2.2.5) through (2.2.10) become:

$$S_1^{(0)} = u_{1,1}^{(0)}, \quad S_2^{(0)} = u_{2,2}^{(0)}, \quad S_3^{(0)} = u_3^{(1)},$$
$$S_4^{(0)} = u_{3,2}^{(0)} + u_2^{(1)}, \quad S_5^{(0)} = u_{3,1}^{(0)} + u_1^{(1)}, \quad S_6^{(0)} = u_{1,2}^{(0)} + u_{2,1}^{(0)}, \tag{2.5.2}$$

$$S_1^{(1)} = u_{1,1}^{(1)}, \quad S_2^{(1)} = u_{2,2}^{(1)}, \quad S_3^{(1)} = 2u_3^{(2)},$$
$$S_4^{(1)} = u_{3,2}^{(1)} + 2u_2^{(2)}, \quad S_5^{(1)} = u_{3,1}^{(1)} + 2u_1^{(2)}, \quad S_6^{(1)} = u_{1,2}^{(1)} + u_{2,1}^{(1)}, \tag{2.5.3}$$

$$S_1^{(2)} = u_{1,1}^{(2)}, \quad S_2^{(2)} = u_{2,2}^{(2)}, \quad S_3^{(2)} = 3u_3^{(3)},$$
$$S_4^{(2)} = u_{3,2}^{(2)} + 3u_2^{(3)}, \quad S_5^{(2)} = u_{3,1}^{(2)} + 3u_1^{(3)}, \quad S_6^{(2)} = u_{1,2}^{(2)} + u_{2,1}^{(2)}, \tag{2.5.4}$$

$$E_1^{(0)} = -\phi_{,1}^{(0)}, \quad E_2^{(0)} = -\phi_{,2}^{(0)}, \quad E_3^{(0)} = -\phi^{(1)}, \tag{2.5.5}$$

$$E_1^{(1)} = -\phi_{,1}^{(1)}, \quad E_2^{(1)} = -\phi_{,2}^{(1)}, \quad E_3^{(1)} = -2\phi^{(2)}, \tag{2.5.6}$$

$$E_1^{(2)} = -\phi_{,1}^{(2)}, \quad E_2^{(2)} = -\phi_{,2}^{(2)}, \quad E_3^{(2)} = 0. \tag{2.5.7}$$

Higher order strains $S_{ij}^{(3)}$ are neglected. From Equation $(2.2.17)_1$ we obtain

$$T_{ab,a}^{(0)} + F_b^{(0)} = 2h\rho\ddot{u}_b^{(0)} + \frac{2h^3}{3}\rho\ddot{u}_b^{(2)}, \quad a,b = 1,2,$$

$$T_{a3,a}^{(0)} + F_3^{(0)} = 2h\rho\ddot{u}_3^{(0)},$$

$$T_{ab,a}^{(1)} - T_{3b}^{(0)} + F_b^{(1)} = \frac{2h^3}{3}\rho\ddot{u}_b^{(1)}, \tag{2.5.8}$$

$$T_{a3,a}^{(1)} - T_{33}^{(0)} + F_3^{(1)} = \frac{2h^3}{3}\rho\ddot{u}_3^{(1)},$$

$$T_{ab,a}^{(2)} - 2T_{3b}^{(1)} + F_b^{(2)} = \frac{2h^3}{3}\rho\ddot{u}_b^{(0)} + \frac{2h^5}{5}\rho\ddot{u}_b^{(2)},$$

where we have truncated the right hand sides by keeping the inertial terms of the displacement components of interest only. Equation $(2.5.8)_1$ is for extension, Equation $(2.5.8)_2$ is for flexure, Equation $(2.5.8)_3$ is for the fundamental thickness-shear, Equation $(2.5.8)_4$ is for the fundamental thickness-stretch, and Equation $(2.5.8)_5$ is for the symmetric thickness-shear. For the charge equations of electrostatics we have, from Equation $(2.2.17)_2$, for an unelectroded plate:

$$D_{a,a}^{(0)} + D^{(0)} = 0,$$

$$D_{a,a}^{(1)} - D_3^{(0)} + D^{(1)} = 0, \tag{2.5.9}$$

$$D_{a,a}^{(2)} - 2D_3^{(1)} + D^{(2)} = 0.$$

The case of an electroded plate will not be examined. In fact, in higher-order theories, when electrodes are present, it is more convenient to use two-dimensional charge equations obtained by the polynomial expansion of ϕ in Equation (2.2.18).

For plate constitutive relations we truncate Equation (2.2.12) as

$$T_{ij}^{(0)} = 2h(c_{ijkl}S_{kl}^{(0)} - e_{kij}E_k^{(0)}) + \frac{2h^3}{3}(c_{ijkl}S_{kl}^{(2)} - e_{kij}E_k^{(2)}),$$

$$D_i^{(0)} = 2h(e_{ijk}S_{jk}^{(0)} + \varepsilon_{ij}E_j^{(0)}) + \frac{2h^3}{3}(e_{ijk}S_{jk}^{(2)} + \varepsilon_{ij}E_j^{(2)}),$$

(2.5.10)

$$T_{ij}^{(1)} = \frac{2h^3}{3}(c_{ijkl}S_{kl}^{(1)} - e_{kij}E_k^{(1)}),$$

$$D_i^{(1)} = \frac{2h^3}{3}(e_{ijk}S_{jk}^{(1)} + \varepsilon_{ij}E_j^{(1)}),$$

(2.5.11)

$$T_{ij}^{(2)} = \frac{2h^3}{3}(c_{ijkl}S_{kl}^{(0)} - e_{kij}E_k^{(0)}) + \frac{2h^5}{5}(c_{ijkl}S_{kl}^{(2)} - e_{kij}E_k^{(2)}),$$

$$D_i^{(2)} = \frac{2h^3}{3}(e_{ijk}S_{jk}^{(0)} + \varepsilon_{ij}E_j^{(0)}) + \frac{2h^5}{5}(e_{ijk}S_{jk}^{(2)} + \varepsilon_{ij}E_j^{(2)}),$$

(2.5.12)

where $S_{ij}^{(3)}$ has been neglected. Since $S_{ij}^{(1)}$ contains $u_3^{(2)}$ and $S_{ij}^{(2)}$ contains $u_3^{(2)}$ and $u_i^{(3)}$, Equations (2.5.10) through (2.5.12) are not yet ready to be used. To obtain the proper constitutive relations we need to make the following stress relaxations:

$$T_{33}^{(1)} = 0,$$

$$T_q^{(2)} = 0, \quad q = 3,4,5,$$

(2.5.13)

from which expression of $S_3^{(1)}$, $S_3^{(2)}$, $S_4^{(2)}$ and $S_5^{(2)}$ in terms of other strain components can be obtained to eliminate $u_3^{(2)}$ and $u_i^{(3)}$ in Equations (2.5.10) through (2.5.12). Then Equations (2.5.8) and (2.5.9) can be written as eleven equations for $u_i^{(0)}$, $u_i^{(1)}$, $u_a^{(2)}$, $\phi^{(0)}$, $\phi^{(1)}$ and $\phi^{(2)}$. In order to predict the exact cutoff frequencies for the two fundamental thickness-shear, the fundamental thickness-stretch, and the two symmetric thickness-shear modes, five correction factors will be needed.

2.5.2 Extension, thickness-stretch and symmetric thickness-shear

To examine the basic behaviors of the second-order equations, we study the special case of an isotropic elastic plate. In this case the second-order equations decouple into two groups. One is for coupled flexure and fundamental thickness-shear, and the other is for coupled extension, thickness-stretch, and symmetric thickness-shear. The former has been discussed in the fourth section of this chapter. We explore the latter below.

It is obvious that when a plate is in extension there is thickness stretch or contraction due to Poisson's effect. At the low frequency range this thickness deformation can be approximately determined from the stress relaxations in the zero-order theory, i.e., Equation (2.3.15) or Equation (2.3.23). When the frequency gets higher or close to the fundamental thickness-stretch frequency, resonance of the thickness-stretch vibration needs to be considered. In addition, high frequency extensional waves become dispersive (see Figure 2.1.4).

First-order, two-dimensional plate equations governing coupled extensional and thickness-stretch motions of plates are due to Kane and Mindlin [33]. This theory includes the resonance of the fundamental thickness-stretch mode and can predict the dispersion of high frequency extensional waves. The equations of the Kane-Mindlin theory can be reduced to the classical equations of extension for low frequencies and long waves [13]. However, the Kane-Mindlin theory has a serious flaw. It cannot predict a complex branch of the dispersion curve that the three-dimensional theory predicts to exist in the frequency range of interest [34]. It was also shown in [34] that in order to capture this complex branch the second order effect of symmetric thickness-shear has to be included. Equations for elastic plates in coupled extension, thickness-stretch and symmetric thickness-shear motions are due to Mindlin and Medick [34] using polynomial expansions. Two-dimensional equations containing extensional, thickness-stretch, and symmetric thickness-shear and higher-order modes were also obtained in [19] using trigonometric expansions. The equations for extensional, thickness-stretch, and symmetric thickness-shear in [19] are equivalent to the Mindlin-Medick theory in that their dispersion curves have the same geometric structure. In the following we will use the equations in [19] for which the correction factors have been determined.

According to the notation used in [19], we orient the plate normal along the x_2 axis (see Fig. 2.5.1)

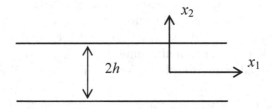

Figure 2.5.1. An isotropic elastic plate and coordinate system.

In the absence of body force, the differential equations for coupled extension, thickness-stretch and symmetric thickness-shear in [19] are

$$\mu u_{a,bb}^{(0)} + (\lambda + \mu)u_{b,ba}^{(0)} + \frac{2\lambda}{h}\alpha_1 u_{2,a}^{(1)} = \rho \ddot{u}_a^{(0)}, \quad a = 1,3,$$

$$\mu\alpha_2 u_{2,bb}^{(1)} - (\lambda + 2\mu)(\frac{\pi}{2h})^2 u_2^{(1)}$$

$$- \frac{2\lambda}{h}\alpha_1 u_{a,a}^{(0)} + \frac{2(\lambda + 4\mu)}{3h}u_{b,b}^{(2)} = \rho \ddot{u}_2^{(1)}, \qquad (2.5.14)$$

$$\mu u_{a,bb}^{(2)} + (\lambda + \mu)u_{b,ba}^{(2)} - \mu(\frac{\pi}{h})^2 u_a^{(2)} - \frac{2(\lambda + 4\mu)}{3h}u_{2,a}^{(1)} = \rho \ddot{u}_a^{(2)},$$

where λ and μ are the Lamè constants of an elastic material. $u_a^{(0)}$ ($a = 1$, 3) are extensional displacements, $u_2^{(1)}$ is thickness-stretch, and $u_a^{(2)}$ is symmetric thickness-shear. α_1 and α_2 are two correction factors given by

$$\alpha_1 = \frac{\pi}{4}, \quad \alpha_2 = \alpha_2(\nu), \qquad (2.5.15)$$

where ν is Poisson's ratio. The three-dimensional displacement field in this theory is given by

$$u_a(x_1, x_2, x_3, t) = u_a^{(0)}(x_1, x_3, t) + u_a^{(2)}(x_1, x_3, t)\cos\pi(1 - \frac{x_2}{h}),$$

$$u_2(x_1, x_2, x_3, t) = u_2^{(1)}(x_1, x_3, t)\cos\frac{\pi}{2}(1 - \frac{x_2}{h}). \qquad (2.5.16)$$

The plate resultants are given below. Those associated with extension are

$$T_{11}^{(0)} = (\lambda + 2\mu)u_{1,1}^{(0)} + \lambda u_{3,3}^{(0)} + \frac{2\lambda}{h}u_2^{(1)},$$

$$T_{33}^{(0)} = (\lambda + 2\mu)u_{3,3}^{(0)} + \lambda u_{1,1}^{(0)} + \frac{2\lambda}{h}u_2^{(1)}, \qquad (2.5.17)$$

$$T_{13}^{(0)} = \mu(u_{1,3}^{(0)} + u_{3,1}^{(0)}).$$

The two self-equilibrating shear resultants $T_{a2}^{(1)}$ for thickness-stretch are

$$T_{a2}^{(1)} = \alpha_2\lambda u_{2,a}^{(1)} + \frac{8\mu}{3h}u_a^{(2)}, \qquad (2.5.18)$$

while the second-order resultants are given by

$$T_{11}^{(2)} = (\lambda + 2\mu)u_{1,1}^{(2)} + \lambda u_{3,3}^{(2)} - \frac{2\lambda}{3h}u_2^{(1)},$$

$$T_{33}^{(2)} = (\lambda + 2\mu)u_{3,3}^{(2)} + \lambda u_{1,1}^{(2)} - \frac{2\lambda}{3h}u_2^{(1)}, \qquad (2.5.19)$$

$$T_{13}^{(2)} = \mu(u_{1,3}^{(2)} + u_{3,1}^{(2)}).$$

We now consider straight-crested waves propagating in the x_1 direction ($\partial / \partial x_3 = 0$). Furthermore, we look for solutions in which $u_3^{(0)} = u_3^{(2)} = 0$. Then, Equations (2.5.14) reduce to

$$(\lambda + 2\mu)u_{1,11}^{(0)} + \frac{2\lambda}{h}\alpha_1 u_{2,1}^{(1)} = \rho\ddot{u}_1^{(0)},$$

$$\mu\alpha_2 u_{2,11}^{(1)} - (\lambda + 2\mu)(\frac{\pi}{2h})^2 u_2^{(1)}$$

$$\frac{2\lambda}{h}\alpha_1 u_{1,1}^{(0)} + \frac{2(\lambda + 4\mu)}{3h}u_{1,1}^{(2)} = \rho\ddot{u}_2^{(1)}, \qquad (2.5.20)$$

$$(\lambda + 2\mu)u_{1,11}^{(2)} - \mu(\frac{\pi}{h})^2 u_1^{(2)} - \frac{2(\lambda + 4\mu)}{3h}u_{2,1}^{(1)} = \rho\ddot{u}_1^{(2)}.$$

Let

$$u_1^{(0)} = A_1^{(0)}\exp i(\xi x_1 - \omega t),$$

$$u_1^{(2)} = A_1^{(2)}\exp i(\xi x_1 - \omega t), \qquad (2.5.21)$$

$$u_2^{(1)} = -iA_2^{(1)}\exp i(\xi x_1 - \omega t),$$

so that Equations (2.5.20) yield

$$(\lambda + 2\mu)(-A_1^{(0)}\xi^2) + \frac{2\lambda}{h}\alpha_1 A_2^{(1)}\xi = -\rho\omega^2 A_1^{(0)},$$

$$\mu\alpha_2(-A_2^{(1)}\xi^2) - (\lambda + 2\mu)(\frac{\pi}{2h})^2 A_2^{(1)}$$

$$+\frac{2\lambda}{h}\alpha_1 A_1^{(0)}\xi - \frac{2(\lambda + 4\mu)}{3h}A_1^{(2)}\xi = -\rho\omega^2 A_2^{(1)}, \qquad (2.5.22)$$

$$(\lambda + 2\mu)(-A_1^{(2)}\xi^2) - \mu(\frac{\pi}{h})^2 A_1^{(2)}$$

$$-\frac{2(\lambda + 4\mu)}{3h}A_2^{(1)}\xi = -\rho\omega^2 A_1^{(2)}.$$

For nontrivial solutions, we have

$$\begin{vmatrix} k^2 X^2 - \Omega^2 & -(k^2 - 2)X & 0 \\ -(k^2 - 2)X & k^2 + \alpha_2 X^2 - \Omega^2 & \frac{4}{3\pi}(k^2 + 2)X \\ 0 & \frac{4}{3\pi}(k^2 + 2)X & k^2 X^2 + 4 - \Omega^2 \end{vmatrix} = 0, \qquad (2.5.23)$$

where we have used the notation

$$X = \frac{2\xi h}{\pi}, \quad \Omega = \omega/(\frac{\pi v_2}{2h}),$$

$$v_1^2 = \frac{\lambda + 2\mu}{\rho}, \quad v_2^2 = \frac{\mu}{\rho}, \quad k^2 = \frac{v_1^2}{v_2^2} = 2\frac{1-v}{1-2v}. \qquad (2.5.24)$$

v_1 and v_2 are the speeds of longitudinal and transverse plane waves in an unbounded, isotropic elastic body, respectively. Equation (2.5.23) determines the dispersion relations for coupled extension, thickness-stretch, and symmetric thickness-shear modes of the plate. These dispersion curves are shown in Figure 2.5.2 for $v = 0.25$ and $\alpha_2 = 0.845$.

Besides being able to describe the slope and curvature of the dispersion curve for long thickness-stretch waves at frequencies close to the first thickness-stretch frequency, Equation (2.5.23) can also predict the complex branch of the dispersion relation that was missing in the Kane-Mindlin theory. We have the following local, asymptotic

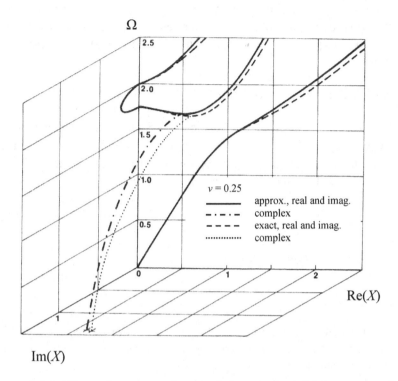

Figure 2.5.2. Dispersion curves for coupled extension, thickness-stretch and symmetric thickness-shear waves [19].

expressions of Equation (2.5.23) for long waves. Near the lowest thickness-stretch cutoff frequency, i.e., near $X = 0$, $\Omega = k$, we have

$$\Omega^2 \cong k^2 - \frac{1}{4-k^2}[\frac{16}{9\pi}(k^2+2)^2 - \alpha_3(4-k^2)]X^2, \qquad (2.5.25)$$

while near the symmetric thickness-shear cutoff frequency ($X = 0$, $\Omega = 2$),

$$\Omega^2 \approx 4 + \frac{1}{4-k^2}[\frac{16}{9\pi^2}(k^2+2)^2 + (4-k^2)k^2]X^2, \qquad (2.5.26)$$

where

$$\alpha_3 = \alpha_2 + \frac{1}{k^2}(k^2-2)^2. \qquad (2.5.27)$$

We note that Equations (2.5.25) and (2.5.26) break down when $k^2 = 4$ or $v = 1/3$; this represents the degenerate case when the lowest thickness-stretch frequency coincides with the symmetric thickness-shear frequency.

2.5.3 Elimination of extension

For long waves at frequencies close to the first thickness-stretch frequency $\Omega = k$, an approximation can be made to eliminate the coupling to extension [35]. Dropping the term quadratic in ξ in Equation $(2.5.22)_1$ for long waves and setting $\Omega = k$, we obtain

$$A_1^{(0)} = -\frac{h}{\pi} \frac{2\lambda}{\lambda + 2\mu} A_2^{(1)} \xi, \qquad (2.5.28)$$

which is equivalent to

$$u_1^{(0)} = -\frac{h}{\pi} \frac{2\lambda}{\lambda + 2\mu} u_{2,1}^{(1)}. \qquad (2.5.29)$$

Substitution of Equation (2.5.29) into Equations $(2.5.20)_{2,3}$ yields the following equations for coupled thickness-stretch and symmetric thickness-shear:

$$\mu \alpha_3 u_{2,11}^{(1)} - (\lambda + 2\mu)(\frac{\pi}{2h})^2 u_2^{(1)} + \frac{2(\lambda + 4\mu)}{3h} u_{1,1}^{(2)} = \rho \ddot{u}_2^{(1)},$$
$$(\lambda + 2\mu) u_{1,11}^{(2)} - \mu(\frac{\pi}{h})^2 u_1^{(2)} - \frac{2(\lambda + 4\mu)}{3h} u_{2,1}^{(1)} = \rho \ddot{u}_1^{(2)}. \qquad (2.5.30)$$

Now, letting

$$u_1^{(2)} = A_1^{(2)} \exp i(\xi x_1 - \omega t),$$
$$u_2^{(1)} = -i A_2^{(1)} \exp i(\xi x_1 - \omega t), \qquad (2.5.31)$$

we obtain the following dispersion relation from Equation (2.5.30):

$$\begin{vmatrix} \alpha_3 X^2 + k^2 - \Omega^2 & \dfrac{4}{3\pi}(k^2 + 2)X \\[2mm] \dfrac{4}{3\pi}(k^2 + 2)X & k^2 X^2 + 4 - \Omega^2 \end{vmatrix} = 0. \qquad (2.5.32)$$

The dispersion curves determined by Equation (2.5.32) for thickness-stretch and symmetric thickness-shear are plotted in Figure 2.5.3 for $v = 0.25$ and $\alpha_2 = 0.845$. A comparison to Figure 2.5.2 shows that Equation (2.5.32) has the same behavior near the first thickness-stretch frequency for the slope, curvature, and complex branch, but the low frequency branch of extension has disappeared. It can be verified that near $X = 0$, $\Omega = k$ and near $X = 0$, $\Omega = 2$, the asymptotic expressions of Equation (2.5.32) are exactly Equations (2.5.25) and (2.5.26). Thus Equation (2.5.30) can be used to study long, coupled thickness-stretch and symmetric thickness-shear waves.

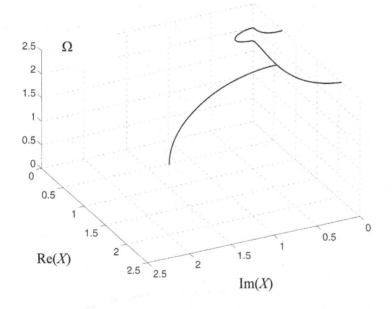

Figure 2.5.3. Dispersion curves for coupled thickness-stretch and symmetric thickness-shear waves [35].

2.5.4 Thickness-stretch approximation

We now further reduce Equation (2.5.30) into a single equation for $u_2^{(1)}$ by eliminating $u_1^{(2)}$ [35]. Similar to the derivation of Equation (2.5.29), for long waves and Ω near k, Equation (2.5.30)$_2$ can be approximated by

$$u_1^{(2)} = \frac{8b}{3\pi^2} \frac{\lambda+4\mu}{\lambda-2\mu} u_{2,1}^{(1)}.$$

(2.5.33)

Substitution of Equation (2.5.33) into Equation (2.5.30)₁ results in

$$\mu r u_{2,11}^{(1)} - (\lambda+2\mu)(\frac{\pi}{2h})^2 u_2^{(1)} = \rho \ddot{u}_2^{(1)},$$

(2.5.34)

where

$$r = \alpha_3 + \frac{16}{9\pi^2} \frac{(\lambda+4\mu)^2}{\mu(\lambda-2\mu)} = \alpha_3 + \frac{16}{9\pi^2} \frac{(k^2+2)^2}{k^2-4}.$$

(2.5.35)

It can be verified that the dispersion relation arising from Equation (2.5.34) is

$$\Omega^2 = k^2 + rX^2$$
$$= k^2 - \frac{1}{4-k^2}[\frac{16}{9\pi}(k^2+2)^2 - \alpha_3(4-k^2)]X^2,$$

(2.5.36)

which is the same as Equation (2.5.25). Therefore Equation (2.5.34) can be used to study long thickness-stretch waves. Equation (2.5.36) is plotted in Figure 2.5.4 for two values of v.

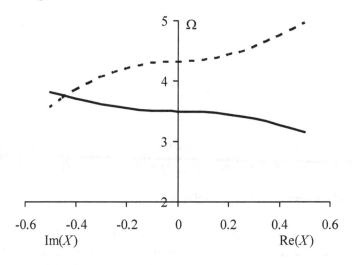

Figure 2.5.4. Dispersion curves for thickness-stretch waves. Solid line: v = 0.3. Dotted line: $v = 0.35$.

In addition to missing the complex branch of the dispersion curves, another flaw of the Kane-Mindlin theory is that it predicts a positive curvature near $X = 0$ for thickness-stretch modes when $v < 1/3$ [13,34,35]. This does not agree with the result of the three-dimensional theory. Equation (2.5.20) or (2.5.30) can predict the correct sign of curvature for the thickness-stretch branch near $X = 0$. Since the dispersion relation of Equation (2.5.34) is asymptotic to that of Equation (2.5.20) or (2.5.30) near $X = 0$, it can be expected that Equation (2.5.34) can predict the correct sign of curvature there. This relies on the sign of r, which must be negative for $v < 1/3$ to yield the correct sign of curvature. Calculations of r are listed in the following where the value of α_2 as a function of v is from [19]:

v	0	0.1	0.2	0.25	0.3	0.35	0.4	
α_2	0.764	0.798	0.830	0.845	0.860	0.874	0.888	(2.5.37)
r	-0.676	-1.03	-1.94	-3.33	-9.39	24.0	9.32	
k^2	2	2.25	2.67	3	3.5	4.33	6	

It is seen that for $k < 2$ or $v < 1/3$, the sign of r is indeed negative and therefore the correct sign of curvature is predicted by Equation (2.5.34) near $X = 0$ for the thickness-stretch branch. We also note the drastic change of r near $k = 2$. In the above analysis we assume that k is sufficiently far away from 2 because the degenerate case of $k = 2$ needs special treatment. Since Equation (2.5.34) cannot predict the complex branch of the dispersion curve, it is valid for small wave numbers before the complex branch comes into play. It is seen from Figure 2.5.2 that the complex branch needs to be considered when X is roughly greater than 0.5. Equation (2.5.24) then implies that Equation (2.5.34) is valid for waves with wavelength $\lambda = 2\pi/\xi > 8h$, or four times the plate thickness. For thickness-stretch vibrations governed by Equation (2.5.34), $u_2^{(1)}$ is the major displacement and the corresponding plate resultant is $T_{12}^{(1)}$. Under the present thickness-stretch approximation, from Equations (2.5.18) and (2.5.33) we obtain

$$T_{12}^{(1)} = (\alpha_2 \lambda + \frac{64\mu}{9\pi^2} \frac{\lambda+4\mu}{\lambda-2\mu})u_{2,1}^{(1)}. \tag{2.5.38}$$

In the fourth section of this chapter it was shown that thickness-shear modes exhibit energy trapping behavior. Since Equation (2.5.34) has the same mathematical structure as that for the thickness-shear approximation, it can be expected that thickness-stretch modes will also exhibit energy trapping. However, due to the fact that r changes sign at k = 2, there may be two types of trapping for thickness-stretch modes. When $k > 2$, the situation is the same as the trapping of thickness-shear modes because in this case the curvature of the dispersion curve of Equation (2.5.34) near $X = 0$ has the same sign as that of the thickness-shear approximation. When $k < 2$, the situation is different and we will discuss this case only. To this end, consider the plate in Figure 2.5.5, which has a small geometric discontinuity with $h_1 < h_2$.

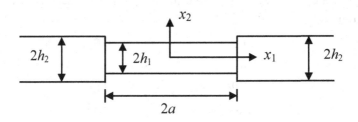

Figure 2.5.5. An elastic plate with slightly different thicknesses in different portions.

Equation (2.5.34) applies to each portion of the plate:

$$v_2^2 r u_{2,11}^{(1)} - \omega_1^2 u_2^{(1)} = \ddot{u}_2^{(1)}, \quad |x_1| < a,$$
$$v_2^2 r u_{2,11}^{(1)} - \omega_2^2 u_2^{(1)} = \ddot{u}_2^{(1)}, \quad |x_1| > a, \tag{2.5.39}$$

where we have denoted

$$\omega_1^2 = \frac{\lambda + 2\mu}{\rho}(\frac{\pi}{2h_1})^2, \quad \omega_2^2 = \frac{\lambda + 2\mu}{\rho}(\frac{\pi}{2h_2})^2. \tag{2.5.40}$$

At a junction of two portions of a plate, the continuity of $u_2^{(1)}$ and $T_{12}^{(1)}$ should be prescribed. In Equation (2.5.40), ω_1 and ω_2 are the thickness-stretch frequencies of infinite plates with thickness $2h_1$ and $2h_2$ respectively. The dispersion curves predicted by Equation (2.5.34) for long thickness-stretch waves in infinite plates with thickness $2h_1$ and $2h_2$ are qualitatively sketched in Figure 2.5.6.

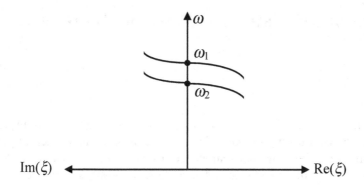

Figure 2.5.6. Dispersion curves for long thickness-stretch waves in infinite plates with thicknesses $2h_1$ and $2h_2$.

Since $h_1 < h_2$, we have $\omega_1 > \omega_2$. We are interested in time-harmonic vibrations with $\omega_2 < \omega < \omega_1$, and for these solutions, we have

$$u_{2,11}^{(1)} + \delta_1^2 u_2^{(1)} = 0, \quad |x_1| < a,$$
$$u_{2,11}^{(1)} - \delta_2^2 u_2^{(1)} = 0, \quad |x_1| > a,$$

(2.5.41)

where

$$\delta_1^2 = \frac{\omega_1^2 - \omega^2}{-v_2^2 r}, \quad \delta_2^2 = \frac{\omega^2 - \omega_2^2}{-v_2^2 r}.$$

(2.5.42)

Since for $k < 2$ or $v < 1/3$, the parameter r is negative, and since $\omega_2 < \omega < \omega_1$, we have $\delta_1^2 > 0$ and $\delta_2^2 > 0$. Equation (2.5.41) then shows that for $|x_1| < a$ the solution is sinusoidal and that for $|x_1| > a$ the solution decays exponentially. Hence, vibration is confined in the central region of $|x_1| < a$. Therefore the vibration energy is trapped. Obviously, when $k > 2$, for energy trapping we must have $h_1 > h_2$, which is similar to the trapping of thickness-shear modes.

Chapter 3
Laminated Plates and Plates on Substrates

Piezoelectric plates are often attached to or embedded in elastic plates for various applications. These structures can be treated in general as laminated plates. In this chapter we first analyze two special structures, i.e., an elastic plate with piezoelectric layers on one or both of its major surfaces. Then a general discussion on laminated piezoelectric plates is presented, followed by the analyses of two problems of a plate on a half-space.

3.1 Elastic Plates with Symmetric Piezoelectric Actuators

In this section, we consider the simplest case of an elastic plate with piezoelectric layers (actuators) symmetrically attached to its major surfaces. In this case, bending of the elastic plate can be produced by anti-symmetric voltages applied to the piezoelectric actuators without coupling to extension. A system of two-dimensional equations for the flexural motion of the combined elastic plate with piezoelectric actuators on it are obtained from the two-dimensional equations for the piezoelectric actuators and the two-dimensional equations for the elastic plate by satisfying the plate-actuator interface mechanical and geometric continuity conditions [36-38]. The piezoelectric actuators can be partially electroded, which has an important consequence of reducing the concentration of the actuating shear stress [37].

3.1.1 Equations for a partially electroded piezoelectric actuator

Consider the thin piezoelectric ceramic plate partially electroded at its major surfaces as shown in Figure 3.1.1.

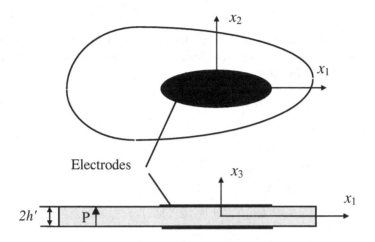

Figure 3.1.1. A partially electroded thin piezoelectric plate.

We summarize the equations for the extensional motion of a piezoelectric plate in the third section of the second chapter below. The major terms in the expansions of the mechanical displacement u_i and electrostatic potential ϕ are

$$u_i = u_i^{(0)}(x_b,t),$$
$$\phi = x_3\phi^{(1)}(x_b,t), \quad a,b = 1,2, \tag{3.1.1}$$

where we have also considered the flexural displacement $u_3^{(0)}$, which will be included in the analysis. Note that in Equation $(3.1.1)_1$ we have not explicitly written the thickness-stretch displacement that accompanies the extension of the plate. In general the expansion of ϕ has a $\phi^{(0)}$ term that is independent of x_3. In the structure to be analyzed $\phi^{(0)}$ has no contribution. Therefore we start from $\phi^{(1)}$. The balance of linear momentum now takes the form

$$T_{ab,a}^{(0)} + F_b^{(0)} = \rho'2h'\ddot{u}_b^{(0)},$$
$$F_3^{(0)} = \rho'2h'\ddot{u}_3^{(0)}, \tag{3.1.2}$$

where ρ' and $2h'$ are the actuator mass density and thickness,

$$T_{ab}^{(0)} = \int_{-h'}^{h'} T_{ab}dx_3, \quad F_j^{(0)} = [T_{3j}]_{-h'}^{h'}. \tag{3.1.3}$$

It is assumed that the plate is very thin and does not resist bending. The surface load is responsible for its motion in the x_3 direction. In the electroded region the electric potential $\phi^{(1)}$ is at most a function of time. In the unelectroded region the following equation of electrostatics is needed to determine $\phi^{(1)}$:

$$D^{(1)}_{a,a} - D^{(0)}_3 + D^{(1)} = 0, \qquad (3.1.4)$$

where

$$D^{(n)}_i = \int_{-h'}^{h'} D_i x_3^n \, dx_3, \quad D^{(1)} = [x_3 D_3]^{h'}_{-h'}. \qquad (3.1.5)$$

The plate constitutive equations are

$$T^{(0)}_r = 2h'(\gamma'_{rs} S^{(0)}_s + \psi_{3r}\phi^{(1)}), \quad r,s = 1,2,6,$$

$$D^{(0)}_3 = 2h'(\psi_{3s} S^{(0)}_s - \varsigma_{33}\phi^{(1)}), \qquad (3.1.6)$$

$$D^{(1)}_a = -\frac{2h'^3}{3}\varsigma_{ab}\phi^{(1)}_{,b},$$

where

$$S^{(0)}_{ab} = \frac{1}{2}(u^{(0)}_{a,b} + u^{(0)}_{b,a}). \qquad (3.1.7)$$

Note that Equation $(3.1.6)_3$ is taken from Equation (2.4.21) by setting $S^{(1)}_{ij} = 0$. For a ceramic plate poled in the x_3 direction, from Equations (2.3.35) and (2.3.36), in the notation of this section

$$[\gamma'_{rs}] = \begin{bmatrix} c^P_{11} & c^P_{12} & 0 \\ c^P_{12} & c^P_{11} & 0 \\ 0 & 0 & c^P_{66} \end{bmatrix}, \quad \begin{array}{l} c^P_{11} = c^E_{11} - (c^E_{13})^2/c^E_{33}, \\ c^P_{12} = c^E_{12} - (c^E_{13})^2/c^E_{33}, \\ c^P_{66} = c^E_{66}, \end{array} \qquad (3.1.8)$$

$$[\psi_{ks}] = \begin{bmatrix} 0 & 0 & 0 \\ 0 & 0 & 0 \\ e^P_{31} & e^P_{31} & 0 \end{bmatrix}, \quad \begin{array}{l} e^P_{31} = e_{31} - e_{33}c^E_{13}/c^E_{33}, \\ k = 1,2,3, \end{array} \qquad (3.1.9)$$

$$[\varsigma_{kj}] = \begin{bmatrix} \varepsilon^P_{11} & 0 & 0 \\ 0 & \varepsilon^P_{11} & 0 \\ 0 & 0 & \varepsilon^P_{33} \end{bmatrix}, \quad \begin{array}{l} \varepsilon^P_{11} = \varepsilon^S_{11} + e^2_{15}/c^E_{44}, \\ \varepsilon^P_{33} = \varepsilon^S_{33} + e^2_{33}/c^E_{33}. \end{array} \qquad (3.1.10)$$

With Equations (3.1.8) through (3.1.10) we can write Equation (3.1.6) as

$$T_{11}^{(0)} = 2h'(c_{11}^P u_{1,1}^{(0)} + c_{12}^P u_{2,2}^{(0)} + e_{31}^P \phi^{(1)}),$$

$$T_{22}^{(0)} = 2h'(c_{12}^P u_{1,1}^{(0)} + c_{11}^P u_{2,2}^{(0)} + e_{31}^P \phi^{(1)}), \qquad (3.1.11)$$

$$T_{12}^{(0)} = 2h' c_{66}^P (u_{1,2}^{(0)} + u_{2,1}^{(0)}),$$

$$D_3^{(0)} = 2h'(e_{31}^P u_{a,a}^{(0)} - \varepsilon_{33}^P \phi^{(1)}),$$

$$D_a^{(1)} = -\frac{2h'^3}{3} \varepsilon_{11}^P \phi_{,a}^{(1)}. \qquad (3.1.12)$$

Substitution of Equation (3.1.11) into Equation (3.1.2)$_1$ yields the equations for the mechanical displacement $u_a^{(0)}$

$$c_{11}^P u_{1,11}^{(0)} + c_{66}^P u_{1,22}^{(0)} + (c_{12}^P + c_{66}^P) u_{2,12}^{(0)} + e_{31}^P \phi_{,1}^{(1)} + \frac{1}{2h'} F_1^{(0)} = \rho' \ddot{u}_1^{(0)},$$

$$(c_{12}^P + c_{66}^P) u_{1,12}^{(0)} + c_{66}^P u_{2,11}^{(0)} + c_{11}^P u_{2,22}^{(0)} + e_{31}^P \phi_{,2}^{(1)} + \frac{1}{2h'} F_2^{(0)} = \rho' \ddot{u}_2^{(0)}. \qquad (3.1.13)$$

Substituting Equation (3.1.12) into Equation (3.1.4) we obtain the following equation for $\phi^{(1)}$ in the unelectroded region:

$$-\frac{2h'^3}{3} \varepsilon_{11}^P \phi_{,aa}^{(1)} - 2h' e_{31}^P u_{a,a}^{(0)} + 2h' \varepsilon_{33}^P \phi^{(1)} = 0, \qquad (3.1.14)$$

where we have taken $D^{(1)} = 0$.

3.1.2 Equations for an elastic plate

A schematic diagram of an elastic plate with thickness $2h$ and density ρ is shown in Figure 3.1.2.

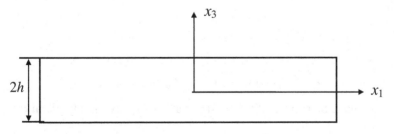

Figure 3.1.2. An elastic plate.

The flexural equations of motion for an elastic plate with the effects of shear deformation and rotatory inertia can be obtained from the equations in the fourth section of the second chapter.

$$T_{a3,a}^{(0)} + F_3^{(0)} = 2\rho h \ddot{u}_3^{(0)},$$

$$T_{ba,b}^{(1)} - T_{3a}^{(0)} + F_a^{(1)} = \frac{2}{3}\rho h^3 \ddot{u}_a^{(1)},$$

$$(3.1.15)$$

where

$$T_{ij}^{(n)} = \int_{-h}^{h} T_{ij} x_3^n dx_3, \quad n = 0,1,$$

$$F_3^{(0)} = T_{33}(h) - T_{33}(-h),$$

$$(3.1.16)$$

$$F_b^{(1)} = h[T_{3b}(h) + T_{3b}(-h)].$$

We consider an orthotropic elastic plate with the following constitutive relations:

$$T_{13}^{(0)} = 2h\kappa_1^2 \bar{c}_{55}(u_{3,1}^{(0)} + u_1^{(1)}),$$

$$T_{23}^{(0)} = 2h\kappa_2^2 \bar{c}_{44}(u_{3,2}^{(0)} + u_2^{(1)}),$$

$$(3.1.17)$$

$$T_{11}^{(1)} = \frac{2}{3}h^3(\gamma_{11}u_{1,1}^{(1)} + \gamma_{12}u_{2,2}^{(1)}),$$

$$T_{22}^{(1)} = \frac{2}{3}h^3(\gamma_{21}u_{1,1}^{(1)} + \gamma_{22}u_{2,2}^{(1)}),$$

$$(3.1.18)$$

$$T_{12}^{(1)} = \frac{2}{3}h^3\gamma_{66}(u_{1,2}^{(1)} + u_{2,1}^{(1)}),$$

where

$$\bar{c}_{44} = c_{44}, \quad \bar{c}_{55} = c_{55},$$

$$\gamma_{11} = c_{11} - c_{13}^2 / c_{33}, \quad \gamma_{22} = c_{22} - c_{23}^2 / c_{33},$$

$$(3.1.19)$$

$$\gamma_{12} = c_{12} - c_{13}c_{32} / c_{33}, \quad \gamma_{66} = c_{66}.$$

The displacement field of the elastic plate is approximately given by

$$u_a = x_3 u_a^{(1)}, \quad u_3 = u_3^{(0)}.$$

$$(3.1.20)$$

3.1.3 Equations for an elastic plate with symmetric actuators

Next we consider an elastic plate with partially electroded piezoelectric actuators symmetrically attached to its major surfaces (see Figure 3.1.3).

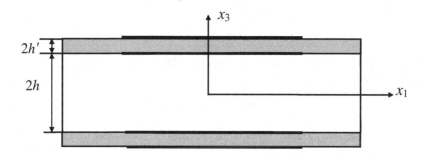

Figure 3.1.3. An elastic plate with symmetric piezoelectric actuators.

We combine the equations for a thin piezoelectric plate (actuator) and the equations for an elastic plate into a set of equations for the flexure of the elastic plate with actuators on it. This is done by satisfying the geometric and mechanical continuity conditions between the elastic plate and the piezoelectric actuators. From this point forward we use superscripts T or B for fields associated with the top or the bottom actuator. Fields without such superscripts are for the elastic plate. From Equation (3.1.3) we obtain the following forces on the top actuator:

$$F_b^{(0)T} = -T_{3b}(h), \quad F_3^{(0)T} = -T_{33}(h). \tag{3.1.21}$$

From the displacement fields in Equations (3.1.1) and (3.1.20), the continuity of mechanical displacement between the top actuator and the elastic plate can be written in the following form:

$$u_a^{(0)T} = hu_a^{(1)}, \quad u_3^{(0)T} = u_3^{(0)}. \tag{3.1.22}$$

Similarly, for the bottom actuator we have

$$F_b^{(0)B} = T_{3b}(-h), \quad F_3^{(0)B} = T_{33}(-h), \tag{3.1.23}$$

$$u_a^{(0)B} = -hu_a^{(1)}, \quad u_3^{(0)B} = u_3^{(0)}. \tag{3.1.24}$$

From Equations (3.1.21), (3.1.23) and (3.1.2) the equations of motion for the top and bottom actuators are

$$T_{ab,a}^{(0)T} - T_{3b}(h) = \rho'2h'\ddot{u}_b^{(0)T},$$

$$- T_{33}(h) = \rho'2h'\ddot{u}_3^{(0)T},$$

(3.1.25)

$$T_{ab,a}^{(0)B} + T_{3b}(-h) = \rho'2h'\ddot{u}_b^{(0)B},$$

$$T_{33}(-h) = \rho'2h'\ddot{u}_3^{(0)B}.$$

(3.1.26)

For the elastic plate, from Equations $(3.1.16)_{2,3}$, (3.1.25) and (3.1.26) we have

$$F_3^{(0)} = -\rho'2h'\ddot{u}_3^{(0)T} - \rho'2h'\ddot{u}_3^{(0)B},$$

(3.1.27)

$$F_b^{(1)} = h[T_{ab,a}^{(0)T} - \rho'2h'\ddot{u}_b^{(0)T} - T_{ab,a}^{(0)B} + \rho'2h'\ddot{u}_b^{(0)B}].$$

(3.1.28)

Substitution of Equations (3.1.27) and (3.1.28) into Equation (3.1.15) yields

$$T_{a3,a}^{(0)} = 2\rho h\ddot{u}_3^{(0)} + \rho'2h'\ddot{u}_3^{(0)T} + \rho'2h'\ddot{u}_3^{(0)B},$$

$$(T_{ba}^{(1)} + hT_{ba}^{(0)T} - hT_{ba}^{(0)B})_{,b} - T_{3a}^{(0)}$$

$$= \frac{2}{3}\rho h^3\ddot{u}_a^{(1)} + h[\rho'2h'\ddot{u}_a^{(0)T} - \rho'2h'\ddot{u}_a^{(0)B}].$$

(3.1.29)

With the continuity conditions in Equations (3.1.22) and (3.1.24), we can write Equations (3.1.29) as

$$T_{a3,a}^{(0)} = \hat{m}\ddot{u}_3^{(0)},$$

$$\hat{T}_{ba,b}^{(1)} - T_{3a}^{(0)} = \hat{I}\ddot{u}_a^{(1)},$$

(3.1.30)

where

$$\hat{T}_{ba}^{(1)} = T_{ba}^{(1)} + hT_{ba}^{(0)T} - hT_{ba}^{(0)B},$$

$$\hat{m} = 2\rho h + 4\rho'h',$$

(3.1.31)

$$\hat{I} = \frac{2}{3}\rho h^3 + 4\rho'h'h^2.$$

$\hat{T}_{ab}^{(1)}$ has the clear physical meaning of total moments consisting of the moments due to the bending of the elastic plate and the extension of the actuators. $\hat{m}$ is the mass per unit area and $\hat{I}$ the rotatory inertia of the plate with the actuators. In the following we assume that the applied voltages on the top and bottom actuators are of equal magnitude and opposite signs and denote

$$\phi^{(1)T} = -\phi^{(1)B} = \phi^{(1)}. \tag{3.1.32}$$

Then for constitutive relations, from Equations $(3.1.31)_1$, $(3.1.11)$, $(3.1.18)$, $(3.1.22)$ and $(3.1.24)$ we obtain

$$\hat{T}_{11}^{(1)} = \hat{\gamma}_{11} u_{1,1}^{(1)} + \hat{\gamma}_{12} u_{2,2}^{(1)} + 4hh'e_{31}^{P}\phi^{(1)},$$

$$\hat{T}_{22}^{(1)} = \hat{\gamma}_{12} u_{1,1}^{(1)} + \hat{\gamma}_{22} u_{2,2}^{(1)} + 4hh'e_{31}^{P}\phi^{(1)}, \tag{3.1.33}$$

$$\hat{T}_{12}^{(1)} = \hat{T}_{21}^{(1)} = \hat{\gamma}_{66} (u_{1,2}^{(1)} + u_{2,1}^{(1)}),$$

where

$$\hat{\gamma}_{11} = \frac{2}{3}h^3\gamma_{11} + 4h'h^2c_{11}^{P}, \quad \hat{\gamma}_{12} = \frac{2}{3}h^3\gamma_{12} + 4h'h^2c_{12}^{P},$$

$$\hat{\gamma}_{22} = \frac{2}{3}h^3\gamma_{22} + 4h'h^2c_{11}^{P}, \quad \hat{\gamma}_{66} = \frac{2}{3}h^3\gamma_{66} + 4h'h^2c_{66}^{P}. \tag{3.1.34}$$

Expressions for $T_{a3}^{(0)}$ in terms of $u_3^{(0)}$ and $u_a^{(1)}$ are given by Equation $(3.1.17)$. Substitution of Equations $(3.1.17)$ and $(3.1.33)$ into Equation $(3.1.30)$ gives the equation for the flexural and thickness-shear motions of the whole plate including the actuators

$$2h\kappa_1^2\bar{c}_{55}(u_{3,11}^{(0)} + u_{1,1}^{(1)}) + 2h\kappa_2^2\bar{c}_{44}(u_{3,22}^{(0)} + u_{2,2}^{(1)}) = \hat{m}\ddot{u}_3^{(0)},$$

$$\hat{\gamma}_{11} u_{1,11}^{(1)} + \hat{\gamma}_{12} u_{2,21}^{(1)} + 4h'he_{31}^{P}\phi_{,1}^{(1)}$$

$$+ \hat{\gamma}_{66} (u_{1,22}^{(1)} + u_{2,12}^{(1)}) - 2h\kappa_1^2\bar{c}_{55}(u_{3,1}^{(0)} + u_1^{(1)}) = \hat{I}\ddot{u}_1^{(1)}, \tag{3.1.35}$$

$$\hat{\gamma}_{22} u_{2,22}^{(1)} + \hat{\gamma}_{12} u_{1,12}^{(1)} + 4h'he_{31}^{P}\phi_{,2}^{(1)}$$

$$+ \hat{\gamma}_{66} (u_{1,21}^{(1)} + u_{2,11}^{(1)}) - 2h\kappa_2^2\bar{c}_{44}(u_{3,2}^{(0)} + u_2^{(1)}) = \hat{I}\ddot{u}_2^{(1)}.$$

In the unelectroded portions of the actuators where $\phi^{(1)}$ is unknown, Equation $(3.1.14)$ is needed which now takes the following form with the use of Equation $(3.1.22)_1$:

$$-\frac{2h'^2}{3}\varepsilon_{11}^{P}\phi_{,aa}^{(1)} - 2h'he_{31}^{P}u_{a,a}^{(1)} + \varepsilon_{33}^{P}2h'\phi^{(1)} = 0. \tag{3.1.36}$$

At the boundary of a finite plate with a unit normal n_a and a unit tangent s_b we need to prescribe

$$n_a \hat{T}_{ab}^{(1)} n_b \quad \text{or} \quad u_a^{(1)} n_a,$$

$$n_a \hat{T}_{ab}^{(1)} s_b \quad \text{or} \quad u_a^{(1)} s_a, \qquad (3.1.37)$$

$$T_{n3}^{(0)} \quad \text{or} \quad u_3^{(0)}.$$

For electric boundary conditions we need to prescribe

$$n_a D_a^{(1)} \quad \text{or} \quad \phi^{(1)}. \qquad (3.1.38)$$

Although the equations derived in this section include shear deformation and rotatory inertia and can be used in situations when the elastic plate is not very thin, the piezoelectric actuators are still assumed to be very thin compared to the elastic plate and are treated as thin films that are in extensional motions and do not resist bending.

3.1.4 Reduction to classical flexure

When the elastic plate is very thin, the effect of shear deformation and rotatory inertia can be neglected. Equations for this case were given in [37], and can be reduced from the equations derived above. For thin plates we neglect shear deformation and rotatory inertia, which results in

$$\hat{I} = 0, \quad u_a^{(1)} = -u_{3,a}^{(0)}. \qquad (3.1.39)$$

Then, similar to the reduction in the fourth section of the second chapter, Equation (3.1.33) and Equations (3.1.35) through (3.1.37) reduce to

$$\hat{T}_{11}^{(1)} = -\hat{\gamma}_{11} u_{3,11}^{(0)} - \hat{\gamma}_{12} u_{3,22}^{(0)} + 4h'he_{31}^P \phi^{(1)},$$

$$\hat{T}_{22}^{(1)} = -\hat{\gamma}_{12} u_{3,11}^{(0)} - \hat{\gamma}_{22} u_{3,22}^{(0)} + 4h'he_{31}^P \phi^{(1)}, \qquad (3.1.40)$$

$$\hat{T}_{12}^{(1)} - \hat{T}_{21}^{(1)} = -2\hat{\gamma}_{66} u_{3,12}^{(0)},$$

$$\hat{T}_{ab,ab}^{(1)} = -\hat{\gamma}_{11} u_{3,1111}^{(0)} - 2(\hat{\gamma}_{12} + 2\hat{\gamma}_{66}) u_{3,1122}^{(0)} - \hat{\gamma}_{22} u_{3,2222}^{(0)}$$

$$+ 4h'he_{31}^P (\phi_{,11}^{(1)} + \phi_{,22}^{(1)}) = \hat{m} \ddot{u}_3^{(0)}, \qquad (3.1.41)$$

$$-\frac{2h'^2}{3} \varepsilon_{11}^P \phi_{,aa}^{(1)} + 2h'he_{31}^P u_{3,aa}^{(0)} + 2h' \varepsilon_{33}^P \phi^{(1)} = 0,$$

$$n_a \hat{T}_{ab}^{(1)} n_b \quad \text{or} \quad \frac{\partial u_3^{(0)}}{\partial n},$$

$$T_{n3}^{(0)} + \frac{\partial \hat{T}_{ns}^{(1)}}{\partial s} \quad \text{or} \quad u_3^{(0)}, \tag{3.1.42}$$

where

$$T_{a3}^{(0)} = \hat{T}_{ba,b}^{(1)}, \quad \hat{T}_{ns}^{(1)} = n_a \hat{T}_{ab}^{(1)} s_b. \tag{3.1.43}$$

3.1.5 Dispersion relations

To see the difference between Equation (3.1.35) and Equation $(3.1.41)_1$, we consider the one-dimensional case of fully electroded actuators with shorted electrodes. Then

$$u_2^{(1)} = 0, \quad \phi^{(1)} = 0, \quad \partial_2 = 0. \tag{3.1.44}$$

From Equation (3.1.41) we obtain

$$-\hat{\gamma}_{11} u_{3,1111}^{(0)} = \hat{m} \ddot{u}_3^{(0)}. \tag{3.1.45}$$

The substitution of the wave solution $u_3^{(0)} = \exp i(\xi x_1 - \omega t)$ into Equation (3.1.45) leads to the following dispersion relation for flexural waves:

$$\omega^2 = \frac{\hat{\gamma}_{11}}{\hat{m}} \xi^4. \tag{3.1.46}$$

Under Equation (3.1.44), from Equation (3.1.35) we have

$$2h\kappa_1^2 \bar{c}_{55} (u_{3,11}^{(0)} + u_{1,1}^{(1)}) = \hat{m} \ddot{u}_3^{(0)},$$

$$\hat{\gamma}_{11} u_{1,11}^{(1)} - 2h\kappa_1^2 \bar{c}_{55} (u_{3,1}^{(0)} + u_1^{(1)}) = \hat{I} \ddot{u}_1^{(1)}. \tag{3.1.47}$$

Consider the following waves:

$$u_3^{(0)} = A \exp i(\xi x_1 - \omega t),$$

$$u_1^{(1)} = B \exp i(\xi x_1 - \omega t). \tag{3.1.48}$$

Substitution of Equation (3.1.48) into Equation (3.1.47) results in

$$2h\kappa_1^2 \bar{c}_{55} (-\xi^2 A + i\xi B) = -\hat{m}\omega^2 A,$$

$$-\hat{\gamma}_{11}\xi^2 B - 2h\kappa_1^2 \bar{c}_{55} (i\xi A + B) = -\hat{I}\omega^2 B, \tag{3.1.49}$$

which is a homogeneous system for A and B. For nontrivial solutions the determinant of the coefficient matrix has to vanish, which yields the dispersion relation

$$\begin{vmatrix} \hat{m}\omega^2 - 2h\kappa_1^2\bar{c}_{55}\xi^2 & 2h\kappa_1^2\bar{c}_{55}i\xi \\ -2h\kappa_1^2\bar{c}_{55}i\xi & \hat{I}\omega^2 - \hat{\gamma}_{11}\xi^2 - 2h\kappa_1^2\bar{c}_{55} \end{vmatrix} = 0. \qquad (3.1.50)$$

Normalized dispersion curves predicted by Equations (3.1.46) and (3.1.50) for the case of $h = 20h'$ are plotted in Figure 3.1.4 in which

$$X = \xi h, \quad Y = \frac{2h}{\pi}\sqrt{\frac{\rho}{\bar{c}_{55}}}\omega. \qquad (3.1.51)$$

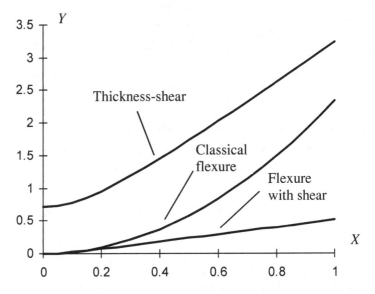

Figure 3.1.4. Dispersion curves of flexural and thickness-shear waves.

The correction factor is simply taken to be 1 in the calculation, as an approximation. One basic difference between Equations (3.1.50) and (3.1.46) is that Equation (3.1.46) has only one flexural branch of the dispersion curve, whereas Equation (3.1.50) has two branches for flexural and thickness-shear waves. For long waves with vanishing ξ, Equation (3.1.46) implies a vanishing ω. The flexural branch of Equation (3.1.50) has a vanishing ω near $\xi = 0$. The thickness-shear branch has a finite cutoff frequency at $\xi = 0$. For flexural motions, Equation (3.1.46) and the flexural branch of Equation (3.1.50) agree only for very small values of ξh. Since in the flexural problem of a finite plate the wave number ξ is of the order of the inverse of the dimension L of the plate, Equation (3.1.46) is valid only for plates with small values of h/L, or very thin plates. When h/L is not very small, the equations with thickness-shear deformation and rotatory inertia should be used.

3.2 Elastic Plates with Piezoelectric Actuators on One Side

In some applications an elastic plate carries a piezoelectric layer only on one side, either as an actuator or a sensor. In this case the structure is sometimes called a unimorph. For unimorphs, due to the asymmetry about the middle plane of the structure, bending and extension of the middle plane are inherently coupled. This causes complications in modeling. In this section we derive two-dimensional equations for a plate piezoelectric unimorph and use the equations obtained to analyze circular disk unimorphs [39]. A unimorph is a special case of laminated plates. Two-dimensional equations for laminated piezoelectric plates can be obtained by combining equations for each layer like in the previous section. They can also be obtained directly by substituting approximate expressions of the global displacement and potential fields into the variational formulation. In the variational approach, since a laminated plate is a body with piecewise constant material parameters that do not have derivatives across an interface, the integrations in the variational formulation have to be performed layer by layer. Interface continuity conditions on traction and electric displacement are part of the stationary conditions of the variational procedure. These conditions can be satisfied exactly by three-dimensional solutions, or approximately by two-dimensional solutions. Another approach for developing equations of single-layered or laminated plates is to calculate moments of various orders of the three-dimensional equations and fields by integrating them through the plate thickness [40,41]. This is the approach we follow in this section.

3.2.1 Classical theory

Consider the two-layer plate in Figure 3.2.1, which is asymmetric about its middle surface.

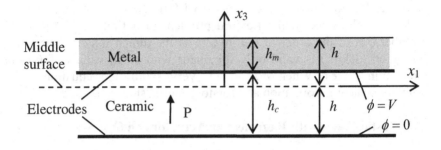

Figure 3.2.1. A piezoelectric unimorph.

Instead of combining separate plate equations for the elastic and piezoelectric layers by the interface continuity conditions, we begin with an approximation of the global displacement and potential fields for both layers. For coupled extension and classical flexure, the major terms of the displacement field are:

$$u_3(x_1,x_2,x_3,t) \cong u_3^{(0)}(x_1,x_2,t),$$
$$u_1(x_1,x_2,x_3,t) \cong u_1^{(0)}(x_1,x_2,t) - x_3 u_{3,1}^{(0)}, \qquad (3.2.1)$$
$$u_2(x_1,x_2,x_3,t) \cong u_2^{(0)}(x_1,x_2,t) - x_3 u_{3,2}^{(0)},$$

where $u_a^{(0)}$ are the middle surface extensional displacements, and $u_3^{(0)}$ is the middle surface flexural displacement. The strains corresponding to Equation (3.2.1) are

$$S_1 = u_{1,1}^{(0)} - x_3 u_{3,11}^{(0)}, \quad S_2 = u_{2,2}^{(0)} - x_3 u_{3,22}^{(0)},$$
$$2S_{12} = S_6 = u_{1,2}^{(0)} + u_{2,1}^{(0)} - 2x_3 u_{3,12}^{(0)}. \qquad (3.2.2)$$

For the ceramic plate which is transversely isotropic, we have

$$S_1 = s_{11}T_1 + s_{12}T_2 + s_{13}T_3 + d_{31}E_3,$$
$$S_2 = s_{12}T_1 + s_{11}T_2 + +s_{13}T_3 + d_{31}E_3, \qquad (3.2.3)$$
$$S_6 = s_{66}T_6,$$

where $s_{66} = 2(s_{11}\text{-}s_{12})$ and $E_3 = -V/h_c$. For a thin plate we make the stress relaxation

$$T_3 = 0 . \tag{3.2.4}$$

Under Equation (3.2.4), Equation (3.2.3) can be inverted to give

$$
\begin{aligned}
T_1 &= c_{11}^p S_1 + c_{12}^p S_2 - e_{31}^p E_3, \\
T_2 &= c_{12}^p S_1 + c_{11}^p S_2 - e_{31}^p E_3, \\
T_{12} &= c_{66}^p S_6,
\end{aligned}
\tag{3.2.5}
$$

where

$$
\begin{aligned}
&c_{11}^p = s_{11} / \Delta, \quad c_{12}^p = -s_{12} / \Delta, \quad \Delta = s_{11}^2 - s_{12}^2, \\
&e_{31}^p = d_{31} / (s_{11} + s_{12}), \quad c_{66}^p = 1 / s_{66} = (c_{11}^p - c_{12}^p) / 2.
\end{aligned}
\tag{3.2.6}
$$

Then the relevant electric constitutive relation takes the form

$$
\begin{aligned}
D_3 &= d_{31}(T_1 + T_2) + \varepsilon_{33} E_3 \\
&= e_{31}^p (S_1 + S_2) + \varepsilon_{33}^p E_3,
\end{aligned}
\tag{3.2.7}
$$

where

$$\varepsilon_{33}^p = \varepsilon_{33} - 2 d_{31} e_{31}^p . \tag{3.2.8}$$

Equation (3.2.7) is useful when the charge or current on the electrodes at the major surfaces of the ceramic plate needs to be calculated.

For the elastic metal layer that is isotropic, we use a hat to indicate its material parameters. Using a method similar to the derivation of Equation (3.2.5), we have

$$
\begin{aligned}
T_1 &= \hat{c}_{11}^p S_1 + \hat{c}_{12}^p S_2, \\
T_2 &= \hat{c}_{12}^p S_1 + \hat{c}_{11}^p S_2, \\
T_{12} &= \hat{c}_{66}^p S_6,
\end{aligned}
\tag{3.2.9}
$$

where

$$
\begin{aligned}
&\hat{c}_{11}^p = \hat{s}_{11} / \hat{\Delta}, \quad \hat{c}_{12}^p = -\hat{s}_{12} / \hat{\Delta}, \quad \hat{\Delta} = \hat{s}_{11}^2 - \hat{s}_{12}^2, \\
&\hat{c}_{66}^p = 1 / \hat{s}_{66} = (\hat{c}_{11}^p - \hat{c}_{12}^p) / 2.
\end{aligned}
\tag{3.2.10}
$$

With Equations (3.2.5) and (3.2.9), we calculate the plate extensional resultants as

$$T_{11}^{(0)} = \int_{-h}^{h} T_{11} dx_3$$
$$= a_{11} u_{1,1}^{(0)} + a_{12} u_{2,2}^{(0)} + b_{11} u_{3,11}^{(0)} + b_{12} u_{3,22}^{(0)} - h_c e_{31}^p E_3,$$

$$T_{22}^{(0)} = \int_{-h}^{h} T_{22} dx_3$$
$$= a_{12} u_{1,1}^{(0)} + a_{11} u_{2,2}^{(0)} + b_{12} u_{3,11}^{(0)} + b_{11} u_{3,22}^{(0)} - h_c e_{31}^p E_3, \qquad (3.2.11)$$

$$T_{12}^{(0)} = \int_{-h}^{h} T_{12} dx_3$$
$$= a_{66} (u_{1,2}^{(0)} + u_{2,1}^{(0)}) + 2 b_{66} u_{3,12}^{(0)},$$

where

$$a_{11} = c_{11}^p h_c + \hat{c}_{11}^p h_m, \quad a_{12} = c_{12}^p h_c + \hat{c}_{12}^p h_m,$$
$$a_{66} = c_{66}^p h_c + \hat{c}_{66}^p h_m = (a_{11} - a_{12})/2,$$
$$b_{11} = \frac{1}{2} \left(c_{11}^p - \hat{c}_{11}^p \right) h_c h_m, \quad b_{12} = \frac{1}{2} \left(c_{12}^p - \hat{c}_{12}^p \right) h_c h_m, \qquad (3.2.12)$$
$$b_{66} = \frac{1}{2} \left(c_{66}^p - \hat{c}_{66}^p \right) h_c h_m = (b_{11} - b_{12})/2.$$

In obtaining Equation (3.2.11) we have made the usual approximation that in the thin ceramic layer the electric field E_3 is essentially independent of x_3 [3]. Similarly, the plate bending and twisting moments are given by

$$T_{11}^{(1)} = \int_{-h}^{h} T_{11} x_3 dx_3$$
$$= -b_{11} u_{1,1}^{(0)} - b_{12} u_{2,2}^{(0)} - \gamma_{11} u_{3,11}^{(0)} - \gamma_{12} u_{3,22}^{(0)} + h_c h_m e_{31}^p E_3,$$

$$T_{22}^{(1)} = \int_{-h}^{h} T_{22} x_3 dx_3$$
$$= -b_{12} u_{1,1}^{(0)} - b_{11} u_{2,2}^{(0)} - \gamma_{12} u_{3,11}^{(0)} - \gamma_{11} u_{3,22}^{(0)} + h_c h_m e_{31}^p E_3, \qquad (3.2.13)$$

$$T_{12}^{(1)} = \int_{-h}^{h} T_{12} x_3 dx_3$$
$$= -b_{66} (u_{1,2}^{(0)} + u_{2,1}^{(0)}) - 2 \gamma_{66} u_{3,12}^{(0)},$$

where

$$\gamma_{11} = \frac{c_{11}^p}{12}\left(h_c^3 + 3h_c h_m^2\right) + \frac{\hat{c}_{11}^p}{12}\left(h_m^3 + 3h_c^2 h_m\right),$$

$$\gamma_{12} = \frac{c_{12}^p}{12}\left(h_c^3 + 3h_c h_m^2\right) + \frac{\hat{c}_{12}^p}{12}\left(h_m^3 + 3h_c^2 h_m\right), \tag{3.2.14}$$

$$\gamma_{66} = \frac{c_{66}^p}{12}\left(h_c^3 + 3h_c h_m^2\right) + \frac{\hat{c}_{66}^p}{12}\left(h_m^2 + 3h_c^2 h_m\right) = (\gamma_{11} - \gamma_{12})/2.$$

For the plate equations of motion for extension and flexure, instead of using the variational principle, we integrate the three-dimensional equations of motion through the plate thickness

$$T_{ab,a}^{(0)} + [T_{3b}]_{-h}^h = (\rho_c h_c + \rho_m h_m)\ddot{u}_b^{(0)} + \frac{1}{2}(\rho_c - \rho_m)h_c h_m \ddot{u}_{3,b}^{(0)},$$

$$T_{a3,a}^{(0)} + [T_{33}]_{-h}^h = (\rho_c h_c + \rho_m h_m)\ddot{u}_3^{(0)}. \tag{3.2.15}$$

The second term on the right hand side of Equation $(3.2.15)_1$ is in fact due to the first-order displacement $u_a^{(1)} = -u_{3,a}^{(0)}$ and is usually smaller than the first term because it depends on the difference of the mass densities and it has a spatial derivative. The classical theory for flexure is for long waves with small wave numbers. A spatial derivative is effectively a multiplication by the small wave number. The plate moment equations are obtained by integrating the product of the three-dimensional equations of motion and x_3 through the plate thickness. This gives the plate transverse shear resultants $T_{a3}^{(0)}$ in terms of the moment resultants

$$T_{ab,a}^{(1)} - T_{3b}^{(0)} + [x_3 T_{3b}]_{-h}^h = -\frac{1}{2}(\rho_c - \rho_m)h_c h_m \ddot{u}_b^{(0)}. \tag{3.2.16}$$

In Equation (3.2.16) the rotatory inertia has been neglected for classical flexure. The surface load in this equation is neglected below.

In summary, we have the two-dimensional equations of extension and flexure (3.2.15), constitutive relations (3.2.11) and (3.2.13), shear force-moment relation (3.2.16), and strain-displacement relations (3.2.2). With successive substitutions, Equation (3.2.15) can be written as the following equations for $u_1^{(0)}$, $u_2^{(0)}$, and $u_3^{(0)}$:

$$a_{66}\nabla^2\mathbf{u}^{(0)} + \frac{a_{11}+a_{12}}{2}\nabla(\nabla\cdot\mathbf{u}^{(0)}) + b_{11}\nabla(\nabla^2 u_3^{(0)})$$

$$-h_c e_{31}^p \nabla E_3 + \left[T_{31}\mathbf{e}_1 + T_{32}\mathbf{e}_2\right]_{-h}^h$$

$$= \left(\rho_c h_c + \rho_m h_m\right)\ddot{\mathbf{u}}^{(0)} + \frac{1}{2}(\rho_c - \rho_m)h_c h_m \nabla\ddot{u}_3^{(0)}, \qquad (3.2.17)$$

$$-\gamma_{11}\nabla^2\nabla^2 u_3^{(0)} - b_{11}\nabla^2(\nabla\cdot\mathbf{u}^{(0)}) + h_c h_m e_{31}^p \nabla^2 E_3 + \left[T_{33}\right]_{-h}^h$$

$$= \left(\rho_c h_c + \rho_m h_m\right)\ddot{u}_3^{(0)} - \frac{1}{2}(\rho_c - \rho_m)h_c h_m \nabla\cdot\ddot{\mathbf{u}}^{(0)},$$

where

$$\mathbf{u}^{(0)} = u_1^{(0)}\mathbf{i}_1 + u_2^{(0)}\mathbf{i}_2,$$

$$\nabla = \mathbf{i}_1\partial_1 + \mathbf{i}_2\partial_2, \quad \nabla^2 = \partial_1^2 + \partial_2^2. \qquad (3.2.18)$$

At the two-dimensional boundary of a plate with an in-plane unit exterior normal $\mathbf{n}$ and an in-plane unit tangent $\mathbf{s}$, we may prescribe

$$\begin{aligned} T_{nn}^{(0)} &\quad \text{or} \quad u_n^{(0)}, \quad T_{ns}^{(0)} \quad \text{or} \quad u_s^{(0)}, \\ T_{n3}^{(0)} + T_{ns,s}^{(1)} &\quad \text{or} \quad u_3^{(0)}, \quad T_{nn}^{(1)} \quad \text{or} \quad u_{3,n}^{(0)}. \end{aligned} \qquad (3.2.19)$$

3.2.2 Stress function formulation for static problems

For static problems, it is often convenient to use a stress function for the extensional part of the problem. Consider the case of $\left[T_{3b}\right]_{-h}^h = 0$. The static form of Equation $(3.2.15)_1$ becomes

$$T_{ab,a}^{(0)} = 0, \qquad (3.2.20)$$

which can be identically satisfied by the introduction of a stress function ψ through

$$T_{11}^{(0)} = \psi_{,22}, \quad T_{22}^{(0)} = \psi_{,11}, \quad T_{12}^{(0)} = -\psi_{,12}. \qquad (3.2.21)$$

Then, from Equation (3.2.11)

$$\psi_{,22} = a_{11}u_{1,1}^{(0)} + a_{12}u_{2,2}^{(0)} + b_{11}u_{3,11}^{(0)} + b_{12}u_{3,22}^{(0)} - h_c e_{31}^p E_3,$$

$$\psi_{,11} = a_{12}u_{1,1}^{(0)} + a_{11}u_{2,2}^{(0)} + b_{12}u_{3,11}^{(0)} + b_{11}u_{3,22}^{(0)} - h_c e_{31}^p E_3, \qquad (3.2.22)$$

$$-\psi_{,12} = a_{66}(u_{1,2}^{(0)} + u_{2,1}^{(0)}) + 2b_{66}u_{3,12}^{(0)}.$$

Equation (3.2.22) can be inverted for

$$u_{1,1}^{(0)} = -\frac{a_{12}}{a_{11}^2 - a_{12}^2}\psi_{,11} + \frac{a_{11}}{a_{11}^2 - a_{12}^2}\psi_{,22}$$

$$-\frac{a_{11}b_{11} - a_{12}b_{12}}{a_{11}^2 - a_{12}^2}u_{3,11}^{(0)} - \frac{a_{11}b_{12} - a_{12}b_{11}}{a_{11}^2 - a_{12}^2}u_{3,22}^{(0)} + \frac{h_c e_{31}^p}{a_{11} + a_{12}}E_3,$$

$$u_{2,2}^{(0)} = \frac{a_{11}}{a_{11}^2 - a_{12}^2}\psi_{,11} - \frac{a_{12}}{a_{11}^2 - a_{12}^2}\psi_{,22}$$

$$-\frac{a_{11}b_{12} - a_{12}b_{11}}{a_{11}^2 - a_{12}^2}u_{3,11}^{(0)} - \frac{a_{11}b_{11} - a_{12}b_{12}}{a_{11}^2 - a_{12}^2}u_{3,22}^{(0)} + \frac{h_c e_{31}^p}{a_{11} + a_{12}}E_3,$$

$$(3.2.23)$$

$$u_{1,2}^{(0)} + u_{2,1}^{(0)} = -\frac{1}{a_{66}}\psi_{,12} - \frac{2b_{66}}{a_{66}}u_{3,12}^{(0)}$$

$$= -\frac{2}{a_{11} - a_{12}}\psi_{,12} - \frac{2(b_{11} - b_{12})}{a_{11} - a_{12}}u_{3,12}^{(0)}.$$

Substituting Equation (3.2.23) into the following compatibility equation:

$$(u_{1,1}^{(0)})_{,22} + (u_{2,2}^{(0)})_{,11} = (u_{1,2}^{(0)} + u_{2,1}^{(0)})_{,12} , \qquad (3.2.24)$$

we obtain

$$\bar{a}_{11}\nabla^4\psi - \bar{b}_{11}\nabla^4 u_3^{(0)} + \frac{h_c e_{31}^p}{a_{11} + a_{12}}\nabla^2 E_3 = 0, \qquad (3.2.25)$$

where

$$\bar{a}_{11} = \frac{a_{11}}{a_{11}^2 - a_{12}^2}, \quad \bar{b}_{11} = \frac{a_{11}b_{12} - a_{12}b_{11}}{a_{11}^2 - a_{12}^2}. \qquad (3.2.26)$$

Substitution of Equation (3.2.23) into the moment expressions in Equation (3.2.13) gives

$$T_{11}^{(1)} = -\bar{\gamma}_{11}u_{3,11}^{(0)} - \bar{\gamma}_{12}u_{3,22}^{(0)} - \bar{b}_{11}\psi_{,11} - \bar{b}_{12}\psi_{,22} + \bar{h}_m h_c e_{31}^p E_3,$$

$$T_{22}^{(1)} = -\bar{\gamma}_{12}u_{3,11}^{(0)} - \bar{\gamma}_{11}u_{3,22}^{(0)} - \bar{b}_{12}\psi_{,11} - \bar{b}_{11}\psi_{,22} + \bar{h}_m h_c e_{31}^p E_3, \qquad (3.2.27)$$

$$T_{12}^{(1)} = -2\bar{\gamma}_{66}u_{3,12}^{(0)} - \bar{b}_{66}\psi_{,12},$$

where

$$\overline{\gamma}_{11} = \gamma_{11} - \frac{a_{11}(b_{11}^2 + b_{12}^2) - 2a_{12}b_{11}b_{12}}{a_{11}^2 - a_{12}^2},$$

$$\overline{\gamma}_{12} = \gamma_{12} - \frac{2a_{11}b_{11}b_{12} - a_{12}(b_{11}^2 + b_{12}^2)}{a_{11}^2 - a_{12}^2},$$

(3.2.28)

$$2\overline{\gamma}_{66} = \gamma_{11} - \gamma_{12} - \frac{(b_{11} - b_{12})^2}{a_{11} - a_{12}}, \quad \overline{h}_m = h_m - \frac{b_{11} + b_{12}}{a_{11} + a_{12}},$$

$$\overline{b}_{12} = -\frac{a_{12}b_{12} - a_{11}b_{11}}{a_{11}^2 - a_{12}^2}, \quad \overline{b}_{66} = -\frac{b_{11} - b_{12}}{a_{11} - a_{12}}.$$

Then Equation $(3.2.15)_2$ takes the following form:

$$-\overline{\gamma}_{11}\nabla^4 u_3^{(0)} - \overline{b}_{11}\nabla^4\psi + \overline{h}_m h_c e_{31}^p \nabla^2 E_3 + [T_{33}]_{-h}^h = 0. \tag{3.2.29}$$

In this stress function formulation the equations that govern ψ and $u_3^{(0)}$ are Equations (3.2.25) and (3.2.29).

3.2.3 A circular plate under a uniform load

As an example we consider circular plates under axi-symmetric loads. For axi-symmetric problems in polar coordinates, we have

$$\nabla^4 = \frac{\partial^4}{\partial r^4} + \frac{2}{r}\frac{\partial^3}{\partial r^3} - \frac{1}{r^2}\frac{\partial^2}{\partial r^2} + \frac{1}{r^3}\frac{\partial}{\partial r},$$

$$T_{rr}^{(0)} = \frac{1}{r}\frac{\partial\psi}{\partial r},$$

$$T_{rr}^{(1)} = -\overline{\gamma}_{11}u_{3,rr}^{(0)} - \frac{\overline{\gamma}_{12}}{r}u_{3,r}^{(0)} - \overline{b}_{11}\psi_{,rr} - \frac{\overline{b}_{12}}{r}\psi_{,r} + \overline{h}_m h_c e_{31}^p E_3, \tag{3.2.30}$$

$$T_{\theta\theta}^{(1)} = -\frac{\overline{\gamma}_{11}}{r}u_{3,r}^{(0)} - \overline{\gamma}_{12}u_{3,rr}^{(0)} - \frac{\overline{b}_{11}}{r}\psi_{,r} - \overline{b}_{12}\psi_{,rr} + \overline{h}_m h_c e_{31}^p E_3,$$

$$T_{3r}^{(0)} = \frac{\partial T_{rr}^{(1)}}{\partial r} + \frac{T_{rr}^{(1)} - T_{\theta\theta}^{(1)}}{r}.$$

First consider a circular plate of a radius R under a uniform pressure $[T_{33}]_{-h}^{h} = q_0$ (see Figure 3.2.2). The plate is simply supported at the edge. In the radial direction it is allowed to undergo extension freely.

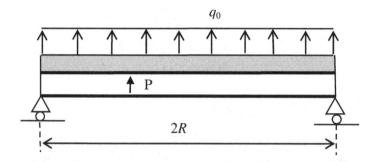

q_0

P

$2R$

Figure 3.2.2. A circular unimorph under a uniform load p.

The problem is axi-symmetric. The boundary value problem is

$$-\bar{\gamma}_{11}\nabla^4 u_3^{(0)} - \bar{b}_{11}\nabla^4 \psi + q_0 = 0, \quad r < R,$$
$$\bar{a}_{11}\nabla^4 \psi - \bar{b}_{11}\nabla^4 u_3^{(0)} = 0, \quad r < R,$$
$$T_{rr}^{(0)} = 0, \quad r = R,$$
$$u_3^{(0)} = 0, \quad r = R, \qquad\qquad (3.2.31)$$
$$T_{rr}^{(1)} = 0, \quad r = R,$$
all fields are finite, $\quad r = 0.$

Note that E_3 appears in Equation $(3.2.31)_5$ through Equation $(3.2.30)_3$. The solution is determined as

$$u_3^{(0)} = \frac{\bar{a}_{11}}{4}Qr^4 + QR^2(A_1 r^2 + A_2 R^2) + A_3 E_3(R^2 - r^2), \qquad (3.2.32)$$

$$\psi = \frac{\bar{b}_{11}}{4}Q(r^4 - 2R^2 r^2), \qquad (3.2.33)$$

where

$$Q = \frac{q_0}{16(\bar{a}_{11}\bar{\gamma}_{11} + \bar{b}_{11}^2)}.$$

$$A_1 = -\frac{3\bar{a}_{11}\bar{\gamma}_{11} + \bar{a}_{11}\bar{\gamma}_{12} + 2\bar{b}_{11}^2 + 2\bar{b}_{11}\bar{b}_{12}}{2(\bar{\gamma}_{11} + \bar{\gamma}_{12})},$$

$$A_2 = \frac{5\bar{a}_{11}\bar{\gamma}_{11} + \bar{a}_{11}\bar{\gamma}_{12} + 4\bar{b}_{11}^2 + 4\bar{b}_{11}\bar{b}_{12}}{4(\bar{\gamma}_{11} + \bar{\gamma}_{12})},$$

$$A_3 = -\frac{\bar{h}_m h_c e_{31}^p}{2(\bar{\gamma}_{11} + \bar{\gamma}_{12})}.$$

(3.2.34)

For numerical results we consider PZT-5H for the ceramic plate. For the metal plate we consider stainless steel with Young's modulus $E = 20 \times 10^{10}$ Pa, shear modulus $G = 74$ GPa, mass density $\rho_m = 8.06 \times 10^{10}$ kg/m^3, and Poisson's ratio $v = 0.313$. Figure 3.2.3 shows the deflection under p for different E_3. By applying a different electric field, the deflection of the plate can be controlled. The dotted lines are from a finite element numerical analysis by ANSYS for comparison.

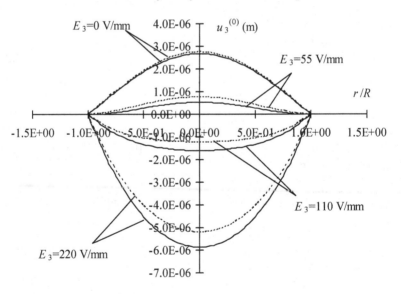

Figure 3.2.3. Deflection under p for different E_3. $q_0\pi R^2 = 2$ N, $h_c = 0.4$ mm, $h_m = 0.15$ mm, $R = 12.7$ mm.

3.2.4 A circular plate under a concentrated load

Next consider a simply supported circular plate under a concentrated load P at the center (see Figure 3.2.4).

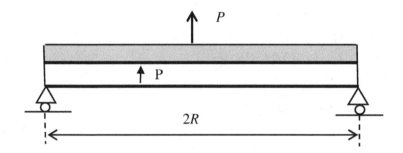

Figure 3.2.4. A circuit unimorph under a concentrated load P.

The boundary value problem is

$$-\bar{\gamma}_{11}\nabla^4 u_3^{(0)} - \bar{b}_{11}\nabla^4 \psi + P\delta(r) = 0, \quad r < R,$$
$$\bar{a}_{11}\nabla^4 \psi - \bar{b}_{11}\nabla^4 u_3^{(0)} = 0, \quad r < R,$$
$$T_{rr}^{(0)} = 0, \quad r = R,$$
$$u_3^{(0)} = 0, \quad r = R, \tag{3.2.35}$$
$$T_{rr}^{(1)} = 0, \quad r = R,$$
$$\text{all fields are finite}, \quad r = 0,$$

where $\delta(r)$ is the Dirac delta function. The solution is found to be

$$u_3^{(0)} = B_1 \bar{P} r^2 \ln r + \bar{P}(B_2 r^2 + B_3 R^2) + B_4 E_3 (R^2 - r^2),$$
$$\psi = B_5 \bar{P} r^2 \ln r + B_6 \bar{P} r^2, \tag{3.2.36}$$

where

$$B_1 = 2(\bar{b}_{11} + \bar{b}_{12}),$$

$$B_2 = -[\bar{\gamma}_{11}(3 + 2\ln R) + \bar{\gamma}_{12}(1 + 2\ln R)]\frac{\bar{b}_{11} + \bar{b}_{12}}{\bar{\gamma}_{11} + \bar{\gamma}_{12}}$$
$$- [\bar{b}_{11}(3 + 2\ln R) + \bar{b}_{12}(1 + 2\ln R)] + (1 + 2\ln R)(\bar{b}_{11} + \bar{b}_{12}),$$

$$B_3 = -2\ln R\left(\bar{b}_{11} + \bar{b}_{12}\right) + [\bar{\gamma}_{11}\left(3 + 2\ln R\right) + \bar{\gamma}_{12}\left(1 + 2\ln R\right)]\frac{\bar{b}_{11} + \bar{b}_{12}}{\bar{\gamma}_{11} + \bar{\gamma}_{12}}$$

$$+ [\bar{b}_{11}\left(3 + 2\ln R\right) + \bar{b}_{12}\left(1 + 2\ln R\right)] - \left(1 + 2\ln R\right)(\bar{b}_{11} + \bar{b}_{12}),$$

$$B_4 = -\frac{\bar{h}_m h_c e_{31}^p}{2\left(\bar{\gamma}_{11} + \bar{\gamma}_{12}\right)},$$

$$B_5 = 2\left(\bar{\gamma}_{11} + \bar{\gamma}_{12}\right),$$

$$B_6 = -\left(1 + 2\ln R\right)\left(\bar{\gamma}_{11} + \bar{\gamma}_{12}\right),$$

$$\bar{P} = \frac{P}{16\pi\left(\bar{b}_{12}\bar{\gamma}_{11} - \bar{b}_{11}\bar{\gamma}_{12}\right)}.$$

$$(3.2.37)$$

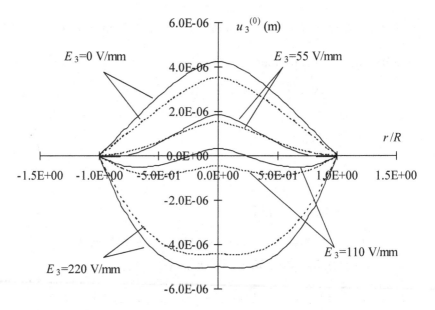

Figure 3.2.5 Deflection under P for different E_3. $P = 1$ N, $h_c = 0.4$ mm, h_m = 0.15 mm, $R = 12.7$ mm.

Figure 3.2.5 shows the deflection under P for different E_3. The behaviors of the curves in Figures 3.2.3 and 3.2.5 are similar. The deflection produced under a concentrated load $P = 1$ N is comparable to the deflection under a uniform load of $\pi R^2 q_0 = 2$ N. This is as expected because a concentrated load at the center is more effective in producing deflection.

3.3 Laminated Piezoelectric Plates

In this section we study motions of multi-layered piezoelectric plate laminates in general. Two-dimensional equations for laminated plates are derived by power series expansions, and are truncated to a first-order theory for coupled extension, flexure and thickness-shear.

3.3.1 Power series expansion

Consider an N-layer piezoelectric plate of total thickness $2h$ with the x_3 axis normal to the plate (see Figure 3.3.1). The two plate major surfaces and the N-1 interfaces are sequentially determined by $x_3 = -h = h_0, h_1, \ldots, h_{N-1}$, and $h_N = h$.

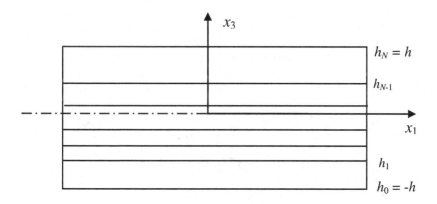

Figure 3.3.1. A laminated piezoelectric plate.

First we expand the mechanical displacement and electric potential into power series in x_3:

$$u_i = \sum_n x_3^n u_i^{(n)}, \quad \phi = \sum_n x_3^n \phi^{(n)} . \qquad (3.3.1)$$

Then

$$S_{ij} = \sum_n x_3^n S_{ij}^{(n)}, \quad E_i = \sum_n x_3^n E_i^{(n)}, \tag{3.3.2}$$

where

$$S_{ij}^{(n)} = \frac{1}{2}[u_{j,i}^{(n)} + u_{i,j}^{(n)} + (n+1)(\delta_{i3}u_j^{(n+1)} + \delta_{3j}u_i^{(n+1)})],$$
$$E_i^{(n)} = -\phi_{,i}^{(n)} - (n+1)\delta_{3i}\phi^{(n+1)}. \tag{3.3.3}$$

For the two-dimensional equations of motion and charge, we multiply

$$T_{ji,j} = \rho\ddot{u}_i, \quad D_{i,i} = 0 \tag{3.3.4}$$

by x_3^n, integrate the resulting equations across the thickness of the I-th layer from h_{I-1} to h_I, sum over I, and make use of the interface continuity conditions of the traction vector and the normal component of the electric displacement. We then obtain the following n-th order field equations

$$T_{ij,i}^{(n)} - nT_{3j}^{(n-1)} + F_j^{(n)} = \sum_m \rho^{(m+n)}\ddot{u}_j^{(m)},$$
$$D_{i,i}^{(n)} - nD_3^{(n-1)} + D^{(n)} = 0, \tag{3.3.5}$$

where

$$\{T_{ij}^{(n)}, D_i^{(n)}\} = \int_{-h}^h \{T_{ij}, D_i\}x_3^n dx_3 = \sum_{I=1}^N \int_{h_{I-1}}^{h_I} \{T_{ij}, D_i\}x_3^n dx_3,$$
$$F_j^{(n)} = [x_3^n T_{3j}]_{-h}^h, \quad D^{(n)} = [x_3^n D_3]_{-h}^h, \tag{3.3.6}$$
$$\rho^{(m+n)} = \sum_{I=1}^N \int_{h_{I-1}}^{h_I} \rho^I x_3^n x_3^m dx_3.$$

Substitution of the three-dimensional constitutive relations of the I-th layer

$$T_{ij} = c_{ijkl}^I S_{kl} - e_{kij}^I E_k,$$
$$D_i = e_{ikl}^I S_{kl} + \varepsilon_{ij}^I E_j \tag{3.3.7}$$

into Equation (3.3.6)$_1$ gives the plate constitutive equations of order n as follows

$$T_{ij}^{(n)} = \sum_m (c_{ijkl}^{(m+n)} S_{kl}^{(m)} - e_{kij}^{(m+n)} E_k^{(m)}),$$

$$D_i^{(n)} = \sum_m (e_{ijk}^{(m+n)} S_{jk}^{(m)} - \varepsilon_{ij}^{(m+n)} E_j^{(m)}),$$

(3.3.8)

where

$$c_{ijkl}^{(m+n)} = \sum_{I=1}^{N} \int_{h_{I-1}}^{h_I} c_{ijkl}^I x_3^n x_3^m dx_3,$$

$$e_{kij}^{(m+n)} = \sum_{I=1}^{N} \int_{h_{I-1}}^{h_I} e_{kij}^I x_3^n x_3^m dx_3,$$

(3.3.9)

$$\varepsilon_{ij}^{(m+n)} = \sum_{I=1}^{N} \int_{h_{I-1}}^{h_I} \varepsilon_{ij}^I x_3^n x_3^m dx_3.$$

3.3.2 First-order theory

For a first-order plate theory of coupled extensional, flexural and thickness-shear motions we make the following truncation of Equation (3.3.1):

$$u_i \cong u_i^{(0)}(x_1, x_2, t) + x_3 u_i^{(1)}(x_1, x_2, t) + x_3^2 u_i^{(2)}(x_1, x_2, t),$$

$$\phi \cong \phi^{(0)}(x_1, x_2, t) + x_3 \phi^{(1)}(x_1, x_2, t),$$

(3.3.10)

where $u_3^{(1)}$ and $u_j^{(2)}$ will be eliminated by stress relaxations. The strains and electric fields are

$$S_p \cong S_p^{(0)} + x_3 S_p^{(1)}, \quad p = 1, 2, \cdots, 6,$$

$$E_i \cong E_i^{(0)} + x_3 E_i^{(1)},$$

(3.3.11)

where the zero- and first-order strains are

$$S_1^{(0)} = u_{1,1}^{(0)}, \quad S_2^{(0)} = u_{2,2}^{(0)}, \quad S_3^{(0)} = u_3^{(1)},$$

$$S_4^{(0)} = u_{3,2}^{(0)} + u_2^{(1)}, \quad S_5^{(0)} = u_{3,1}^{(0)} + u_1^{(1)}, \quad S_6^{(0)} = u_{1,2}^{(0)} + u_{2,1}^{(0)},$$

(3.3.12)

and

$$S_1^{(1)} = u_{1,1}^{(1)}, \quad S_2^{(1)} = u_{2,2}^{(1)}, \quad S_3^{(1)} = 2u_3^{(2)},$$

$$S_4^{(1)} = u_{3,2}^{(1)} + 2u_2^{(0)} \cong 0, \quad S_5^{(1)} = u_{3,1}^{(1)} + 2u_1^{(2)} \cong 0, \qquad (3.3.13)$$

$$S_6^{(1)} = u_{1,2}^{(1)} + u_{2,1}^{(1)}.$$

We note that $S_3^{(0)}$ and $S_3^{(1)}$ are involved with $u_3^{(1)}$ and $u_3^{(2)}$, which are to be eliminated. $S_4^{(1)}$ and $S_5^{(1)}$ are neglected as an approximation. The zero- and first-order electric fields are given by

$$E_1^{(0)} = -\phi_{,1}^{(0)}, \quad E_2^{(0)} = -\phi_{,2}^{(0)}, \quad E_3^{(0)} = -\phi^{(1)}, \qquad (3.3.14)$$

and

$$E_1^{(1)} = -\phi_{,1}^{(1)}, \quad E_2^{(1)} = -\phi_{,2}^{(1)}, \quad E_3^{(1)} = 0. \qquad (3.3.15)$$

The zero- and first-order equations of motion and charge are

$$T_{ab,a}^{(0)} + F_b^{(0)} = \sum_{I=1}^{N} \rho^I [(h_I - h_{I-1})\ddot{u}_b^{(0)} + \frac{h_I^2 - h_{I-1}^2}{2}\ddot{u}_b^{(1)}],$$

$$T_{a3,a}^{(0)} + F_3^{(0)} = \sum_{I=1}^{N} \rho^I [(h_I - h_{I-1})\ddot{u}_3^{(0)}],$$

$$T_{ab,a}^{(1)} - T_{3b}^{(0)} + F_b^{(1)} = \sum_{I=1}^{N} \rho^I [\frac{h_I^2 - h_{I-1}^2}{2}\ddot{u}_b^{(0)} + \frac{h_I^3 - h_{I-1}^3}{3}\ddot{u}_b^{(1)}], \qquad (3.3.16)$$

$$D_{a,a}^{(0)} + D^{(0)} = 0,$$

$$D_{a,a}^{(1)} - D_3^{(0)} + D^{(1)} = 0.$$

The zero-order plate constitutive relations are

$$T_{ij}^{(0)} = (c_{ijkl}^{(0)} S_{kl}^{(0)} - e_{kij}^{(0)} E_k^{(0)}) + (c_{ijkl}^{(1)} S_{kl}^{(1)} - e_{kij}^{(1)} E_k^{(1)}),$$

$$D_i^{(0)} = (e_{ijk}^{(0)} S_{jk}^{(0)} + \varepsilon_{ij}^{(0)} E_j^{(0)}) + (e_{ijk}^{(1)} S_{jk}^{(1)} + \varepsilon_{ij}^{(1)} E_j^{(1)}), \qquad (3.3.17)$$

where

$$c_{ijkl}^{(0)} = \sum_{I=1}^{N}(h_I - h_{I-1})\bar{c}_{ijkl}^I, \quad e_{kij}^{(0)} = \sum_{I=1}^{N}(h_I - h_{I-1})\bar{e}_{kij}^I,$$

$$\varepsilon_{ij}^{(0)} = \sum_{I=1}^{N}(h_I - h_{I-1})\bar{\varepsilon}_{ij}^I,$$

(3.3.18)

and

$$c_{ijkl}^{(1)} = \sum_{I=1}^{N}(\frac{h_I^2 - h_{I-1}^2}{2})\bar{c}_{ijkl}^I, \quad e_{kij}^{(1)} = \sum_{I=1}^{N}(\frac{h_I^2 - h_{I-1}^2}{2})\bar{e}_{kij}^I,$$

$$\varepsilon_{ij}^{(1)} = \sum_{I=1}^{N}(\frac{h_I^2 - h_{I-1}^2}{2})\bar{\varepsilon}_{ij}^I.$$

(3.3.19)

The first-order constitutive relations are

$$T_{ij}^{(1)} = (c_{ijkl}^{(1)}S_{kl}^{(0)} - e_{kij}^{(1)}E_k^{(0)}) + (c_{ijkl}^{(2)}S_{kl}^{(1)} - e_{kij}^{(2)}E_k^{(1)}),$$

$$D_i^{(1)} = (e_{ijk}^{(1)}S_{jk}^{(0)} + \varepsilon_{ij}^{(1)}E_j^{(0)}) + (e_{ijk}^{(2)}S_{jk}^{(1)} + \varepsilon_{ij}^{(2)}E_j^{(1)}),$$

(3.3.20)

where

$$c_{ijkl}^{(2)} = \sum_{I=1}^{N}(\frac{h_I^3 - h_{I-1}^3}{3})\bar{c}_{ijkl}^I, \quad e_{kij}^{(2)} = \sum_{I=1}^{N}(\frac{h_I^3 - h_{I-1}^3}{3})\bar{e}_{kij}^I,$$

$$\varepsilon_{ij}^{(2)} = \sum_{I=1}^{N}(\frac{h_I^3 - h_{I-1}^3}{3})\bar{\varepsilon}_{ij}^I.$$

(3.3.21)

Note that in Equations (3.3.18), (3.3.19) and (3.3.21) we have used the following material constants after the simpler version of stress relaxation (see Equations (2.4.13) through (2.4.15)):

$$\bar{c}_{ijkl}^I = c_{ijkl}^I - c_{ij33}^I c_{33kl}^I / c_{3333}^I,$$

$$\bar{e}_{kij}^I = e_{kij}^I - e_{k33}^I c_{33ij}^I / c_{3333}^I,$$

$$\bar{\varepsilon}_{ij}^I = \varepsilon_{ij}^I + e_{i33}^I e_{j33}^I / c_{3333}^I.$$

(3.3.22)

Then $u_3^{(1)}$ and $u_3^{(2)}$ will not appear in the constitutive relations in Equations (3.3.17) and (3.3.20). A more sophisticated stress relaxation using $T_{33}^{(0)} = 0$ and $T_{3j}^{(1)} = 0$ to eliminate $S_{33}^{(0)}$ and $S_{3j}^{(1)}$ (and hence $u_3^{(1)}$ and $u_j^{(2)}$) can be performed. Two shear correction factors κ_1 and κ_2 can be introduced by replacing the following zero-order strains:

$$S_{31}^{(0)} \rightarrow \kappa_1 S_{31}^{(0)}, \quad S_{32}^{(0)} \rightarrow \kappa_2 S_{32}^{(0)} \tag{3.3.23}$$

to correct the two fundamental thickness-shear resonant frequencies.

In summary, we have the two-dimensional equations of motion and charge (3.3.16), constitutive relations (3.3.17) and (3.3.20), and strains and electric fields (3.3.12) through (3.3.15). With successive substitutions, seven equations for $u_i^{(0)}$, $u_a^{(1)}$, $\phi^{(0)}$ and $\phi^{(1)}$ can be obtained. At the boundary of a plate with an in-plane unit exterior normal $\mathbf{n}$ and an in-plane unit tangent $\mathbf{s}$, we may prescribe

$$
\begin{aligned}
&T_{nn}^{(0)} \quad \text{or} \quad u_n^{(0)}, \quad T_{ns}^{(0)} \quad \text{or} \quad u_s^{(0)}, \quad T_{n3}^{(0)} \quad \text{or} \quad u_3^{(0)}, \\
&T_{nn}^{(1)} \quad \text{or} \quad u_n^{(1)}, \quad T_{ns}^{(1)} \quad \text{or} \quad u_s^{(1)}, \\
&D_n^{(0)} \quad \text{or} \quad \phi^{(0)}, \quad D_n^{(1)} \quad \text{or} \quad \phi^{(1)}.
\end{aligned}
\tag{3.3.24}
$$

For a laminated plate, extension and flexure are usually coupled unless the plate is laminated symmetrically about its middle surface.

3.4 A Plate on a Substrate

A thin film on the surface of a half-space or at the interface between two half-spaces is a common structure in device applications. In this case, for long waves, the film can be modeled by the two-dimensional equations [42], and the half-space(s) by the three-dimensional equations. In this section we analyze two problems of this type.

3.4.1 A Piezoelectric film on an elastic half-space

First consider a piezoelectric film on an elastic half-space (see Figure 3.4.1) [43]. This problem is useful for understanding the behavior of piezoelectric actuators. The film is modeled by the equations for the extensional deformation of thin plates, and the substrate is governed by the equations of elasticity. The film is of ceramics with thickness poling, and is partially electroded. The thick lines represent electrodes. The substrate is isotropic.

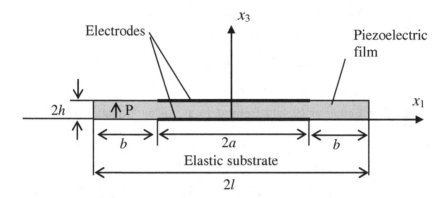

Figure 3.4.1. A partially electroded piezoelectric film on an elastic substrate.

We study plane strain deformations with $u_2 = 0$ and $\partial/\partial x_2 = 0$. The top and bottom electrodes on the film are under given applied electric potentials $\pm V/2$. Then for the electroded portion we have $\phi^{(0)} = 0$ and $\phi^{(1)} = V/2h$. In the unelectroded portions, since $\phi^{(0)}(\pm a) = 0$ and $\phi_{,1}^{(0)}(\pm l) = 0$, it can be concluded that $\phi^{(0)} = 0$. However, $\phi^{(1)}$ remains unknown in the unelectroded portions and is coupled to the mechanical fields. Denote the shear stress between the film and the substrate by $\tau(x_1)$ for $|x_1| < l$. Since the film is assumed to be very thin, the continuity of the tangential displacement at the interface implies that the tangential surface displacement of the substrate is also $u_1^{(0)}(x_1)$. Then we have the following relation for the surface strain due to a surface distribution of shear stress τ [43]

$$u_{1,1}^{(0)}(x_1) = -\frac{2(1-v^2)}{\pi E} \int_{-l}^{l} \frac{\tau(s)}{x_1 - s} ds , \qquad (3.4.1)$$

where E is the Young's modulus and v is the Poisson's ratio of the substrate. Equation (3.4.1) is from the theory of elasticity. For the extensional force $T_{11}^{(0)}$ in the film we have

$$T_{11}^{(0)} = 2h(c_{11}^P u_{1,1}^{(0)} + e_{31}^P \phi^{(1)}) . \qquad (3.4.2)$$

$T_{11}^{(0)}$ must satisfy the following edge conditions

$$T_{11}^{(0)}(\pm l) = 0 .$$ (3.4.3)

The equation of motion of the film is

$$\tau = T_{11,1}^{(0)} .$$ (3.4.4)

Integrating Equation (3.4.4) from $-l$ to x_1, using Equation (3.4.3), we obtain

$$\int_{-l}^{x_1} \tau(s)ds = T_{11}^{(0)}(x_1) .$$ (3.4.5)

Substituting Equation (3.4.1) into Equation (3.4.2) and the resulting expression into Equation (3.4.5), we have the following integral equation for τ :

$$\int_{-l}^{x_1} \tau(s)ds + \frac{2hc_{11}^P 2(1-v^2)}{\pi E} \int_{-l}^{l} \frac{\tau(s)}{x_1 - s} ds$$

$$= 2he_{31}^P \phi^{(1)}(x_1), \quad |x_1| < l.$$ (3.4.6)

For the electric potential we have

$$\phi^{(1)} = \begin{cases} V/2h, & |x_1| < a, \\ \text{unknown}, & a < |x_1| < l, \end{cases}$$ (3.4.7)

where $\phi^{(1)}$ in the unelectroded portions is part of the unknowns and has to be obtained by solving simultaneous equations. The additional equation for determining $\phi^{(1)}$ takes the following form with the substitution of Equation (3.4.1) into Equation (3.1.14) and h' replaced by h:

$$\frac{2he_{31}^P 2(1-v^2)}{\pi E} \int_{-l}^{l} \frac{\tau(s)}{x_1 - s} ds$$

$$= \frac{2h^3}{3} \varepsilon_{11}^P \phi_{,11}^{(1)} - 2h\varepsilon_{33}^P \phi^{(1)}, \quad a < |x_1| < l,$$ (3.4.8)

which is coupled to τ. Equation (3.4.8) is an integro-differential equation to which we need to impose the following boundary conditions:

$$\phi^{(1)}(\pm a) = V/2h, \quad \phi_{,1}^{(1)}(\pm l) = 0 .$$ (3.4.9)

We have obtained a pair of coupled integro-differential equations (3.4.6) and (3.4.8) for determining $\tau(x_1)$ and $\phi^{(1)}(x_1)$, with boundary conditions (3.4.9). Equation (3.4.8) is over the region of the unelectroded portions of the film only, where $\phi^{(1)}$ is unknown.

3.4.1.1 Rigid substrate

First consider the special case of a fully electroded film on a rigid substrate. We have $u_1^{(0)} = 0$. Because of symmetry we only consider half of the film with $0 < x_1 < l$. For a fully electroded film with $a = l$, we have $\phi^{(1)} = V / 2h$ over the whole film. Equation (3.4.2) then implies that the film extensional force $T_{11}^{(0)}$ has a constant value of $e_{31}^P V$ along the film. Then Equation (3.4.4) further implies that the shear stress τ is zero under the film. However, Equation (3.4.3) requires that $T_{11}^{(0)}$ must vanish at both ends of the film. This is possible only when two concentrated shear forces of equal magnitude $e_{31}^P V$ and opposite in direction are present at both ends of the film. Mathematically, the shear stress distribution can be represented by

$$\tau(x_1) = -e_{31}^P V [\delta(x_1 - l) - \delta(x_1 + l)], \qquad (3.4.10)$$

where $\delta(x_1)$ is the Dirac delta function.

A slightly more general case is when a partially electroded film is on a rigid substrate. In this case from Equations (3.4.6) and (3.4.8) we have

$$e_{31}^P 2h\phi_{,1}^{(1)} - \tau = 0,$$
$$-\frac{2h^3}{3}\varepsilon_{11}^P \phi_{,11}^{(1)} + 2h\varepsilon_{33}^P \phi^{(1)} = 0, \qquad (3.4.11)$$

for which the boundary conditions in Equation (3.4.9) still apply. From Equation (3.4.11)$_2$ and Equation (3.4.9), the following solution for $\phi^{(1)}$ can be obtained

$$\phi^{(1)} = \begin{cases} V / 2h, & 0 < x_1 < a, \\ G\exp[-\xi(x_1 - a)] + H\exp[\xi(x_1 - a)], & a < x_1 < l, \end{cases} \qquad (3.4.12)$$

where

$$\xi = \frac{2}{h}\sqrt{3\frac{\varepsilon_{33}^P}{\varepsilon_{11}^P}}, \quad G = \frac{V}{2h[1+\exp(-2\xi b)]},$$

$$H = \frac{V}{2h[1+\exp(-2\xi b)]}\exp(-2\xi b),$$

(3.4.13)

and $b = l\text{-}a$. Substitution of Equation (3.4.12) into Equation (3.4.11)$_1$ gives the shear stress distribution

$$\tau = \begin{cases} 0, & 0 < x_1 < a, \\ \dfrac{e_{31}^P \xi V}{1+\exp(-2\xi b)}\{-\exp[-\xi(x_1 - a] \\ \quad + \exp(-2\xi b)\exp[\xi(x_1 - a]\}, & a < x_1 < l. \end{cases}$$

(3.4.14)

Under the electroded portion there is no shear stress. In the unelectroded portion the shear stress decays exponentially with $\tau(l) = 0$. The important difference between Equations (3.4.14) and (3.4.10) is that one predicts a distribution of finite shear stresses while the other has a singular distribution of the delta function. This shows the advantage of partially electroded actuators in reducing stress concentration. The total shear force predicted by Equation (3.4.14) can be obtained by integrating the shear stress over the unelectroded region (a, l), which yields

$$Q = \frac{e_{31}^P V}{1+\exp(-2\xi b)}[-1 + 2\exp(-\xi b) - \exp(-2\xi b)].$$

(3.4.15)

For the equations of a thin film to be valid in the unelectroded portion, the length of the unelectroded portion b has to be much larger than the thickness $2h$. Then it can be seen from Equation (3.4.13)$_1$ that in this case $\exp(-\xi b) \ll 1$. If we neglect $\exp(-\xi b)$ compared to one in the above expressions, we obtain the following simpler and physically more revealing expressions:

$$\phi^{(1)} \cong \begin{cases} V/2h, & 0 < x_1 < a, \\ \dfrac{V}{2h}\exp[-\xi(x_1 - a)], & a < x_1 < l, \end{cases}$$

(3.4.16)

$$\tau \cong \begin{cases} 0, & 0 < x_1 < a, \\ -e_{31}^p \xi V \exp[-\xi(x_1 - a)], & a < x_1 < l, \end{cases} \tag{3.4.17}$$

$$Q \cong -e_{31}^p V. \tag{3.4.18}$$

The shear stress distributions given by Equations (3.4.10) and (3.4.17) are shown qualitatively in Figure 3.4.2 where a minus sign has been dropped. The figure shows that the shear stress distribution under a partially electroded film is much less concentrated than a fully electroded film. From Equation (3.4.18) we can see that the total shear force is about the same as Equation (3.4.10), being equal to the tension (or contraction) force in the unelectroded portion of the film.

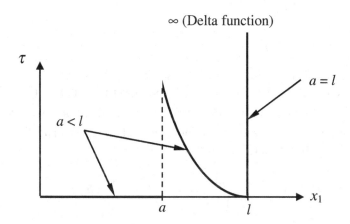

Figure 3.4.2. Shear stress distribution under a piezoelectric film on a rigid substrate. $a = l$: fully electroded. $a < l$: partially electroded.

3.4.1.2 Compliant substrate

Next consider the special case when the piezoelectric film is fully electroded and is much stiffer than the substrate. With the introduction of the following small, dimensionless parameter:

$$\lambda = \frac{E}{c_{11}^p 2(1 - v^2)}, \tag{3.4.19}$$

and the change of the integration variable $s = lt$, Equation (3.4.6) can be written as

$$\frac{1}{\pi}\int_{-1}^{1}\frac{f(t)}{\varsigma - t}dt = \lambda e_{31}^{p}\frac{V}{2h} - \lambda\frac{l}{2h}\int_{-1}^{\varsigma}f(t)dt, \qquad (3.4.20)$$

where we have denoted $f(t) = \tau(lt)$, and $\varsigma = x_1/l$. We now seek the following perturbation solution:

$$f(t) = f^{(0)}(t) + \lambda f^{(1)}(t) + \cdots. \qquad (3.4.21)$$

Substituting Equation (3.4.21) into (3.4.20), we obtain the following zero- and first-order problems:

$$\frac{1}{\pi}\int_{-1}^{1}\frac{f^{(0)}(t)}{\varsigma - t}dt = 0,$$

$$\frac{1}{\pi}\int_{-1}^{1}\frac{f^{(1)}(t)}{\varsigma - t}dt = e_{31}^{p}\frac{V}{2h} - \frac{l}{2h}\int_{-1}^{\varsigma}f^{(0)}(t)dt. \qquad (3.4.22)$$

These equations are analytically solvable. Equation $(3.4.22)_1$ is homogeneous. A physically meaningful solution to Equation $(3.3.22)_1$ is $f^{(0)}(t) \equiv 0$, which represents the situation that the substrate is mechanically not resisting the deformation of the film. The first-order solution from Equation $(3.4.22)_2$ takes the form

$$f^{(1)}(\varsigma) = \frac{1}{\pi}\int_{-1}^{1}\sqrt{\frac{1-t^2}{1-\varsigma^2}}\frac{1}{\varsigma - t}(-e_{31}^{p}\frac{V}{2h})dt, \qquad (3.4.23)$$

or

$$\tau(x_1) \cong -\frac{E}{c_{11}^{p}2(1-v^2)}e_{31}^{p}\frac{V}{2h}\frac{1}{\pi}\int_{-l}^{l}\sqrt{\frac{l^2-s^2}{l^2-x_1^2}}\frac{1}{x_1 - s}ds. \qquad (3.4.24)$$

It can be verified that Equation (3.4.24) is an odd function and is singular at $x_1 = \pm l$. The integration of τ over $(0,l)$ is still finite to give the total shear force. $f^{(1)}(\varsigma)$ in Equation (3.4.23) is normalized and shown in Figure 3.4.3. Comparison of Equations (3.4.24) and (3.4.10) shows that for the case of fully electroded actuators, the shear stress distribution is more concentrated when the substrate is stiffer than the film.

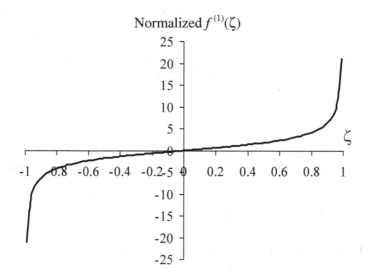

Figure 3.4.3. Normalized shear stress $\pi f^{(1)}(\varsigma)/(-e_{31}^{P}V/2h)$.

3.4.1.3 Elastic substrate in general

In general we need to solve coupled integro-differential equations (3.4.6) and (3.4.8). We divide the domain into elements and solve the equations numerically. First we re-write Equations (3.4.6) and (3.4.8) as follows:

$$\int_{-l}^{x} \tau(s)ds + A\int_{-l}^{l} \frac{\tau(s)}{x-s}ds = e_{31}^{P}\Phi(x), \quad |x|<l, \tag{3.4.25}$$

and

$$B\int_{-l}^{l} \frac{\tau(s)}{x-s}ds = C\Phi_{,11}(x) - \varepsilon_{33}^{P}\Phi(x), \quad a<|x|<l, \tag{3.4.26}$$

where we have introduced the following for simplicity:

$$A = \frac{2hc_{11}^{P}2(1-v^{2})}{\pi E}, \quad B = \frac{2he_{31}^{P}2(1-v^{2})}{\pi E},$$

$$C = \frac{h^{2}}{3}\varepsilon_{11}^{P}, \quad \Phi = 2h\phi^{(1)}, \quad x = x_{1}. \tag{3.4.27}$$

Then, we divide the whole interval $(-l, l)$ into small segments with length Δ_i, $i = 1, 2, 3, ..., N$, where N $(= 2N_a + 2N_b)$ is the total number of the segments, $2N_a$ is the number for the electroded portion, and $2N_b$ is the number for the unelectroded portions (see Figure 3.4.4). One node is placed on each element.

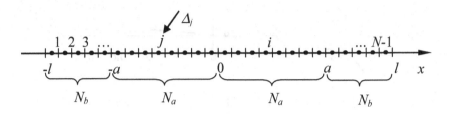

Figure 3.4.4. Discretization of the domain $(-l, l)$ into N $(= 2N_a + 2N_b)$ small segments (elements).

The shear stress and electrostatic potential are approximated by undetermined constants over each element. Let $x = x_i$ (i.e., at node i) for $i = 1, 2, 3, ..., N$, then the two integrals in Equation (3.4.25) can be discretized as:

$$\int_{-l}^{x_i} \tau(s)ds = \sum_{j=1}^{i} \int_{\Delta_j} \tau(s)ds = \sum_{j=1}^{i} \tau_j \Delta_j , \qquad (3.4.28)$$

$$\int_{-l}^{l} \frac{\tau(s)}{x_i - s}ds = \sum_{j=1}^{N} \int_{\Delta_j} \frac{\tau(s)}{x_i - s}ds = \sum_{j=1}^{N} \int_{\Delta_j} \frac{1}{x_i - s}ds\tau_j$$

$$= \sum_{\substack{j=1 \\ j \neq i}}^{N} \ln\left|\frac{1}{x_i - s}\right|_{x_j - \frac{1}{2}\Delta_j}^{x_j + \frac{1}{2}\Delta_j} \tau_j = \sum_{\substack{j=1 \\ j \neq i}}^{N} \ln\left|\frac{x_i - (x_j - \frac{1}{2}\Delta_j)}{x_i - (x_j + \frac{1}{2}\Delta_j)}\right| \tau_j , \qquad (3.4.29)$$

in which $\tau_i = \tau(x_i)$. The singular integral $\int_{\Delta_j} \frac{1}{x_i - s}ds$ can be shown to vanish in the Cauchy-principal value sense and thus is excluded from the summation. Using the above results, we can write the discretized form of Equation (3.4.25) as:

$$\Delta\sum_{j=1}^{i}\tau_j + A\sum_{\substack{j=1\\j\neq i}}^{N}\ln\left|\frac{x_i - (x_j - \frac{1}{2}\Delta)}{x_i - (x_j + \frac{1}{2}\Delta)}\right|\tau_j = e_{31}^p\,\Phi_i,$$

$$i = 1, 2, 3, ..., N,$$

(3.4.30)

where $\Phi_i = \Phi(x_i)$ and a uniform length for all the elements has been assumed (i.e., $\Delta_j = \Delta$, $j = 1, 2, 3, ..., N$). Notice that $\Phi_{N_b+1} = \Phi_{N_b+2}$ $= ... = \Phi_{N_b+2N_a} = V$, as given in Equation (3.4.7). There are a total of N equations in Equation (3.4.30). Applying the central difference scheme, we have

$$\Phi_{,11}(x_i) = \frac{\Phi_{i+1} - 2\Phi_i + \Phi_{i-1}}{\Delta^2},$$

$$i = 2, 3, ..., N_b, N_b + 2N_a + 1, ..., N - 1.$$

(3.4.31)

Thus, the discretized form of Equation (3.4.26) is found to be

$$B\sum_{\substack{j=1\\j\neq i}}^{N}\ln\left|\frac{x_i - (x_j - \frac{1}{2}\Delta)}{x_i - (x_j + \frac{1}{2}\Delta)}\right|\tau_j = \frac{C}{\Delta^2}(\Phi_{i+1} - 2\Phi_i + \Phi_{i-1}) - \varepsilon_{33}^p\,\Phi_i,$$

$$i = 2, 3, ..., N_b, N_b + N_a + 1, ..., N - 1,$$

(3.4.32)

in which

$$\Phi_{N_b+1} = \Phi_{N_b+2N_a} = V,$$

(3.4.33)

from condition (3.4.7) or (3.4.9)$_1$ should be applied. There are only $2N_b - 2$ equations in Equation (3.4.32). Two more equations are found by enforcing the zero-slope boundary condition in Equation (3.4.9)$_2$ using the finite difference, that is,

$$\Phi_1 = \Phi_2, \quad \Phi_{N-1} = \Phi_N.$$

(3.4.34)

There are a total of $N + 2N_b$ equations in Equations (3.4.30), (3.4.32) and (3.4.34), which are used to solve for the N unknowns τ_i ($i = 1, 2, 3, ..., N$) and $2N_b$ unknowns Φ_i, ($i = 1, 2, 3, ..., N_b$, $N_b+2N_a+1, ..., N-1, N$). Equations (3.4.30), (3.4.32) and (3.4.34) can be written in a matrix form and solved for τ and Φ together or solved for τ and Φ separately using partitioning of the whole matrix. For numerical results, we consider PZT-7A for the piezoelectric film. For geometric parameters, we choose $l = 0.1$ m and $h = 0.0025$ m. The substrate is made of steel ($E = 2.0 \times 10^{11}$ Pa, $v = 0.3$). Numerical tests show that $N = 200$ is sufficient and is thus used for all the following calculations.

Figures 3.4.5 and 3.4.6 show the results for the shear stresses and electric potentials, respectively, with the electroded region of three different sizes. It is observed that the location of the peak value of shear stress moves towards the edges of the film when a increases, as expected. Since the shear stresses decay so rapidly, they essentially do not feel the ends of the film in the cases shown with almost the same shear stress distributions concentrated at different locations. Figure 3.4.5 suggests that if a partially electroded actuator is used for the purpose of reducing shear stress concentration, only very small portions in the order of about five times the film thickness near the ends of the actuator need to be left unelectroded. This provides some guidance for design.

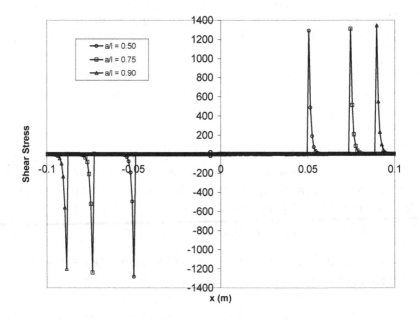

Figure 3.4.5. Shear stress distribution for different sizes of the electroded area.

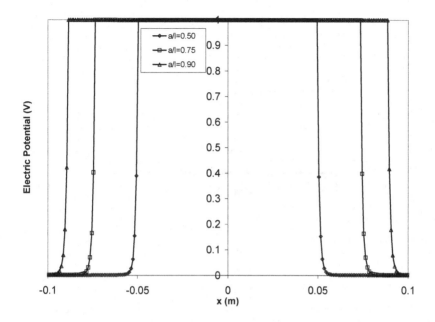

Figure 3.4.6. Electric potential distributiosn for different sizes of the electroded area.

From the numerical data of Figure 3.4.5, it is observed that the shear stresses under the electrodes are not zero, but much smaller in value compared with the peak values near the edges of the electroded area. To illustrate this, the distributions of the shear stresses under the electrode are plotted in Figure 3.4.7 for three substrates with decreasing Young's moduli. The shear stresses under the electrode are larger as the substrate becomes softer, as also suggested by the comparison of Figures 3.4.2 and 3.4.3. However, the values of the shear stress under the electrode are still several orders smaller than the overall peak values of the shear stress right outside the edges of the electrode. Practically the shear stress under the electrode can be neglected for many purposes. Since the shear stress under the electrode is much smaller than the peak values right outside the electrode, the shear stress is discontinuous at the ends of the electrode. The discontinuity in the shear stress distribution is related to the thin film model used.

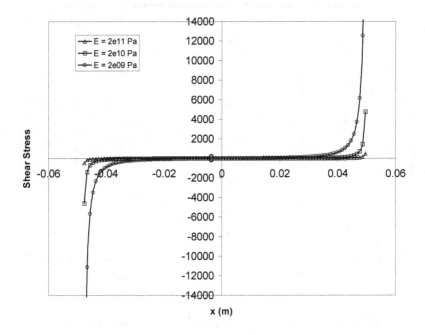

Figure 3.4.7. Distributions of the shear stresses ($\times 10^9$) under the electrodes.

3.4.2 Piezoelectric surface waves guided by a thin elastic film

Consider the propagation of long, anti-plane (shear horizontal or SH) surface waves over a piezoelectric half-space carrying a thin, non-piezoelectric dielectric film (see Figure 3.4.8). The ceramic half-space is poled in the x_3 direction.

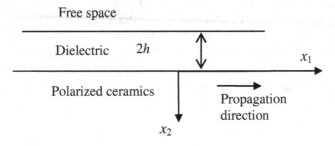

Figure 3.4.8. A ceramic half-space with a dielectric film.

For the ceramic half-space, for motions of $u_3(x_1, x_2, t)$ and $\phi(x_1, x_2, t)$, the governing equations are [12,9]

$$\bar{c}_{44} \nabla^2 u_3 = \rho \ddot{u}_3,$$
$$\nabla^2 \psi = 0,$$

(3.4.35)

$$\psi = \phi - \frac{e_{15}}{\varepsilon_{11}} u_3,$$

(3.4.36)

and

$$T_{23} = \bar{c}_{44} u_{3,2} + e_{15} \psi_{,2},$$
$$T_{31} = \bar{c}_{44} u_{3,1} + e_{15} \psi_{,1},$$
$$D_1 = -\varepsilon_{11} \psi_{,1},$$
$$D_2 = -\varepsilon_{11} \psi_{,2},$$

(3.4.37)

where the piezoelectrically-stiffened shear constant is defined by

$$\bar{c}_{44} = c_{44} + \frac{e_{15}^2}{\varepsilon_{11}} = c_{44}(1 + k_{15}^2), \quad k_{15}^2 = \frac{e_{15}^2}{\varepsilon_{11} c_{44}}.$$

(3.4.38)

For a surface wave solution we must have

$$u_3, \phi \to 0, \quad x_2 \to +\infty.$$

(3.4.39)

Consider the possibility of solutions in the following form:

$$u_3 = A \exp(-\xi_2 x_2) \exp[i(\xi_1 x_1 - \omega t)],$$
$$\psi = B \exp(-\xi_1 x) \exp[i(\xi_1 x_1 - \omega t)],$$

(3.4.40)

where A and B are undetermined constants, and ξ_2 should be positive for decaying behavior away from the surface. Equation $(3.4.40)_2$ already satisfies Equation $(3.4.35)_2$. For Equation $(3.4.40)_1$ to satisfy Equation $(3.4.35)_1$ we must have

$$\bar{c}_{44}(\xi_1^2 - \xi_2^2) = \rho \omega^2,$$

(3.4.41)

which leads to the following expression for ξ_2:

$$\xi_2^2 = \xi_1^2 - \frac{\rho \omega^2}{\bar{c}_{44}} = \xi_1^2 (1 - \frac{v^2}{v_T^2}) > 0,$$

(3.4.42)

where

$$v^2 = \frac{\omega^2}{\xi_1^2}, \quad v_T^2 = \frac{\bar{c}_{44}}{\rho}. \tag{3.4.43}$$

The following are needed for prescribing boundary and continuity conditions:

$$\phi = [B\exp(-\xi_1 x_2) + \frac{e_{15}}{\varepsilon_{11}} A\exp(-\xi_2 x_2)]\exp[i(\xi_1 x_1 - \omega t)],$$

$$T_{23} = -[A\bar{c}_{44}\xi_2 \exp(-\xi_2 x_2) \tag{3.4.44}$$
$$+ e_{15}B\xi_1 \exp(-\xi_1 x_2)]\exp[i(\xi_1 x_1 - \omega t)],$$

$$D_2 = \varepsilon_{11}B\xi_1 \exp(-\xi_1 x_2)\exp[i(\xi_1 x_1 - \omega t)].$$

Electric fields can also exist in the free space of $x_2 < -2h \cong 0$, which is governed by

$$\nabla^2 \phi = 0, \quad x_2 < 0,$$
$$\phi \to 0, \quad x_2 \to -\infty. \tag{3.4.45}$$

A surface wave solution to Equation (3.4.45) is

$$\phi = C\exp(\xi_1 x_2)\exp[i(\xi_1 x_1 - \omega t)], \tag{3.4.46}$$

where C is an undetermined constant. From Equation (3.4.46), in the free space

$$D_2 = -\varepsilon_0 \xi_1 C \exp(\xi_1 x_2)\exp[i(\xi_1 x_1 - \omega t)]. \tag{3.4.47}$$

The film is assumed to be very thin in the sense that its thickness is much smaller than the wavelength of the waves we are interested in. We use a prime to indicate the elastic and dielectric constants as well as the mass density of the film. Consider a film of a cubic crystal with m3m symmetry. The elastic and dielectric constants are given by

$$\begin{pmatrix} c'_{11} & c'_{12} & c'_{12} & 0 & 0 & 0 \\ c'_{12} & c'_{11} & c_{12} & 0 & 0 & 0 \\ c'_{12} & c'_{12} & c'_{11} & 0 & 0 & 0 \\ 0 & 0 & 0 & c'_{44} & 0 & 0 \\ 0 & 0 & 0 & 0 & c'_{44} & 0 \\ 0 & 0 & 0 & 0 & 0 & c'_{44} \end{pmatrix}, \begin{pmatrix} \varepsilon'_{11} & 0 & 0 \\ 0 & \varepsilon'_{11} & 0 \\ 0 & 0 & \varepsilon'_{11} \end{pmatrix}. \tag{3.4.48}$$

For the film, we use the two-dimensional plate equations. This approach was used in [42]. The lowest order effects of the film are governed by the zero-order equations for extension. We have

$$T_{ab,a}^{(0)} + T_{2b}(x_2 = 0) - T_{2b}(x_2 = -2h) = 2\rho'h\ddot{u}_b^{(0)},$$
$$D_{a,a}^{(0)} + D_2(x_2 = 0) - D_2(x_2 = -2h) = 0, \quad a,b = 1,3. \tag{3.4.49}$$

The constitutive relations for the film are

$$T_r^{(0)} = c_{rs}^p S_s^{(0)},$$
$$D_i^{(0)} = \varepsilon_{ij}' E_j^{(0)}, \quad r,s = 1,3,5, \tag{3.4.50}$$

where

$$c_{rs}^p = c_{rs}' - c_{rv}' c_{vw}'^{-1} c_{ws}', \quad v,w = 2,4,6. \tag{3.4.51}$$

The fields in the film are

$$u_3 = A \exp[i(\xi_1 x_1 - \omega t)],$$
$$\phi = C \exp[i(\xi_1 x_1 - \omega t)]. \tag{3.4.52}$$

Equation (3.4.52) already satisfies the continuity of the mechanical displacement between the film and the ceramic half-space, and the continuity of the electric potential between the film and the free space. From Equation (3.4.50) we obtain:

$$T_{13}^{(0)} = 2hc_{55}^p S_{13}^{(0)} = 2hc_{44}^p u_{3,1}^{(0)}$$
$$= 2hc_{44}^p i\xi_1 A \exp[i(\xi_1 x_1 - \omega t)],$$
$$D_1^{(0)} = 2h\varepsilon_{11}' E_1 = -2h\varepsilon_{11}'\phi_{,1} \tag{3.4.53}$$
$$= -2h\varepsilon_{11}' i\xi_1 C \exp[i(\xi_1 x_1 - \omega t)].$$

Substitution of Equations (3.4.44), (3.4.46), (3.4.47) and (3.4.53) into the continuity condition of the electric potential between the ceramic half-space and the film, Equation (3.4.49)$_1$ for $b = 3$, and Equation (3.4.49)$_2$ yields

$$B + \frac{e_{15}}{\varepsilon_{11}} A = C,$$

$$-c_{44}^p \xi_1^2 A - \frac{1}{2h}(A\bar{c}_{44}\xi_2 + e_{15}B\xi_1) = -\rho'\omega^2 A, \qquad (3.4.54)$$

$$\varepsilon_{11}' \xi_1^2 C + \frac{1}{2h}(\varepsilon_{11}B\xi_1 + \varepsilon_0\xi_1 C) = 0,$$

which is a system of linear, homogeneous equations for A, B and C. For nontrivial solutions the determinant of the coefficient matrix has to vanish

$$\begin{vmatrix} \dfrac{e_{15}}{\varepsilon_{11}} & 1 & -1 \\[2mm] \rho'\omega^2 - c_{44}^p \xi_1^2 - \dfrac{\bar{c}_{44}\xi_2}{2h} & -\dfrac{e_{15}\xi_1}{2h} & 0 \\[2mm] 0 & \dfrac{\varepsilon_{11}\xi_1}{2h} & \dfrac{\varepsilon_0\xi_1}{2h} + \varepsilon_{11}'\xi_1^2 \end{vmatrix} = 0, \qquad (3.4.55)$$

which determines the dispersion relation, a relation between ω and ξ_1, of the surface wave. In terms of the surface wave speed $v = \omega / \xi_1$, Equation (3.4.55) can be written in the following form:

$$\left(\frac{v^2}{v_T'^2} - 1\right)\frac{c_{44}^p}{\bar{c}_{44}} 2h\xi_1 - \sqrt{1 - \frac{v^2}{v_T^2}} + \bar{k}_{15}^2 = \frac{\bar{k}_{15}^2}{1 + \dfrac{\varepsilon_0}{\varepsilon_{11}} + \dfrac{\varepsilon_{11}'}{\varepsilon_{11}}\xi_1 2h}, \qquad (3.4.56)$$

where Equation (3.4.42) has been used, and

$$v_T'^2 = \frac{c_{44}^p}{\rho'}, \qquad \bar{k}_{15}^2 = \frac{e_{15}^2}{\varepsilon_{11}\bar{c}_{44}}. \qquad (3.4.57)$$

When $h = 0$, i.e., the dielectric film does not exist, Equation (3.4.56) reduces to

$$v^2 = v_T^2\left[1 - \frac{\bar{k}_{15}^4}{(1 + \varepsilon_{11}/\varepsilon_0)^2}\right], \qquad (3.4.58)$$

which is the speed of the well-known Bleustein-Gulyaev piezoelectric surface wave. When $\bar{k}_{15}^2 = 0$, i.e., the half-space is nonpiezoelectric, electromechanical coupling disappears and the wave is pure elastic. In this case, Equation (3.4.56) reduces to

$$\left(\frac{v^2}{v_T'^2} - 1 \right) \frac{c_{44}^P}{\bar{c}_{44}} 2h\xi_1 - \sqrt{1 - \frac{v^2}{v_T^2}} = 0 , \qquad (3.4.59)$$

which is the equation that determines the speed of Love wave (an anti-plane surface wave in an elastic half-space carrying an elastic layer) in the limit when the film is very thin compared to the wavelength ($\xi_1 h \ll 1$).

Chapter 4
Nonlinear Effects in Electroelastic Plates

In this chapter we consider the effects of initial or biasing fields in an electroelastic plate. In this case the three-dimensional equations are linear (see Section 3 of Chapter 1), but they can only be derived from the nonlinear equations. We will also study nonlinear effects in an electroelastic plate in thickness-shear vibrations with relatively large amplitude.

4.1 Plates under Biasing Fields

Two-dimensional equations for elastic and electroelastic plates under biasing fields were studied in [44-47]. The derivation below is from [46]. Consider an electroelastic plate in the reference configuration with the X_3 axis along the plate normal, as shown in Figure 4.1.1.

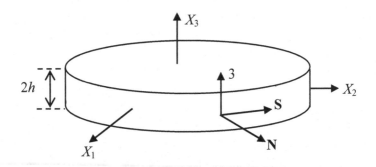

Figure 4.1.1. The reference configuration of an electroelastic plate and coordinate system.

4.1.1 First-order theory

For a first-order theory we make the following expansions of the incremental displacement and electric potential:

$$u_\gamma \cong u_\gamma^{(0)}(X_1,X_2,t) + X_3 u_\gamma^{(1)}(X_1,X_2,t) + X_3^2 u_\gamma^{(2)}(X_1,X_2,t),$$
$$\phi^1 \cong \phi^{(0)}(X_1,X_2,t) + X_3 \phi^{(1)}(X_1,X_2,t),$$

(4.1.1)

where $u_3^{(1)}$ and $u_\gamma^{(2)}$ will be either eliminated by stress relaxation or neglected later. From Equation (4.1.1) we calculate the displacement gradients and electric field as

$$u_{\gamma,L} \cong U_{\gamma L}^{(0)}(X_1,X_2,t) + X_3 U_{\gamma L}^{(1)}(X_1,X_2,t),$$
$$\mathcal{E}_L^1 \cong \mathcal{E}_L^{(0)}(X_1,X_2,t) + X_3 \mathcal{E}_L^{(1)}(X_1,X_2,t),$$

(4.1.2)

where

$$U_{11}^{(0)} = u_{1,1}^{(0)}, \quad U_{12}^{(0)} = u_{1,2}^{(0)}, \quad U_{13}^{(0)} = u_1^{(1)},$$
$$U_{21}^{(0)} = u_{2,1}^{(0)}, \quad U_{22}^{(0)} = u_{2,2}^{(0)}, \quad U_{23}^{(0)} = u_2^{(1)},$$
$$U_{31}^{(0)} = u_{3,1}^{(0)}, \quad U_{32}^{(0)} = u_{3,2}^{(0)}, \quad U_{33}^{(0)} = u_3^{(1)},$$
$$U_{11}^{(1)} = u_{1,1}^{(1)}, \quad U_{12}^{(1)} = u_{1,2}^{(1)}, \quad U_{13}^{(1)} = 2u_1^{(2)} \cong 0,$$
$$U_{21}^{(1)} = u_{2,1}^{(1)}, \quad U_{22}^{(1)} = u_{2,2}^{(1)}, \quad U_{23}^{(1)} = 2u_2^{(2)} \cong 0,$$
$$U_{31}^{(1)} = u_{3,1}^{(1)} \cong 0, \quad U_{32}^{(1)} = u_{3,2}^{(1)} \cong 0, \quad U_{33}^{(1)} = 2u_3^{(2)},$$

(4.1.3)

$$\mathcal{E}_1^{(0)} = -\phi_{,1}^{(0)}, \quad \mathcal{E}_2^{(0)} = -\phi_{,2}^{(0)}, \quad \mathcal{E}_3^{(0)} = -\phi^{(1)},$$
$$\mathcal{E}_1^{(1)} = -\phi_{,1}^{(1)}, \quad \mathcal{E}_2^{(1)} = -\phi_{,2}^{(1)}, \quad \mathcal{E}_3^{(1)} = 0.$$

(4.1.4)

In Equation (4.1.3), as an approximation, we have neglected $u_{3,1}^{(1)}$, $u_{3,2}^{(1)}$, $u_1^{(2)}$ and $u_2^{(2)}$. Substituting Equation (4.1.1) into the variational formulation in Equation (1.3.12), for independent variations of $\delta u_1^{(0)}$, $\delta u_2^{(0)}$, $\delta u_3^{(0)}$, $\delta u_1^{(1)}$, $\delta u_2^{(1)}$, $\delta \phi^{(0)}$ and $\delta \phi^{(1)}$, we obtain the following two-dimensional equations of motion and electrostatics:

$$K_{A\alpha,A}^{(0)} + F_\alpha^{(0)} = 2\rho_0 h \ddot{u}_\alpha^{(0)}, \quad \alpha = 1,2,3,$$
$$K_{A\alpha,A}^{(1)} - K_{3\alpha}^{(0)} + F_\alpha^{(1)} = \frac{2\rho_0 h^3}{3} \ddot{u}_\alpha^{(1)}, \quad \alpha = 1,2,$$
$$\mathcal{D}_{A,A}^{(0)} + \mathcal{D}^{(0)} = 0,$$
$$\mathcal{D}_{A,A}^{(1)} - \mathcal{D}_3^{(0)} + \mathcal{D}^{(1)} = 0,$$

(4.1.5)

where we have introduced an index convention that index A assumes 1 and 2 but not 3. Equations $(4.1.5)_1$ for $\alpha = 1, 2$ are the equations for extension, and for $\alpha = 3$ the equation for flexure. Equations $(4.1.5)_2$ are for shear in the X_1 and X_2 directions. In Equation (4.1.5) the plate resultants and surface loads of various orders are defined by

$$\{K_{K\alpha}^{(n)}, \mathcal{D}_K^{(n)}\} = \int_{-h}^{h} \{K_{K\alpha}^1, \mathcal{D}_K^1\} X_3^n dX_3,$$

$$F_\alpha^{(n)} = [X_3^n K_{3\alpha}^1]_{-h}^h, \quad \mathcal{D}^{(n)} = [X_3^n \mathcal{D}_3^1]_{-h}^h, \quad n = 0, 1, \tag{4.1.6}$$

where $K_{K\alpha}^{(n)}$ represent plate extensional and shearing forces, and bending and twisting moments. For thin plates we make the following stress relaxation:

$$K_{33}^1 = G_{33L\gamma} u_{\gamma,L} - R_{L33}\mathcal{E}_L^1 = 0. \tag{4.1.7}$$

From Equation (4.1.7) we can solve for $u_{3,3}$, with the result

$$u_{3,3} = \frac{-1}{G_{3333}} [G_{33L\gamma} u_{\gamma,L} - G_{3333} u_{3,3} - R_{L33}\mathcal{E}_L^1]. \tag{4.1.8}$$

$u_{3,3}$ has been eliminated from the right hand side of Equation (4.1.8) because when $L = \gamma = 3$ the two terms containing $u_{3,3}$ cancel each other out. Substituting Equation (4.1.8) back into the three-dimensional constitutive relations in Equation (1.3.6), we obtain

$$K_{K\alpha}^1 = \overline{G}_{K\alpha L\gamma} u_{\gamma,L} - \overline{R}_{LK\alpha}\mathcal{E}_L^1,$$

$$\mathcal{D}_K^1 = \overline{R}_{KL\gamma} u_{\gamma,L} + \overline{L}_{KL}\mathcal{E}_L^1, \tag{4.1.9}$$

where the relaxed constants are

$$\overline{G}_{K\alpha L\gamma} = G_{K\alpha L\gamma} - G_{K\alpha 33} G_{33L\gamma} / G_{3333},$$

$$\overline{R}_{KL\gamma} = R_{KL\gamma} - R_{K33} G_{33L\gamma} / G_{3333}, \tag{4.1.10}$$

$$\overline{L}_{KL} = L_{KL} + R_{K33} R_{L33} / G_{3333}.$$

In Equation (4.1.9) $K_{33}^1 = 0$ is satisfied and its right hand side does not contain $u_{3,3}$. From Equation $(4.1.6)_1$, with the substitution of Equations (4.1.9) and (4.1.2), we obtain the plate constitutive relations as

$$K_{K\alpha}^{(0)} = G_{K\alpha L\gamma}^{(0)} U_{\gamma L}^{(0)} + G_{K\alpha L\gamma}^{(1)} U_{\gamma L}^{(1)} - R_{LK\alpha}^{(0)} \mathcal{E}_L^{(0)} - R_{LK\alpha}^{(1)} \mathcal{E}_L^{(1)},$$

$$K_{K\alpha}^{(1)} = G_{K\alpha L\gamma}^{(1)} U_{\gamma L}^{(0)} + G_{K\alpha L\gamma}^{(2)} U_{\gamma L}^{(1)} - R_{LK\alpha}^{(1)} \mathcal{E}_L^{(0)} - R_{LK\alpha}^{(2)} \mathcal{E}_L^{(1)},$$

$$\mathcal{D}_K^{(0)} = R_{KL\gamma}^{(0)} U_{\gamma L}^{(0)} + R_{KL\gamma}^{(1)} U_{\gamma L}^{(1)} + L_{KL}^{(0)} \mathcal{E}_L^{(0)} + L_{KL}^{(1)} \mathcal{E}_L^{(1)},$$

$$\mathcal{D}_K^{(1)} = R_{KL\gamma}^{(1)} U_{\gamma L}^{(0)} + R_{KL\gamma}^{(2)} U_{\gamma L}^{(1)} + L_{KL}^{(1)} \mathcal{E}_L^{(0)} + L_{KL}^{(2)} \mathcal{E}_L^{(1)},$$

(4.1.11)

where

$$\{G_{K\alpha L\gamma}^{(n)}, R_{Kl\gamma}^{(n)}, L_{KL}^{(n)}\}$$
$$= \int_{-h}^{h} \{\overline{G}_{K\alpha L\gamma}, \overline{R}_{KL\gamma}, \overline{L}_{KL}\} X_3^n \mathrm{d}X_3, \quad n = 0,1,2.$$

(4.1.12)

Physically, $G_{K\alpha L\gamma}^{(n)}$ represent the plate flexural and extensional stiffness, etc. Because of the use of the relaxed material constants in Equation (4.1.9), $u_3^{(1)}$ and $u_3^{(2)}$ do not appear on the right hand sides of Equation (4.1.11). A more sophisticated stress relaxation using $K_{33}^{(0)} = 0$ and $K_{3j}^{(1)} = 0$ to eliminate $U_{33}^{(0)}$ and $U_{3L}^{(1)}$ (and hence $u_3^{(1)}$ and $u_3^{(2)}$) can be performed. Equation (4.1.11) shows that extension and bending may be coupled due to biasing fields. In a plate theory only the moments of various orders of the biasing fields matter, not the exact three-dimensional distributions of the biasing fields. Shear correction factors may be introduced but they are not pursued here.

In summary, we have obtained the two-dimensional equations of motion and electrostatics (4.1.5), constitutive relations (4.1.11), and the displacement gradients (4.1.3) and electric fields (4.1.4). With successive substitutions, Equations (4.1.5) can be written as seven equations for the seven unknowns of $u_1^{(0)}$, $u_2^{(0)}$, $u_3^{(0)}$, $u_1^{(1)}$, $u_2^{(1)}$, $\phi^{(0)}$, and $\phi^{(1)}$. To these equations the proper forms of boundary conditions can be determined from the variational formulation in Equation (1.3.12). At the boundary of a plate with a unit exterior normal **N** and a unit tangent **S**, we need to prescribe

$$K_{NN}^{(0)} \quad \text{or} \quad u_N^{(0)}, \quad K_{NS}^{(0)} \quad \text{or} \quad u_S^{(0)}, \quad K_{N3}^{(0)} \quad \text{or} \quad u_3^{(0)},$$

$$K_{NN}^{(1)} \quad \text{or} \quad u_N^{(1)}, \quad K_{NS}^{(1)} \quad \text{or} \quad u_S^{(1)},$$

$$\mathcal{D}_N^{(0)} \quad \text{or} \quad \phi^{(0)}, \quad \mathcal{D}_N^{(1)} \quad \text{or} \quad \phi^{(1)}.$$

(4.1.13)

4.1.2 Buckling of ceramic plates

One application of the equations for a plate under biasing fields is the buckling analysis of thin plates. For the classical description of the buckling phenomenon, the electroelastic counterpart of the initial stress theory in elasticity is sufficient, which is a special case of the theory for small fields superposed on a bias. The effective material constants of such a theory are given in Equation (1.3.20). We perform buckling analysis of a few piezoelectric plates and bimorphs made from polarized ceramics [46]. Three cases corresponding to Figure 4.1.2 are considered. We limit our discussion to plane strain analysis with $u_2 = 0$ and $\partial/\partial X_2 = 0$. For all three cases the two major surfaces of the plate at $X_3 = \pm h$ are traction-free and are unelectroded, with vanishing normal electric displacement. The plates are mechanically simply supported at their end faces at $X_1 = 0$ and $X_1 = l$. The electrical end conditions will be specified later in each specific case. The two ends of the plates are loaded by the following axial force per unit length in the X_2 direction, which is responsible for the biasing deformation:

$$p = 2hK_{11}^0. \tag{4.1.14}$$

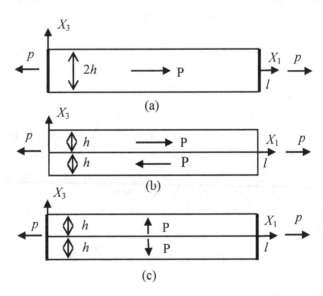

Figure 4.1.2. Simply supported ceramic plates of length l and thickness $2h$.

4.1.2.1 A single-layered plate

Consider a ceramic plate poled in the X_1 direction as shown in Figure 4.1.2(a). The material matrices are in Equation (2.3.45). The plate is electroded at its two end faces at $X_1 = 0$ and $X_1 = l$. The electrodes are shown by thick lines in the figure. When the initial load p is being applied, the end electrodes are shorted to eliminate the initial electric field E_1 which otherwise would exist. Once p is already loaded, there exist initial charges on the end electrodes. The electrodes are then opened during the incremental flexural deformation and there are no incremental charges on these electrodes. Therefore for the incremental fields the electric displacement vanishes at both ends. The governing equations for the incremental fields take the form

$$G^{(0)}_{1313}u^{(0)}_{3,11} + G^{(0)}_{1331}u^{(1)}_{1,1} + R^{(0)}_{313}\phi^{(1)}_{,1} = 0,$$

$$G^{(2)}_{1111}u^{(1)}_{1,11} - G^{(0)}_{3113}u^{(0)}_{3,1} - G^{(0)}_{3131}u^{(1)}_1 + R^{(2)}_{111}\phi^{(1)}_{,11} - R^{(0)}_{331}\phi^{(1)} = 0,$$

$$L^{(0)}_{11}\phi^{(0)}_{,11} = 0,$$
(4.1.15)

$$R^{(2)}_{111}u^{(1)}_{1,11} - R^{(0)}_{313}u^{(0)}_{3,1} - R^{(0)}_{331}u^{(1)}_1 - L^{(2)}_{11}\phi^{(1)}_{,11} + L^{(0)}_{33}\phi^{(1)} = 0,$$

and the boundary conditions are

$$u^{(0)}_3 = 0, \quad X_1 = 0, l,$$

$$K^{(1)}_{11} = G^{(2)}_{1111}u^{(1)}_{1,1} + R^{(2)}_{111}\phi^{(1)}_{,1} = 0, \quad X_1 = 0, l,$$

$$\mathcal{D}^{(0)}_1 = -2h\bar{\varepsilon}_{33}\phi^{(0)}_{,1} = 0, \quad X_1 = 0, l,$$
(4.1.16)

$$\mathcal{D}^{(1)}_1 = R^{(2)}_{111}u^{(1)}_{1,1} - L^{(2)}_{11}\phi^{(1)}_{,1} = 0, \quad X_1 = 0, l,$$

where

$$G^{(0)}_{1313} = 2hc_{44} + p, \quad G^{(0)}_{1331} = G^{(0)}_{3113} = G^{(0)}_{3131} = 2hc_{44},$$

$$G^{(2)}_{1111} = \frac{1}{3}h^2(2h\bar{c}_{33} + p), \quad R^{(0)}_{313} = R^{(0)}_{331} = 2he_{15},$$

$$R^{(2)}_{111} = \frac{2}{3}h^3\bar{e}_{33}, \quad L^{(2)}_{11} = \frac{2}{3}h^3\bar{\varepsilon}_{33}, \quad L^{(0)}_{33} = 2h\varepsilon_{11},$$
(4.1.17)

$$\bar{c}_{33} = c_{33} - \frac{c^2_{13}}{c_{11}}, \quad \bar{e}_{33} = e_{33} - \frac{e_{31}c_{31}}{c_{11}}, \quad \bar{\varepsilon}_{33} = \varepsilon_{33} + \frac{e^2_{31}}{c_{11}}.$$

In Equation $(4.1.16)_2$ the bending moment $K_{11}^{(1)}$ is coupled to $\mathcal{E}_1^{(1)} = -\phi_{,1}^{(1)}$ through $\bar{e}_{33}$ as expected. Furthermore, for the transverse shear force we have $K_{13}^{(0)} = G_{1313}^{(0)} u_{3,1}^{(0)} + G_{1331}^{(0)} u_1^{(1)} + R_{313}^{(0)} \phi^{(1)}$ which is coupled to $\mathcal{E}_3^{(0)} = -\phi^{(1)}$ through e_{15} as expected. Equations $(4.1.15)_3$ and $(4.1.16)_3$ show that $\phi^{(0)}$ is a constant which can be taken to be zero. Let

$$u_3^{(0)} = A \sin \xi X_1, \quad u_1^{(1)} = B \cos \xi X_1,$$
$$\phi^{(1)} = C \cos \xi X_1, \tag{4.1.18}$$

where A, B, and C are undetermined constants and $\xi = \pi/l$. The boundary conditions in Equation (4.1.16) are satisfied automatically. Then the buckling load can be determined from the following equations obtained by substituting Equation (4.1.18) into Equation $(4.1.15)_{1,2,4}$:

$$\xi^2 G_{1313}^{(0)} A + \xi G_{1331}^{(0)} B + \xi R_{313}^{(0)} C = 0,$$
$$\xi G_{3113}^{(0)} A + (\xi^2 G_{1111}^{(2)} + G_{3131}^{(0)}) B + (\xi^2 R_{111}^{(2)} + R_{331}^{(0)}) C = 0, \tag{4.1.19}$$
$$\xi R_{313}^{(0)} A + (\xi^2 R_{111}^{(2)} + R_{331}^{(0)}) B - (\xi^2 L_{11}^{(2)} + L_{33}^{(0)}) C = 0.$$

For non-trivial solutions of A, B, or C, the following condition must be satisfied

$$\begin{vmatrix} \xi^2 G_{1313}^{(0)} & \xi G_{1331}^{(0)} & \xi R_{313}^{(0)} \\ \xi G_{3113}^{(0)} & \xi^2 G_{1111}^{(2)} + G_{3131}^{(0)} & \xi^2 R_{111}^{(2)} + R_{331}^{(0)} \\ \xi R_{313}^{(0)} & \xi^2 R_{111}^{(2)} + R_{331}^{(0)} & -(\xi^2 L_{11}^{(2)} + L_{33}^{(0)}) \end{vmatrix} = 0, \tag{4.1.20}$$

which can be written as

$$a^{(1)} \bar{p}^{(1)2} + b^{(1)} \bar{p}^{(1)} + c^{(1)} = 0, \tag{4.1.21}$$

where

$$\bar{p}^{(1)} = \frac{p}{2h\bar{c}_{33}}, \quad \lambda_0 = \frac{\pi^2}{3}\left(\frac{h}{l}\right)^2, \quad a^{(1)} = \lambda_0,$$

$$b^{(1)} = \lambda_0 + \frac{1}{\bar{c}_{33}}\left[(\lambda_0 + 1)c_{44} + \frac{\lambda_0 e_{15}^2 + (\lambda_0 \bar{e}_{33} + e_{15})^2}{\varepsilon_{11} + \lambda_0 \bar{\varepsilon}_{33}}\right], \tag{4.1.22}$$

$$c^{(1)} = \frac{1}{\bar{c}_{33}}\left[\lambda_0 c_{44} + \frac{\lambda_0 e_{15}^2 + \lambda_0^2 \bar{c}_{33}^{-1} \bar{e}_{33}^2 c_{44}}{\varepsilon_{11} + \lambda_0 \bar{\varepsilon}_{33}}\right].$$

Since $\lambda_0 \ll 1$ for a thin plate, it follows from Equation (4.1.22) that $(b^{(1)})^2 \gg 4a^{(1)}c^{(1)}$ and $b^{(1)} \gg a^{(1)}$ for thin plates. Hence, an approximate solution to Equation (4.1.21) can be found as

$$\overline{p}^{(1)} \approx -\frac{c^{(1)}}{b^{(1)}}\left(1 + \frac{a^{(1)}c^{(1)}}{(b^{(1)})^2}\right). \tag{4.1.23}$$

Our main interest is the effect of piezoelectric coupling on the buckling load. To see this more clearly, we let $c_{44} \to \infty$ in Equation (4.1.23), which effectively eliminates plate shear deformations and the related piezoelectric coupling through e_{15}. We then expand Equation (4.1.23) into a polynomial of λ_0 that is small. This leads to

$$\overline{p}^{(1)} \approx -\lambda_0(1 + \lambda_0\overline{k}^2), \quad \overline{k}^2 = \overline{e}_{33}^2/(\overline{c}_{33}\varepsilon_{11}), \tag{4.1.24}$$

where $\overline{k}^2$ may be considered as an electromechanical coupling factor.

Under our notation $-\lambda_0$ is the dimensionless buckling load from an elastic analysis without considering the piezoelectric effect. The second term on the right hand side of Equation (4.1.24)$_1$ represents the effect of piezoelectric coupling due to $\overline{e}_{33}$. This additional term is proportional to $\overline{k}^2$, which ranges from 0.1 to 0.6 for most polarized ceramics. Since $\overline{k}^2$ is multiplied by λ_0, which is a small number, the piezoelectric modification on the buckling load is a small addition to the elastic buckling load. Hence an elastic analysis without considering the piezoelectric effect gives a conservative estimate of the buckling load.

4.1.2.2 A bimorph with in-plane poling

In this case we consider a ceramic bimorph as shown in Figure 4.1.2(b). The end faces are unelectroded. The governing equations for the incremental fields are

$$\begin{aligned}
& G_{1313}^{(0)}u_{3,11}^{(0)} + G_{1331}^{(0)}u_{1,1}^{(1)} = 0, \\
& G_{1111}^{(2)}u_{1,11}^{(1)} + R_{111}^{(1)}\phi_{,11}^{(0)} - G_{3113}^{(0)}u_{3,1}^{(0)} - G_{3131}^{(0)}u_1^{(1)} = 0, \\
& R_{111}^{(1)}u_{1,11}^{(1)} - L_{11}^{(0)}\phi_{,11}^{(0)} = 0, \\
& L_{11}^{(2)}\phi_{,11}^{(1)} - L_{33}^{(0)}\phi^{(1)} = 0,
\end{aligned} \tag{4.1.25}$$

with the following boundary conditions:

$$u_3^{(0)} = 0, \quad X_1 = 0, l,$$

$$K_{11}^{(1)} = G_{1111}^{(2)} u_{1,1}^{(1)} + R_{111}^{(1)} \phi_{,1}^{(0)} = 0, \quad X_1 = 0, l,$$

$$\mathcal{D}_1^{(0)} = R_{111}^{(1)} u_{1,1}^{(1)} - L_{11}^{(0)} \phi_{,1}^{(0)} = 0, \quad X_1 = 0, l,$$
(4.1.26)

$$\mathcal{D}_1^{(1)} = -L_{11}^{(2)} \phi_{,1}^{(1)} = 0, \quad X_1 = 0, l,$$

where the plate material constants are as in Equation (4.1.17) plus

$$R_{111}^{(1)} = h^2 e_{33}.$$
(4.1.27)

In Equation (4.1.26)$_2$ the bending moment $K_{11}^{(1)}$ is coupled to $\mathcal{E}_1^{(0)} = -\phi_{,1}^{(0)}$ through e_{33} as expected. We note that Equations (4.1.25)$_4$ and (4.1.26)$_4$ imply $\phi^{(1)} = 0$. Let

$$u_3^{(0)} = A \sin \xi X_1, \quad u_1^{(1)} = B \cos \xi X_1,$$

$$\phi^{(0)} = D \cos \xi X_1,$$
(4.1.28)

which satisfy the boundary conditions in Equation (4.1.26) when $\xi = \pi/l$. Eliminating $\phi^{(0)}$ from Equation (4.1.25)$_2$ and Equation (4.1.25)$_3$, and substituting Equations (4.1.28)$_{1,2}$ into the resulting equation and Equation (4.1.25)$_1$, we obtain

$$\xi^2 G_{1313}^{(0)} A + \xi G_{1331}^{(0)} B = 0,$$

$$\xi G_{3113}^{(0)} A + \{\xi^2 [G_{1111}^{(2)} + (L_{11}^{(0)})^{-1} R_{111}^{(1)2}] + G_{3131}^{(0)}\} B = 0.$$
(4.1.29)

For nontrivial solutions of A and B, the following must be true:

$$\begin{vmatrix} \xi \ G_{1313}^{(0)} & G_{1331}^{(0)} \\ \xi G_{3113}^{(0)} & \xi^2 [G_{1111}^{(2)} + (L_{11}^{(0)})^{-1} R_{111}^{(1)2}] + G_{3131}^{(0)} \end{vmatrix} = 0,$$
(4.1.30)

or

$$a^{(2)} \bar{p}^{(2)2} + b^{(2)} \bar{p}^{(2)} + c^{(2)} = 0$$
(4.1.31)

where

$$\bar{p}^{(2)} = \frac{p}{2h\bar{c}_{33}}, \qquad a^{(2)} = \lambda_0,$$

$$b^{(2)} = \lambda_0 + \frac{1}{\bar{c}_{33}}\left[(\lambda_0 + 1)c_{44} + \frac{3}{4}\lambda_0\bar{e}_{33}^2\bar{\varepsilon}_{33}^{-1}\right], \qquad (4.1.32)$$

$$c^{(2)} = \frac{1}{\bar{c}_{33}}\left[\lambda_0 c_{44} + \frac{3}{4}\lambda_0 c_{44}\bar{c}_{33}^{-1}\bar{e}_{33}^2\bar{\varepsilon}_{33}^{-1}\right].$$

An approximate solution to Equation (4.1.31) is given by

$$\bar{p}^{(2)} \approx -\frac{c^{(2)}}{b^{(2)}}\left(1 + \frac{a^{(2)}c^{(2)}}{(b^{(2)})^2}\right). \qquad (4.1.33)$$

Letting $c_{44} \to \infty$ in Equation (4.1.33) and expanding the resulting expression into a polynomial of λ_0, we have

$$\bar{p}^{(2)} \approx -\lambda_0(1 + \frac{3}{4}\bar{k}_{33}^2), \quad \bar{k}_{33}^2 = \bar{e}_{33}^2/(\bar{c}_{33}\varepsilon_{33}). \qquad (4.1.34)$$

Comparing Equation $(4.1.34)_1$ with Equation $(4.1.24)_1$, we notice the important difference that in Equation $(4.1.34)_1$ the piezoelectric modification on the buckling load is not multiplied by the small number λ_0. This is because in Equation (4.1.15) the plate piezoelectric coefficient $R_{111}^{(2)}$ is proportional to h^3. However, in Equation (4.1.25) $R_{111}^{(1)}$ is proportional to h^2 and it is the squares of the piezoelectric coefficients that appear in the buckling loads. Therefore the piezoelectric effect on the buckling load is much stronger in Case (b) than in Case (a).

4.1.2.3 A bimorph with thickness poling

The third case is a ceramic bimorph as shown in Figure 4.1.2(c). The plate is electroded at $X_1 = 0$ and $X_1 = l$, with shorted and grounded electrodes. The governing equations take the following form:

$$G_{1313}^{(0)} u_{3,11}^{(0)} + G_{1331}^{(0)} u_{1,1}^{(1)} + R_{113}^{(1)} \phi_{,11}^{(1)} = 0,$$

$$G_{1111}^{(2)} u_{1,11}^{(1)} - G_{3113}^{(0)} u_{3,1}^{(0)} - G_{3131}^{(0)} u_1^{(1)} + [R_{311}^{(1)} - R_{131}^{(1)}]\phi_{,1}^{(1)} = 0,$$

$$L_{11}^{(0)} \phi_{,11}^{(0)} = 0,$$ (4.1.35)

$$R_{113}^{(1)} u_{3,11}^{(0)} - [R_{311}^{(1)} - R_{131}^{(1)}]u_{1,1}^{(1)} - L_{11}^{(2)} \phi_{,11}^{(1)} + L_{33}^{(0)} \phi^{(1)} = 0.$$

The boundary conditions are

$$u_3^{(0)} = 0, \quad X_1 = 0, l,$$

$$K_{11}^{(1)} = G_{1111}^{(2)} u_{1,1}^{(1)} + h^2 \bar{e}_{31} \phi^{(1)} = 0, \quad X_1 = 0, l,$$ (4.1.36)

$$\phi^{(0)} = 0, \quad X_1 = 0, l,$$

$$\phi^{(1)} = 0, \quad X_1 = 0, l,$$

where

$$G_{1111}^{(2)} = \frac{1}{3} h^2 (2h\bar{c}_{11} + p), \quad G_{1313}^{(0)} = 2hc_{44} + p,$$

$$G_{1331}^{(0)} = G_{3113}^{(0)} = G_{3131}^{(0)} = 2hc_{44}, \quad R_{113}^{(1)} = R_{131}^{(1)} = h^2 e_{15},$$

$$R_{311}^{(1)} = h^2 \bar{e}_{31}, \quad L_{11}^{(0)} = 2h\varepsilon_{11}, \quad L_{11}^{(2)} = \frac{2}{3} h^3 \varepsilon_{11}, \quad L_{33}^{(0)} = 2h\bar{\varepsilon}_{33}, \quad (4.1.37)$$

$$\bar{c}_{11} = c_{11} - c_{13}^2 / c_{33}, \quad \bar{e}_{31} = e_{31} - e_{33}c_{31} / c_{33},$$

$$\bar{\varepsilon}_{33} = \varepsilon_{33} + e_{33}^2 / c_{33}.$$

In Equation (4.1.36)$_2$ the bending moment $K_{11}^{(1)}$ is coupled to $\mathcal{E}_3^{(0)} = -\phi^{(1)}$ through $\bar{e}_{31}$ as expected. In this case $\phi^{(0)}$ is zero. We let

$$u_3^{(0)} = A \sin \xi X_1, \quad u_1^{(1)} = B \cos \xi X_1,$$

$$\phi^{(1)} = C \sin \xi X_1,$$ (4.1.38)

which satisfy the boundary conditions in Equation (4.1.36). Substituting Equation (4.1.38) into Equations (4.1.35)$_{1,2,4}$:

$$\xi^2 G_{1313}^{(0)} A + \xi G_{1331}^{(0)} B + \xi^2 R_{113}^{(1)} C = 0,$$

$$\xi G_{3113}^{(0)} A + (\xi^2 G_{1111}^{(2)} + G_{3131}^{(0)}) B + \xi (R_{131}^{(1)} - R_{311}^{(1)}) C = 0,$$ (4.1.39)

$$\xi^2 R_{113}^{(1)} A + \xi (R_{131}^{(1)} - R_{311}^{(1)}) B - (\xi^2 L_{11}^{(2)} + L_{33}^{(0)}) C = 0.$$

For nontrivial solutions of A, B, or C, the following condition must be satisfied:

$$\begin{vmatrix} \xi^2 G_{1313}^{(0)} & \xi G_{1331}^{(0)} & \xi^2 R_{113}^{(1)} \\ \xi G_{3113}^{(0)} & \xi^2 G_{1111}^{(2)} + G_{3131}^{(0)} & -\xi(R_{311}^{(1)} - R_{131}^{(1)}) \\ \xi^2 R_{113}^{(1)} & -\xi(R_{311}^{(1)} - R_{131}^{(1)}) & -(\xi^2 L_{11}^{(2)} + L_{33}^{(0)}) \end{vmatrix} = 0, \tag{4.1.40}$$

or

$$a^{(3)} \overline{p}^{(3)2} + b^{(3)} \overline{p}^{(3)} + c^{(3)} = 0, \tag{4.1.41}$$

where

$$\overline{p}^{(3)} = \frac{p}{2h\overline{c}_{11}}, \quad a^{(3)} = \lambda_0,$$

$$b^{(3)} = \lambda_0 + \frac{1}{\overline{c}_{11}} \left[(\lambda_0 + 1)c_{44} + \frac{3\lambda_0 [\lambda_0 e_{15}^2 + (e_{15} - \overline{e}_{31})^2]}{4(\lambda_0 \varepsilon_{11} + \overline{\varepsilon}_{33})} \right], \tag{4.1.42}$$

$$c^{(3)} = \frac{1}{\overline{c}_{11}} \left[\lambda_0 c_{44} + \frac{3\lambda_0 [\lambda_0 e_{15}^2 + \overline{c}_{11}^{-1} c_{44} \overline{e}_{31}^2]}{4(\lambda_0 \varepsilon_{11} + \overline{\varepsilon}_{33})} \right].$$

An approximate solution to Equation (4.1.41) is found to be

$$\overline{p}^{(3)} \approx -\frac{c^{(3)}}{b^{(3)}} \left(1 + \frac{a^{(3)} c^{(3)}}{(b^{(3)})^2} \right). \tag{4.1.43}$$

Letting $c_{44} \to \infty$ in Equation (4.1.43) and expanding it into a polynomial of λ_0, we obtain

$$\overline{p}^{(3)} \approx -\lambda_0 (1 + \frac{3}{4} \overline{k}_{31}^2), \quad \overline{k}_{31}^2 = \overline{e}_{31}^2 / (\overline{c}_{11} \overline{\varepsilon}_{33}) \tag{4.1.44}$$

which shows the same behavior as Equation (4.1.34). Buckling analysis of circular piezoelectric plates using two-dimensional equations was performed in [47].

4.2 Large Thickness-Shear Deformations

We are interested in thickness-shear vibrations with relatively large shear deformations [48] for a plate whose reference configuration is shown in Figure 4.2.1, with the X_2 axis as the plate normal. The relevant relatively large displacement gradient components are $u_{1,2}$ and $u_{3,2}$. Up to cubic nonlinearity, the three-dimensional equations are given in the fourth section of the first chapter.

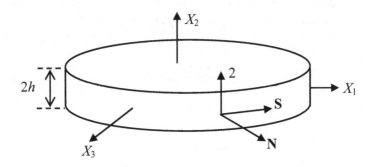

Figure 4.2.1. The reference configuration of an electroelastic plate and coordinate system.

We introduce a convention that subscripts A, B and C assume 1 and 3 only but not 2. Keeping nonlinear terms of $u_{1,2}$ and $u_{3,2}$ only, from Equation (1.4.6)$_1$ we have

$$K_{LM} = c_{LMRS} u_{R,S} - e_{KLM} \mathcal{E}_K$$
$$+ c^e_{LMA2B2} u_{A,2} u_{B,2} + c^e_{LMA2B2C2} u_{A,2} u_{B,2} u_{C,2}. \tag{4.2.1}$$

4.2.1 First-order theory

For a first-order plate theory we make the following expansions of the mechanical displacement and electric potential:

$$u_L \cong u_L^{(0)}(X_1, X_3, t) + X_2 u_L^{(1)}(X_1, X_3, t) + X_2^2 u_L^{(2)}(X_1, X_3, t),$$
$$\phi \cong \phi^{(0)}(X_1, X_3, t) + X_2 \phi^{(1)}(X_1, X_3, t), \tag{4.2.2}$$

where $u_2^{(1)}$ and $u_L^{(2)}$ will be eliminated or neglected later. From Equation (4.2.2) we can write

$$u_{R,S} = U_{RS}^{(0)} + X_2 U_{RS}^{(1)},$$

$$\mathcal{E}_K = -\phi_{,K} = \mathcal{E}_K^{(0)} + X_2 \mathcal{E}_K^{(1)}, \tag{4.2.3}$$

where

$$U_{11}^{(0)} = u_{1,1}^{(0)}, \quad U_{12}^{(0)} = u_1^{(1)}, \quad U_{13}^{(0)} = u_{1,3}^{(0)},$$

$$U_{21}^{(0)} = u_{2,1}^{(0)}, \quad U_{22}^{(0)} = u_2^{(1)}, \quad U_{23}^{(0)} = u_{2,3}^{(0)},$$

$$U_{31}^{(0)} = u_{3,1}^{(0)}, \quad U_{32}^{(0)} = u_3^{(1)}, \quad U_{33}^{(0)} = u_{3,3}^{(0)},$$

$$U_{11}^{(1)} = u_{1,1}^{(1)}, \quad U_{12}^{(1)} = 2u_1^{(2)} \cong 0, \quad U_{13}^{(1)} = u_{1,3}^{(1)},$$

$$U_{21}^{(1)} = u_{2,1}^{(1)} \cong 0, \quad U_{22}^{(1)} = 2u_2^{(2)}, \quad U_{23}^{(1)} = u_{2,3}^{(1)} \cong 0,$$

$$U_{31}^{(1)} = u_{3,1}^{(1)}, \quad U_{32}^{(1)} = 2u_3^{(2)} \cong 0, \quad U_{33}^{(1)} = u_{3,3}^{(1)}, \tag{4.2.4}$$

and

$$\mathcal{E}_3^{(0)} = -\phi_{,3}^{(0)}, \quad \mathcal{E}_1^{(0)} = -\phi_{,1}^{(0)}, \quad \mathcal{E}_2^{(0)} = -\phi^{(1)},$$

$$\mathcal{E}_3^{(1)} = -\phi_{,3}^{(1)}, \quad \mathcal{E}_1^{(1)} = -\phi_{,1}^{(1)}, \quad \mathcal{E}_2^{(1)} = 0. \tag{4.2.5}$$

Substituting Equation (4.2.2) into the variational formulation in Equation (1.1.24), for independent variations of $\delta u_A^{(0)}$, $\delta u_2^{(0)}$, $\delta u_A^{(1)}$, $\delta \phi^{(0)}$ and $\delta \phi^{(1)}$, we obtain the following two-dimensional equations of motion and electrostatics:

$$K_{BM,B}^{(0)} + F_M^{(0)} = 2\rho_0 h \ddot{u}_M^{(0)}, \quad M = 1, 2, 3,$$

$$K_{BA,B}^{(1)} - K_{2A}^{(0)} + F_A^{(1)} = \frac{2\rho_0 h^3}{3} \ddot{u}_A^{(1)}, \quad A = 1, 3,$$

$$\mathcal{D}_{A,A}^{(0)} + \mathcal{D}^{(0)} = 0,$$

$$\mathcal{D}_{A,A}^{(1)} - \mathcal{D}_2^{(0)} + \mathcal{D}^{(1)} = 0. \tag{4.2.6}$$

Equation $(4.2.6)_1$ for $M = 1$ and 3 are the equations for extension, and for $M = 2$ the equation for flexure. Equation $(4.2.6)_2$ is for thickness-shear in the X_1 and X_3 directions. In Equation (4.2.6) the plate resultants and surface as well as body loads of various orders are defined by

$$\{K_{LM}^{(n)}, \mathcal{D}_K^{(n)}\} = \int_{-h}^{h} \{K_{LM}, \mathcal{D}_K\} X_2^n dX_2,$$

$$F_M^{(n)} = [X_2^n K_{2M}]_{-h}^{h} + \int_{-h}^{h} \rho_0 f_M X_2^n dX_2, \qquad (4.2.7)$$

$$\mathcal{D}^{(n)} = [X_2^n \mathcal{D}_2]_{-h}^{h}, \quad n = 0, 1,$$

where $K_{LM}^{(n)}$ represent plate extensional and shearing forces, and bending and twisting moments. Since the plate is assumed to be thin, we make the stress relaxation of $K_{22} = 0$. This implies, through Equation (4.2.1) by setting $L = M = 2$, the following expression for $u_{2,2}$ in terms of other components of the displacement and potential gradients:

$$u_{2,2} = -\frac{1}{c_{2222}}(c_{22RS}u_{R,S} - c_{2222}u_{2,2} - e_{K22}\mathcal{E}_K \qquad (4.2.8)$$
$$+ c_{22A2B2}^e u_{A,2}u_{B,2} + c_{22A2B2C2}^e u_{A,2}u_{B,2}u_{C,2}).$$

In Equation (4.2.8) $u_{2,2}$ has been eliminated from the right hand side. When $R = S = 2$ the two terms containing $u_{2,2}$ cancel each other out. Substituting Equation (4.2.8) back into Equation (4.2.1) and Equation $(1.4.6)_2$, we obtain the following relaxed constitutive relations for thin plates:

$$K_{LM} = \bar{c}_{LMRS}u_{R,S} - \bar{e}_{KLM}\mathcal{E}_K$$
$$+ \bar{c}_{LMA2B2}u_{A,2}u_{B,2} + \bar{c}_{LMA2B2C2}u_{A,2}u_{B,2}u_{C,2}, \qquad (4.2.9)$$
$$\mathcal{D}_K = \bar{e}_{KRS}u_{R,S} + \bar{\varepsilon}_{KL}\mathcal{E}_L,$$

where the plate material constants are defined by

$$\bar{c}_{LMRS} = c_{LMRS} - c_{LM22}c_{22RS} / c_{2222},$$
$$\bar{c}_{LMA2B2} - c_{LMA2B2}^e - c_{LM22}c_{22A2B2}^e / c_{2222},$$
$$\bar{c}_{LMA2B2C2} = c_{LMA2B2C2}^e - c_{LM22}c_{22A2B2C2}^e / c_{2222}, \qquad (4.2.10)$$
$$\bar{e}_{KLM} = e_{KLM} - c_{LM22}e_{K22} / c_{2222},$$
$$\bar{\varepsilon}_{KL} = \varepsilon_{KL} + e_{K22}e_{L22} / c_{2222}.$$

The right hand side of Equation (4.2.9) does not contain $u_{2,2}$ and $K_{22} = 0$ is automatically satisfied by Equation (4.2.9). Integrating Equation (4.2.9) through the plate thickness, we obtain the zero-order two-dimensional plate constitutive relations

$$K_{LM}^{(0)} = 2h(\bar{c}_{LMRS}' U_{RS}^{(0)} - \bar{e}_{KLM}' \mathcal{E}_K^{(0)}$$
$$+ \bar{c}_{LMA2B2} u_A^{(1)} u_B^{(1)} + \bar{c}_{LMA2B2C2} u_A^{(1)} u_B^{(1)} u_C^{(1)}), \qquad (4.2.11)$$
$$\mathcal{D}_K^{(0)} = 2h(\bar{e}_{KRS}' U_{RS}^{(0)} + \bar{\varepsilon}_{KL} \mathcal{E}_L^{(0)}),$$

where we have modified the zero-order linear plate constants by the introduction of two shear correction factors κ_1 and κ_3 as follows:

$$U_{12}^{(0)} \rightarrow \kappa_1 U_{12}^{(0)}, \quad U_{32}^{(0)} \rightarrow \kappa_3 U_{32}^{(0)}. \qquad (4.2.12)$$

Multiplying both sides of Equation (4.2.9) by X_2 and integrating the resulting equation through the plate thickness, we have the first-order plate constitutive relations

$$K_{LM}^{(1)} = \frac{2h^3}{3}(\bar{c}_{LMRS} U_{RS}^{(1)} - \bar{e}_{KLM} \mathcal{E}_K^{(1)}),$$
$$\qquad (4.2.13)$$
$$\mathcal{D}_K^{(1)} = \frac{2h^3}{3}(\bar{e}_{KRS} U_{RS}^{(1)} + \bar{\varepsilon}_{KL} \mathcal{E}_L^{(1)}).$$

A more sophisticated stress relaxation using $K_{22}^{(0)} = 0$ and $K_{2j}^{(1)} = 0$ to eliminate $U_{22}^{(0)}$ and $U_{2j}^{(1)}$ (and hence $u_2^{(1)}$ and $u_2^{(2)}$) can be performed.

In summary, we have obtained two-dimensional equations of motion and electrostatics (4.2.6), constitutive relations (4.2.11) and (4.2.13), and the displacement and potential gradients (4.2.4) and (4.2.5). With successive substitutions, Equation (4.2.6) can be written as seven equations for $u_1^{(0)}$, $u_2^{(0)}$, $u_3^{(0)}$, $u_1^{(1)}$, $u_3^{(1)}$, $\phi^{(0)}$, and $\phi^{(1)}$. To these equations the proper forms of boundary conditions can be determined from the variational formulation in Equation (1.1.24). At the boundary of a plate with a unit exterior normal $\mathbf{N}$ and a unit in-plane tangent $\mathbf{S}$ in the reference configuration, we need to prescribe

$$K_{NN}^{(0)} \quad \text{or} \quad u_N^{(0)}, \quad K_{NS}^{(0)} \quad \text{or} \quad u_S^{(0)}, \quad K_{N2}^{(0)} \quad \text{or} \quad u_2^{(0)},$$
$$K_{NN}^{(1)} \quad \text{or} \quad u_N^{(1)}, \quad K_{NS}^{(1)} \quad \text{or} \quad u_S^{(1)}, \qquad (4.2.14)$$
$$\mathcal{D}_N^{(0)} \quad \text{or} \quad \phi^{(0)}, \quad \mathcal{D}_N^{(1)} \quad \text{or} \quad \phi^{(1)}.$$

4.2.2 Thickness-shear vibration of a quartz plate

As an example, we analyze large thickness-shear vibration in the X_1 direction of a rotated Y-cut quartz plate (see Figure 4.2.2) [48], which is a widely used operating mode of piezoelectric resonators.

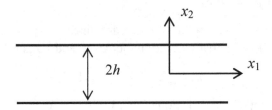

Figure 4.2.2. An electroded quartz plate.

4.2.2.1 Governing equations

The material matrices of rotated Y-cut quartz can be found in Equation (2.1.2). Consider a plate electroded at its two major faces. For thickness-shear in the X_1 direction, $u_1^{(1)}$ is the dominating mechanical displacement which is coupled to $\phi^{(1)}$. For a very thin plate, edge effects can be neglected and the thickness-shear mode does not vary with X_1 and X_3. Then the relevant equations are

$$-K_{21}^{(0)} = \frac{2\rho_0 h^3}{3}\ddot{u}_1^{(1)},$$

$$K_{21}^{(0)} = 2h(\bar{c}'_{2112}U_{12}^{(0)} - \bar{e}'_{221}\mathcal{E}_2^{(0)}$$
$$+ \bar{c}_{211212}u_1^{(1)}u_1^{(1)} + \bar{c}_{21121212}u_1^{(1)}u_1^{(1)}u_1^{(1)}),$$
$$\mathcal{D}_2^{(0)} = 2h(\bar{e}'_{212}U_{12}^{(0)} + \bar{\varepsilon}_{22}\mathcal{E}_2^{(0)}), \qquad (4.2.15)$$
$$U_{12}^{(0)} = u_1^{(1)},$$
$$\mathcal{E}_2^{(0)} = -\phi^{(1)}.$$

From Equations $(4.2.15)_{2,4,5}$ the equation of motion in Equation $(4.2.15)_1$ can be written as

$$\ddot{u} + \omega_\infty^2 u + \beta u^2 + \gamma u^3 = \alpha\phi^{(1)}. \qquad (4.2.16)$$

In Equation (4.2.16) we have denoted

$$u = u_1^{(1)}, \quad \omega_\infty^2 = 3\bar{c}'_{2112}/(\rho_0 h^2), \quad \alpha = -3\bar{e}'_{221}/(\rho_0 h^2),$$

$$\bar{c}'_{2112} = \kappa_1^2 \bar{c}_{2112}, \quad \bar{e}'_{221} = \kappa_1 \bar{e}_{221},$$

$$\kappa_1^2 = \frac{\pi^2}{12}(1 - \frac{8\bar{k}_{26}^2}{\pi^2}), \quad \bar{k}_{26}^2 = \frac{e_{26}^2}{\varepsilon_{22}(c_{66} + e_{26}^2/\varepsilon_{22})}, \tag{4.2.17}$$

$$\beta = 3\bar{c}_{211212}/(\rho_0 h^2), \quad \gamma = 3\bar{c}_{21121212}/(\rho_0 h^2),$$

where the thickness-shear correction factor is taken from [27]. In Equation (4.2.17), ω is the fundamental thickness-shear frequency from a linear solution. We want to study free and forced vibrations near ω.

4.2.2.2 Free vibration

For free vibrations we look for a periodic solution with undetermined frequency ω and amplitude A to the homogeneous form of Equation (4.2.16). For rotated Y-cut quartz, $\beta = 0$ and therefore the quadratic nonlinear term in Equation (4.2.16) disappears. Substituting $u = A\cos\omega t$ into Equation (4.2.16) (with $V = 0$), and neglecting the $\cos3\omega t$ term, we obtain

$$-\omega^2 A + \omega_\infty^2 A + \frac{3}{4}\gamma A^3 = 0. \tag{4.2.18}$$

Equation (4.2.18) yields the following expression for the nonlinear resonant frequency

$$\omega = \sqrt{\omega_\infty^2 + \frac{3}{4}\gamma A^2} \cong \omega_\infty (1 + \frac{3\gamma A^2}{8\omega_\infty^2}), \tag{4.2.19}$$

and the following corresponding free vibration mode

$$u = A\cos[\omega_\infty (1 + \frac{3\gamma A^2}{8\omega_\infty^2})t]. \tag{4.2.20}$$

Equation (4.2.19) shows that for large amplitude vibrations the frequency becomes amplitude dependent.

4.2.2.3 Forced vibration

Next consider electrically forced vibrations under a voltage across the thickness of the plate with $\phi(\pm h) = \pm 0.5V\cos\omega t$. Then Equation

$(4.2.2)_2$ implies that

$$\phi^{(0)} = 0, \quad \phi^{(1)} = \frac{V}{2h}\cos\omega t . \tag{4.2.21}$$

With $\phi^{(1)}$ from Equation (4.2.21), Equation (4.2.16) can be written as

$$\ddot{u} + \omega_\infty^2 u + c\dot{u} + \gamma u^3 = \frac{\alpha V}{2h}\cos\omega t, \tag{4.2.22}$$

where we have also introduced a damping term with a damping coefficient $c = 2\omega_\infty\zeta = \omega_\infty/Q$. ζ is the relative damping coefficient and Q is the quality factor. We look for a solution to Equation (4.2.22) in the following form:

$$u = A\cos(\omega t + \psi), \tag{4.2.23}$$

where A and ψ are undetermined constants. Substituting Equation (4.2.23) into Equation (4.2.22), neglecting the $\cos 3(\omega t + \psi)$ term, and collecting the coefficients of $\sin\omega t$ and $\cos\omega t$, we obtain

$$\begin{cases} [(\omega_\infty^2 - \omega^2)A + \dfrac{3}{4}\gamma A^3]\cos\psi - cA\omega\sin\psi = \dfrac{\alpha V}{2h}, \\[2mm] [(\omega_\infty^2 - \omega^2)A + \dfrac{3}{4}\gamma A^3]\sin\psi + cA\omega\cos\psi = 0. \end{cases} \tag{4.2.24}$$

Multiplying Equations $(4.2.24)_{1,2}$ by $\cos\psi$ and $\sin\psi$ respectively and then adding them, and multiplying $(4.2.24)_{1,2}$ by $\sin\psi$ and $\cos\psi$ respectively and then subtracting one from the other, we have

$$\begin{cases} (\omega_\infty^2 - \omega^2)A + \dfrac{3}{4}\gamma A^3 = \dfrac{\alpha V}{2h}\cos\psi, \\[2mm] cA\omega = -\dfrac{\alpha V}{2h}\sin\psi. \end{cases} \tag{4.2.25}$$

Squaring both sides of Equation $(4.2.25)_{1,2}$ and then adding them yields

$$[(\omega_\infty^2 - \omega^2)A + \dfrac{3}{4}\gamma A^3]^2 + (cA\omega)^2 = (\dfrac{\alpha V}{2h})^2 . \tag{4.2.26}$$

We are interested in resonant behaviors near ω_∞. Therefore we denote $\omega = \omega_\infty + \Delta\omega$. Then $\omega_\infty^2 - \omega^2 \cong -2\omega_\infty\Delta\omega$, and Equation (4.2.26) can be written as

$$(\Delta\omega - \frac{3\gamma A^2}{8\omega_\infty})^2 + \frac{c^2}{4} = (\frac{\alpha V}{4h\omega_\infty A})^2,$$ (4.2.27)

from which we can solve for $\Delta\omega$:

$$\Delta\omega = \frac{3\gamma A^2}{8\omega_\infty} \pm \frac{1}{2}[(\frac{\alpha V}{2h\omega_\infty A})^2 - c^2]^{\frac{1}{2}}.$$ (4.2.28)

We calculate the electric current flowing in or out of the driving electrodes, which is important to resonator design. From Equations (4.2.15)$_3$, (4.2.17)$_1$, and (4.2.23) we have, for the free electric charge per unit undeformed area of the electrode at $X_2 = h$

$$Q_e = -\frac{\mathcal{D}_2^{(0)}}{2h} = -(\kappa_1 \bar{e}_{26} u - \bar{\varepsilon}_{22} \phi^{(1)})$$
$$\cong -\kappa_1 \bar{e}_{26} u = -\kappa_1 \bar{e}_{26} A \cos(\omega t + \psi),$$ (4.2.29)

where, for near resonance behavior, we have neglected the electrostatic term in the expression of Q_e which is much smaller than the piezoelectric term. Then the current flows out of the electrode is

$$-\dot{Q}_e = -\kappa_1 \bar{e}_{26} A\omega \sin(\omega t + \psi) = I \sin(\omega t + \psi),$$
$$I = -\kappa_1 \bar{e}_{26} A\omega \cong -\kappa_1 \bar{e}_{26} A\omega_\infty.$$ (4.2.30)

From Equations (4.2.30)$_2$ and (4.2.28), we obtain the frequency-current amplitude relation

$$\Delta\omega = \frac{3\gamma}{8\omega_\infty} \left(\frac{I}{\kappa_1 \bar{e}_{26}\omega_\infty} \right)^2 \pm \frac{1}{2}\left[\left(\frac{\alpha V \kappa_1 \bar{e}_{26}}{2hI} \right)^2 - c^2 \right]^{1/2}.$$ (4.2.31)

Quartz is a material with very little damping. We choose the quality factor to be $Q = 10^5$. Consider a 1 MHz fundamental mode resonator with $h = 0.8273$ mm. The frequency-current amplitude relation predicted by Equation (4.2.31) is plotted in Figure 4.2.3, which is typical for nonlinear resonance.

The vertices of the curves in Figure 4.2.3 fall on a parabola as predicted by an approximate analysis from the three-dimensional equations [49]. The Equation for the corresponding parabola predicted by the two-dimensional equations here is Equation (4.2.31), without the last term on the right hand side. In Figure 4.2.4 we plot the two parabolas from the two- and three-dimensional solutions as a comparison. It can be seen that the two parabolas are very close to each other.

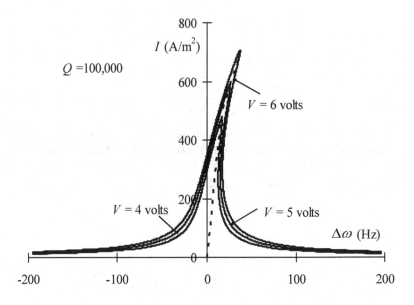

Figure 4.2.3. Nonlinear amplitude-frequency behavior near resonance.

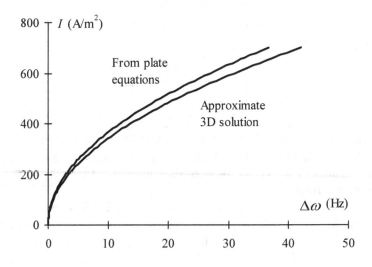

Figure 4.2.4. Amplitude-frequency behavior predicted by two- and three-dimensional equations.

Chapter 5
Piezoelectric Shells

As a natural continuation of piezoelectric plates, we next study motions of thin electroelastic shells, which are common structures for devices.

5.1 First-Order Theory

Consider an element of a thin shell (see Figure 5.1.1).

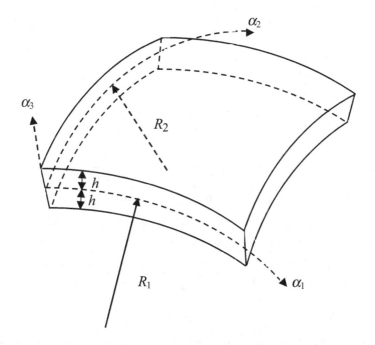

Figure 5.1.1. A shell element and coordinate system.

α_1 and α_2 are the middle surface principal coordinates, and α_3 is the thickness coordinate. $(\alpha_1,\alpha_2,\alpha_3)$ is an orthogonal curvilinear coordinate system. The thickness of the shell, $2h$, is much smaller than the radii of curvature R_1 and R_2 of the middle surface. Let A_1 and A_2 be the Lamè coefficients corresponding to α_1 and α_2 at the middle surface. The metric in $(\alpha_1,\alpha_2,\alpha_3)$ is given by $A_1(1+\alpha_3/R_1)$, $A_2(1+\alpha_3/R_2)$, and 1, which determines the tensor operations in $(\alpha_1,\alpha_2,\alpha_3)$. The strain-displacement relations take the following form [50]:

$$S_{11} = \frac{1}{A_1(1+\dfrac{\alpha_3}{R_1})}\left[\frac{\partial u_1}{\partial \alpha_1} + \frac{u_2}{A_2}\frac{\partial A_1}{\partial \alpha_2} + \frac{A_1 u_3}{R_1}\right],$$

$$S_{22} = \frac{1}{A_2(1+\dfrac{\alpha_3}{R_2})}\left[\frac{\partial u_2}{\partial \alpha_2} + \frac{u_1}{A_1}\frac{\partial A_2}{\partial \alpha_1} + \frac{A_2 u_3}{R_2}\right], \quad S_{33} = \frac{\partial u_3}{\partial \alpha_3},$$

(5.1.1)

$$2S_{23} = A_2(1+\frac{\alpha_3}{R_2})\frac{\partial}{\partial \alpha_3}\left[\frac{u_2}{A_2(1+\dfrac{\alpha_3}{R_2})}\right] + \frac{1}{A_2(1+\dfrac{\alpha_3}{R_2})}\frac{\partial u_3}{\partial \alpha_2},$$

$$2S_{31} = A_1(1+\frac{\alpha_3}{R_1})\frac{\partial}{\partial \alpha_3}\left[\frac{u_1}{A_1(1+\dfrac{\alpha_3}{R_1})}\right] + \frac{1}{A_1(1+\dfrac{\alpha_3}{R_1})}\frac{\partial u_3}{\partial \alpha_1},$$

$$2S_{12} = \frac{A_1(1+\dfrac{\alpha_3}{R_1})}{A_2(1+\dfrac{\alpha_3}{R_2})}\frac{\partial}{\partial \alpha_2}\left[\frac{u_1}{A_1(1+\dfrac{\alpha_3}{R_1})}\right] + \frac{A_2(1+\dfrac{\alpha_3}{R_2})}{A_1(1+\dfrac{\alpha_3}{R_1})}\frac{\partial}{\partial \alpha_1}\left[\frac{u_2}{A_2(1+\dfrac{\alpha_3}{R_2})}\right].$$

(5.1.2)

The electric field-potential relation can be obtained from the gradient operation on a scalar in curvilinear coordinates

$$E_1 = -\frac{1}{A_1(1+\frac{\alpha_3}{R_1})}\frac{\partial\phi}{\partial\alpha_1}, \quad E_2 = -\frac{1}{A_2(1+\frac{\alpha_3}{R_2})}\frac{\partial\phi}{\partial\alpha_2},$$

$$E_3 = -\frac{\partial\phi}{\partial\alpha_3}. \tag{5.1.3}$$

The charge equation of electrostatics is given by the divergence operation on a vector in curvilinear coordinates

$$\nabla\cdot\mathbf{D} = \frac{1}{A_1(1+\frac{\alpha_3}{R_1})A_2(1+\frac{\alpha_3}{R_2})}\left\{\frac{\partial}{\partial\alpha_1}\left[A_2(1+\frac{\alpha_3}{R_2})D_1\right]\right.$$

$$+\frac{\partial}{\partial\alpha_2}\left[A_1(1+\frac{\alpha_3}{R_1})D_2\right] \tag{5.1.4}$$

$$\left.+\frac{\partial}{\partial\alpha_3}\left[A_1(1+\frac{\alpha_3}{R_1})A_2(1+\frac{\alpha_3}{R_2})D_3\right]\right\} = 0.$$

For thin shells, if the dependence of $A_1(1+\alpha_3/R_1)$ and $A_2(1+\alpha_3/R_2)$ on α_3 is neglected after their derivatives in the above formulas with respect to α_3 have been carried out, Equations (5.1.1) through (5.1.4) reduce to

$$S_{11} = \frac{1}{A_1}\left[\frac{\partial u_1}{\partial\alpha_1}+\frac{u_2}{A_2}\frac{\partial A_1}{\partial\alpha_2}+\frac{A_1 u_3}{R_1}\right],$$

$$S_{22} = \frac{1}{A_2}\left[\frac{\partial u_2}{\partial\alpha_2}+\frac{u_1}{A_1}\frac{\partial A_2}{\partial\alpha_1}+\frac{A_2 u_3}{R_2}\right], \quad S_{33} = \frac{\partial u_3}{\partial\alpha_3},$$

$$2S_{23} = \frac{\partial u_2}{\partial\alpha_3}-\frac{u_2}{R_2}+\frac{1}{A_2}\frac{\partial u_3}{\partial\alpha_2}, \quad 2S_{31} = \frac{\partial u_1}{\partial\alpha_3}-\frac{u_1}{R_1}+\frac{1}{A_1}\frac{\partial u_3}{\partial\alpha_1}, \tag{5.1.5}$$

$$2S_{12} = \frac{A_1}{A_2}\frac{\partial}{\partial\alpha_2}\left[\frac{u_1}{A_1}\right]+\frac{A_2}{A_1}\frac{\partial}{\partial\alpha_1}\left[\frac{u_2}{A_2}\right],$$

$$E_1 = -\frac{1}{A_1}\frac{\partial\phi}{\partial\alpha_1}, \quad E_2 = -\frac{1}{A_2}\frac{\partial\phi}{\partial\alpha_2}, \quad E_3 = -\frac{\partial\phi}{\partial\alpha_3}, \tag{5.1.6}$$

$$\nabla \cdot \mathbf{D} = \frac{1}{A_1 A_2} \left\{ \frac{\partial}{\partial \alpha_1} [A_2 D_1] + \frac{\partial}{\partial \alpha_2} [A_1 D_2] + \frac{\partial}{\partial \alpha_3} [A_1 A_2 D_3] \right.$$

$$\left. + A_1 A_2 \left(\frac{1}{R_1} + \frac{1}{R_2} \right) D_3 \right\} = 0.$$

(5.1.7)

The power series expansion method for plates also applies to shells [51]. For a first-order shear deformation theory we make the following expansions of the displacement and electric potential in $(\alpha_1, \alpha_2, \alpha_3)$:

$$u_j \cong u_j^{(0)}(\alpha_1, \alpha_2, t) + \alpha_3 u_j^{(1)}(\alpha_1, \alpha_2, t) + \alpha_3^2 u_j^{(2)}(\alpha_1, \alpha_2, t),$$

$$\phi \cong \phi^{(0)}(\alpha_1, \alpha_2, t) + \alpha_3 \phi^{(1)}(\alpha_1, \alpha_2, t),$$

(5.1.8)

where $u_3^{(1)}$ and $u_j^{(2)}$ will be eliminated or neglected later. With Equation (5.1.8), the strains and electric field for thin shells can be written as:

$$S_{kl} = S_{kl}^{(0)} + \alpha_3 S_{kl}^{(1)}, \quad E_k = E_k^{(0)} + \alpha_3 E_k^{(1)},$$

(5.1.9)

where

$$S_{11}^{(0)} = \frac{1}{A_1} \left[\frac{\partial u_1^{(0)}}{\partial \alpha_1} + \frac{u_2^{(0)}}{A_2} \frac{\partial A_1}{\partial \alpha_2} + \frac{A_1 u_3^{(0)}}{R_1} \right],$$

$$S_{22}^{(0)} = \frac{1}{A_2} \left[\frac{\partial u_2^{(0)}}{\partial \alpha_2} + \frac{u_1^{(0)}}{A_1} \frac{\partial A_2}{\partial \alpha_1} + \frac{A_2 u_3^{(0)}}{R_2} \right], \quad S_{33}^{(0)} = u_3^{(1)},$$

$$2 S_{23}^{(0)} = u_2^{(1)} - \frac{u_2^{(0)}}{R_2} + \frac{1}{A_2} \frac{\partial u_3^{(0)}}{\partial \alpha_2},$$

(5.1.10)

$$2 S_{31}^{(0)} = u_1^{(1)} - \frac{u_1^{(0)}}{R_1} + \frac{1}{A_1} \frac{\partial u_3^{(0)}}{\partial \alpha_1},$$

$$2 S_{12}^{(0)} = \frac{A_1}{A_2} \frac{\partial}{\partial \alpha_2} \left[\frac{u_1^{(0)}}{A_1} \right] + \frac{A_2}{A_1} \frac{\partial}{\partial \alpha_1} \left[\frac{u_2^{(0)}}{A_2} \right],$$

$$S_{11}^{(1)} = \frac{1}{A_1}\left[\frac{\partial u_1^{(1)}}{\partial \alpha_1} + \frac{u_2^{(1)}}{A_2}\frac{\partial A_1}{\partial \alpha_2} + \frac{A_1 u_3^{(1)}}{R_1}\right] \cong \frac{1}{A_1}\left[\frac{\partial u_1^{(1)}}{\partial \alpha_1} + \frac{u_2^{(1)}}{A_2}\frac{\partial A_1}{\partial \alpha_2}\right],$$

$$S_{22}^{(1)} = \frac{1}{A_2}\left[\frac{\partial u_2^{(1)}}{\partial \alpha_2} + \frac{u_1^{(1)}}{A_1}\frac{\partial A_2}{\partial \alpha_1} + \frac{A_2 u_3^{(1)}}{R_2}\right] \cong \frac{1}{A_2}\left[\frac{\partial u_2^{(1)}}{\partial \alpha_2} + \frac{u_1^{(1)}}{A_1}\frac{\partial A_2}{\partial \alpha_1}\right],$$

$$S_{33}^{(1)} = 2u_3^{(2)}, \quad 2S_{23}^{(1)} = 2u_2^{(2)} + \frac{1}{A_2}\frac{\partial u_3^{(1)}}{\partial \alpha_2} - \frac{u_2^{(1)}}{R_2}, \tag{5.1.11}$$

$$S_{31}^{(1)} = 2u_1^{(2)} + \frac{1}{A_1}\frac{\partial u_3^{(1)}}{\partial \alpha_1} - \frac{u_1^{(1)}}{R_1},$$

$$2S_{12}^{(1)} = \frac{A_1}{A_2}\frac{\partial}{\partial \alpha_2}\left[\frac{u_1^{(1)}}{A_1}\right] + \frac{A_2}{A_1}\frac{\partial}{\partial \alpha_1}\left[\frac{u_2^{(1)}}{A_2}\right],$$

$$E_1^{(0)} = -\frac{1}{A_1}\frac{\partial \phi^{(0)}}{\partial \alpha_1}, \quad E_2^{(0)} = -\frac{1}{A_2}\frac{\partial \phi^{(0)}}{\partial \alpha_2}, \quad E_3^{(0)} = -\phi^{(1)},$$

$$\tag{5.1.12}$$

$$E_1^{(1)} = -\frac{1}{A_1}\frac{\partial \phi^{(1)}}{\partial \alpha_1}, \quad E_2^{(1)} = -\frac{1}{A_2}\frac{\partial \phi^{(1)}}{\partial \alpha_2}, \quad E_3^{(1)} = 0.$$

The two-dimensional equations of motion are obtained by substituting Equation (5.1.8) into the variational formulation in Equation (1.2.26) for independent variations of $\delta u_1^{(0)}$, $\delta u_2^{(0)}$, $\delta u_3^{(0)}$, $\delta u_1^{(1)}$ and $\delta u_2^{(1)}$. The results are [50,52]

$$\frac{\partial(N_{11}A_2)}{\partial \alpha_1} + \frac{\partial(N_{21}A_1)}{\partial \alpha_2}$$

$$+ N_{12}\frac{\partial A_1}{\partial \alpha_2} - N_{22}\frac{\partial A_2}{\partial \alpha_1} + Q_{13}A_1A_2\frac{1}{R_1} + F_1^{(0)} = 2\rho h A_1 A_2 \ddot{u}_1^{(0)},$$

$$\frac{\partial(N_{12}A_2)}{\partial \alpha_1} + \frac{\partial(N_{22}A_1)}{\partial \alpha_2}$$

$$+ N_{21}\frac{\partial A_2}{\partial \alpha_1} - N_{11}\frac{\partial A_1}{\partial \alpha_2} + Q_{23}A_1A_2\frac{1}{R_2} + F_2^{(0)} = 2\rho h A_1 A_2 \ddot{u}_2^{(0)},$$

$$\frac{\partial}{\partial \alpha_1}(Q_{13}A_2) + \frac{\partial}{\partial \alpha_2}(Q_{23}A_1)$$

$$- N_{11}\frac{A_1A_2}{R_1} - N_{22}\frac{A_1A_2}{R_2} + F_3^{(0)} = 2\rho h A_1 A_2 \ddot{u}_3^{(0)}, \tag{5.1.13}$$

$$\frac{\partial(M_{11}A_2)}{\partial \alpha_1} + \frac{\partial(M_{21}A_1)}{\partial \alpha_2}$$

$$+ M_{12}\frac{\partial A_1}{\partial \alpha_2} - M_{22}\frac{\partial A_2}{\partial \alpha_1} - Q_{31}A_1A_2 + F_1^{(1)} = \frac{2\rho h^3}{3}A_1A_2\ddot{u}_1^{(1)},$$

$$\frac{\partial(M_{12}A_2)}{\partial \alpha_1} + \frac{\partial(M_{22}A_1)}{\partial \alpha_2} \tag{5.1.14}$$

$$+ M_{21}\frac{\partial A_2}{\partial \alpha_1} - M_{11}\frac{\partial A_1}{\partial \alpha_2} - Q_{32}A_1A_2 + F_2^{(1)} = \frac{2\rho h^3}{3}A_1A_2\ddot{u}_2^{(1)}.$$

The two-dimensional charge equations are obtained by substituting Equation (5.1.8) into the variational formulation in Equation (1.2.26) for independent variations of $\delta\phi^{(0)}$ and $\delta\phi^{(1)}$, or by moment operations on the three-dimensional charge equation in Equation (5.1.7). The results are

$$\frac{\partial\left(D_1^{(0)}A_2\right)}{\partial \alpha_1} + \frac{\partial\left(D_2^{(0)}A_1\right)}{\partial \alpha_2} + (\frac{1}{R_1} + \frac{1}{R_2})A_1A_2D_3^{(0)} + D^{(0)} = 0,$$

$$\frac{\partial\left(D_1^{(1)}A_2\right)}{\partial \alpha_1} + \frac{\partial\left(D_2^{(1)}A_1\right)}{\partial \alpha_2} + (\frac{1}{R_1} + \frac{1}{R_2})A_1A_2D_3^{(1)} \tag{5.1.15}$$

$$- A_1A_2D_3^{(0)} + D^{(1)} = 0.$$

The shell resultants and mechanical surface loads are defined by

$$\{N_{ab}, Q_{3c}, Q_{c3}\} = \{T_{ab}^{(0)}, T_{3c}^{(0)}, T_{c3}^{(0)}\} = \int_{-h}^{h}\{T_{ab}, T_{3c}, T_{c3}\}d\alpha_3,$$

$$M_{ab} = T_{ab}^{(1)} = \int_{-h}^{h} T_{ab}\alpha_3 d\alpha_3, \tag{5.1.16}$$

$$F_j^{(n)} = \left[\alpha_3^n T_{3j} A_1 A_2\right]_{-h}^{h}, \quad n = 0,1, \quad a,b,c = 1,2.$$

N_{ab} are the extensional and shear forces in the tangent plane. $Q_{3c} = Q_{c3}$ are transverse shear forces. M_{ab} are bending and twisting moments. The resultants of the electric dispalcement and surface charge are

$$D_k^{(n)} = \int_{-h}^{h} D_k \alpha_3^n d\alpha_3, \quad D^{(n)} = \left[\alpha_3^n D_3 A_1 A_2\right]_{-h}^{h}. \tag{5.1.17}$$

Equations (5.1.13) are for extension and flexure, and Equations (5.1.14) are for shear deformations in the α_1 and α_2 directions. For shell constitutive relations we substitute the three-dimensional constitutive relations from Equation (1.2.7) into Equations (5.1.16) and (5.1.17), and perform the following stress relaxations:

$$T_{33}^{(0)} = \int_{-h}^{h} T_{33} d\alpha_3 = 0,$$
$$T_{3j}^{(1)} = \int_{-h}^{h} T_{33} \alpha_3 d\alpha_3 = 0, \quad j = 1,2,3, \tag{5.1.18}$$

so that $S_{33}^{(0)}$ and $S_{3j}^{(1)}$, and hence $u_3^{(1)}$ and $u_j^{(2)}$ can be eliminated. Shear correction factors can be introduced in the manner of Equation (2.4.16). Then the following two-dimensional constitutive relations will result:

$$T_{ij}^{(0)} = 2h(\bar{c}_{ijkl}' S_{kl}^{(0)} - \bar{e}_{kij}' E_k^{(0)}),$$
$$D_k^{(0)} = 2h(\bar{e}_{kij}' S_{ij}^{(0)} + \bar{\varepsilon}_{kl} E_l^{(0)}), \tag{5.1.19}$$

$$T_r^{(1)} = \frac{2h^3}{3}(\gamma_{rs} S_s^{(1)} - \psi_{kr} E_k^{(1)}),$$
$$D_i^{(1)} = \frac{2h^3}{3}(\psi_{is} S_s^{(1)} + \zeta_{ij} E_j^{(1)}), \quad r,s = 1,2,6, \tag{5.1.20}$$

where the relaxed material constants are defined in Equations (2.3.20), (2.4.18), (2.4.19), and (2.4.15).

In summary, we have obtained two-dimensional equations of motion (5.1.13) and (5.1.14), charge equations (5.1.15), constitutive relations (5.1.19) and (5.1.20), strain-displacement relations (5.1.10) and (5.1.11), and electric field-potential relations (5.1.12). With successive substitutions, Equations (5.1.13), (5.1.14) and (5.1.15) can be written as seven equations for $u_1^{(0)}$, $u_2^{(0)}$, $u_3^{(0)}$, $u_1^{(1)}$, $u_2^{(1)}$, $\phi^{(0)}$ and $\phi^{(1)}$. At the boundary of a shell with a unit exterior normal **n** and a unit tangent **s**, we may prescribe

$$N_{nn} \quad \text{or} \quad u_n^{(0)}, \quad N_{ns} \quad \text{or} \quad u_s^{(0)}, \quad Q_{n3}^{(0)} \quad \text{or} \quad u_3^{(0)},$$

$$M_{nn} \quad \text{or} \quad u_n^{(1)}, \quad M_{ns} \quad \text{or} \quad u_s^{(1)}, \qquad (5.1.21)$$

$$D_n^{(0)} \quad \text{or} \quad \phi^{(0)}, \quad D_n^{(1)} \quad \text{or} \quad \phi^{(1)}.$$

Equations for laminated piezoelectric shells can be found in [53]. Equations for electroelastic shells under biasing fields can be found in [54,55] along with buckling analysis [55]. Nonlinear equations for electroelastic shells in large thickness-shear deformations can be found in [56].

5.2 Classical Theory

For the classical theory of a shell in coupled extension and flexure without shear deformations, we set the relevant shear strains to zero:

$$2S_{31}^{(0)} = u_1^{(1)} - \frac{u_1^{(0)}}{R_1} + \frac{1}{A_1}\frac{\partial u_3^{(0)}}{\partial \alpha_1} = 0,$$

$$2S_{23}^{(0)} = u_2^{(1)} - \frac{u_2^{(0)}}{R_2} + \frac{1}{A_2}\frac{\partial u_3^{(0)}}{\partial \alpha_2} = 0. \qquad (5.2.1)$$

This allows us to express $u_1^{(1)}$ and $u_2^{(1)}$ in terms of the zero-order displacements for extension and flexure. Furthermore, we ignore the rotatory inertia in Equation (5.1.14) and obtain

$$\frac{\partial\left(M_{11}A_2\right)}{\partial \alpha_1} + \frac{\partial\left(M_{21}A_1\right)}{\partial \alpha_2}$$

$$+ M_{12}\frac{\partial A_1}{\partial \alpha_2} - M_{22}\frac{\partial A_2}{\partial \alpha_1} - Q_{31}A_1A_2 + F_1^{(1)} = 0,$$

$$\frac{\partial\left(M_{12}A_2\right)}{\partial \alpha_1} + \frac{\partial\left(M_{22}A_1\right)}{\partial \alpha_2} \qquad (5.2.2)$$

$$+ M_{21}\frac{\partial A_2}{\partial \alpha_1} - M_{11}\frac{\partial A_1}{\partial \alpha_2} - Q_{32}A_1A_2 + F_2^{(1)} = 0,$$

which yields expressions for transverse shear forces Q_{31} and Q_{32} in terms of the bending and twisting moments. In summary, the equations for the classical theory are

$$\frac{\partial(N_{11}A_2)}{\partial\alpha_1} + \frac{\partial(N_{21}A_1)}{\partial\alpha_2}$$

$$+ N_{12}\frac{\partial A_1}{\partial\alpha_2} - N_{22}\frac{\partial A_2}{\partial\alpha_1} + Q_{13}A_1A_2\frac{1}{R_1} + F_1^{(0)} = 2\rho h A_1 A_2 \ddot{u}_1^{(0)},$$

$$\frac{\partial(N_{12}A_2)}{\partial\alpha_1} + \frac{\partial(N_{22}A_1)}{\partial\alpha_2}$$

$$+ N_{21}\frac{\partial A_2}{\partial\alpha_1} - N_{11}\frac{\partial A_1}{\partial\alpha_2} + Q_{23}A_1A_2\frac{1}{R_2} + F_2^{(0)} = 2\rho h A_1 A_2 \ddot{u}_2^{(0)},$$

$$\frac{\partial}{\partial\alpha_1}(Q_{13}A_2) + \frac{\partial}{\partial\alpha_2}(Q_{23}A_1)$$

$$- N_{11}\frac{A_1A_2}{R_1} - N_{22}\frac{A_1A_2}{R_2} + F_3^{(0)} = 2\rho h A_1 A_2 \ddot{u}_3^{(0)},$$

(5.2.3)

$$\frac{\partial(D_1^{(0)}A_2)}{\partial\alpha_1} + \frac{\partial(D_2^{(0)}A_1)}{\partial\alpha_2}$$

$$+ (\frac{1}{R_1} + \frac{1}{R_2})A_1A_2D_3^{(0)} + D^{(0)} = 0,$$

$$\frac{\partial(D_1^{(1)}A_2)}{\partial\alpha_1} + \frac{\partial(D_2^{(1)}A_1)}{\partial\alpha_2}$$

$$+ (\frac{1}{R_1} + \frac{1}{R_2})A_1A_2D_3^{(1)} - A_1A_2D_3^{(0)} + D^{(1)} = 0,$$

(5.2.4)

where

$$N_r^{(0)} = 2h(\bar{c}_{rs}'S_s^{(0)} - \bar{e}_{kr}'E_k^{(0)}),$$

$$D_i^{(0)} = 2h(\bar{e}_{is}'S_s^{(0)} + \bar{\varepsilon}_{ij}E_j^{(0)}), \quad r,s = 1,2,6,$$

(5.2.5)

$$T_r^{(1)} = \frac{2h^3}{3}(\gamma_{rs}S_s^{(1)} - \psi_{kr}E_k^{(1)}),$$

$$D_i^{(1)} = \frac{2h^3}{3}(\psi_{is}S_s^{(1)} + \zeta_{ij}E_j^{(1)}), \quad r,s = 1,2,6, \tag{5.2.6}$$

$$Q_{31}A_1A_2 = \frac{\partial(M_{11}A_2)}{\partial\alpha_1} + \frac{\partial(M_{21}A_1)}{\partial\alpha_2}$$

$$+ M_{12}\frac{\partial A_1}{\partial\alpha_2} - M_{22}\frac{\partial A_2}{\partial\alpha_1} + F_1^{(1)},$$

$$Q_{32}A_1A_2 = \frac{\partial(M_{12}A_2)}{\partial\alpha_1} + \frac{\partial(M_{22}A_1)}{\partial\alpha_2} \tag{5.2.7}$$

$$+ M_{21}\frac{\partial A_2}{\partial\alpha_1} - M_{11}\frac{\partial A_1}{\partial\alpha_2} + F_2^{(1)},$$

$$S_{11}^{(0)} = \frac{1}{A_1}\left[\frac{\partial u_1^{(0)}}{\partial\alpha_1} + \frac{u_2^{(0)}}{A_2}\frac{\partial A_1}{\partial\alpha_2} + \frac{A_1 u_3^{(0)}}{R_1}\right],$$

$$S_{22}^{(0)} = \frac{1}{A_2}\left[\frac{\partial u_2^{(0)}}{\partial\alpha_2} + \frac{u_1^{(0)}}{A_1}\frac{\partial A_2}{\partial\alpha_1} + \frac{A_2 u_3^{(0)}}{R_2}\right], \tag{5.2.8}$$

$$2S_{12}^{(0)} = \frac{A_1}{A_2}\frac{\partial}{\partial\alpha_2}\left[\frac{u_1^{(0)}}{A_1}\right] + \frac{A_2}{A_1}\frac{\partial}{\partial\alpha_1}\left[\frac{u_2^{(0)}}{A_2}\right],$$

$$S_{11}^{(1)} = \frac{1}{A_1}\left[\frac{\partial u_1^{(1)}}{\partial\alpha_1} + \frac{u_2^{(1)}}{A_2}\frac{\partial A_1}{\partial\alpha_2}\right],$$

$$S_{22}^{(1)} = \frac{1}{A_2}\left[\frac{\partial u_2^{(1)}}{\partial\alpha_2} + \frac{u_1^{(1)}}{A_1}\frac{\partial A_2}{\partial\alpha_1}\right], \tag{5.2.9}$$

$$S_{12}^{(1)} = \frac{A_1}{A_2}\frac{\partial}{\partial\alpha_2}\left[\frac{u_1^{(1)}}{A_1}\right] + \frac{A_2}{A_1}\frac{\partial}{\partial\alpha_1}\left[\frac{u_2^{(1)}}{A_2}\right],$$

$$u_1^{(1)} = \frac{u_1^{(0)}}{R_1} - \frac{1}{A_1} \frac{\partial u_3^{(0)}}{\partial \alpha_1},$$

$$u_2^{(1)} = \frac{u_2^{(0)}}{R_2} - \frac{1}{A_2} \frac{\partial u_3^{(0)}}{\partial \alpha_2}, \tag{5.2.10}$$

$$E_1^{(0)} = -\frac{1}{A_1} \frac{\partial \phi^{(0)}}{\partial \alpha_1}, \quad E_2^{(0)} = -\frac{1}{A_2} \frac{\partial \phi^{(0)}}{\partial \alpha_2}, \quad E_3^{(0)} = -\phi^{(1)},$$

$$E_1^{(1)} = -\frac{1}{A_1} \frac{\partial \phi^{(1)}}{\partial \alpha_1}, \quad E_2^{(1)} = -\frac{1}{A_2} \frac{\partial \phi^{(1)}}{\partial \alpha_2}, \quad E_3^{(1)} = 0. \tag{5.2.11}$$

With successive substitutions, Equations (5.2.3) and (5.2.4) can be written as five equations for $u_1^{(0)}$, $u_2^{(0)}$, $u_3^{(0)}$, $\phi^{(0)}$ and $\phi^{(1)}$. At the boundary of a shell with unit exterior normal **n** and unit tangent **s**, we may prescribe

$$N_{nn} \quad \text{or} \quad u_n^{(0)}, \quad N_{ns} \quad \text{or} \quad u_s^{(0)},$$

$$Q_{n3}^{(0)} + M_{ns,s} \quad \text{or} \quad u_3^{(0)}, \quad M_{nn} \quad \text{or} \quad u_{3,n}^{(0)}, \tag{5.2.12}$$

$$D_n^{(0)} \quad \text{or} \quad \phi^{(0)}, \quad D_n^{(1)} \quad \text{or} \quad \phi^{(1)}.$$

5.3 Membrane Theory

A thin piezoelectric shell can sometimes be treated as a membrane that does not resist bending, without bending moments and transverse shear forces. In this case we have the following membrane theory:

$$\frac{\partial(N_{11}A_2)}{\partial \alpha_1} + \frac{\partial(N_{21}A_1)}{\partial \alpha_2} + N_{12}\frac{\partial A_1}{\partial \alpha_2} - N_{22}\frac{\partial A_2}{\partial \alpha_1} + F_1^{(0)} = 2\rho h A_1 A_2 \ddot{u}_1^{(0)},$$

$$\frac{\partial(N_{12}A_2)}{\partial \alpha_1} + \frac{\partial(N_{22}A_1)}{\partial \alpha_2} + N_{21}\frac{\partial A_2}{\partial \alpha_1} - N_{11}\frac{\partial A_1}{\partial \alpha_2} + F_2^{(0)} = 2\rho h A_1 A_2 \ddot{u}_2^{(0)}, \tag{5.3.1}$$

$$-N_{11}\frac{1}{R_1} - N_{22}\frac{1}{R_2} + F_3^{(0)} = 2\rho h \ddot{u}_3^{(0)},$$

$$\frac{\partial \left(D_1^{(0)} A_2 \right)}{\partial \alpha_1} + \frac{\partial \left(D_2^{(0)} A_1 \right)}{\partial \alpha_2} + (\frac{1}{R_1} + \frac{1}{R_2}) A_1 A_2 D_3^{(0)} + D^{(0)} = 0,$$

$$\frac{\partial \left(D_1^{(1)} A_2 \right)}{\partial \alpha_1} + \frac{\partial \left(D_2^{(1)} A_1 \right)}{\partial \alpha_2} + (\frac{1}{R_1} + \frac{1}{R_2}) A_1 A_2 D_3^{(1)}$$

$$- A_1 A_2 D_3^{(0)} + D^{(1)} = 0,$$

(5.3.2)

$$N_r^{(0)} = 2h(\gamma_{rs} S_s^{(0)} - \psi_{kr} E_k^{(0)}),$$

$$D_i^{(0)} = 2h(\psi_{is} S_s^{(0)} + \zeta_{ij} E_j^{(0)}), \quad r,s = 1,2,6,$$

(5.3.3)

$$D_i^{(1)} = \frac{2h^3}{3} \varepsilon_{ij} E_j^{(1)},$$

(5.3.4)

$$S_{11}^{(0)} = \frac{1}{A_1} \left[\frac{\partial u_1^{(0)}}{\partial \alpha_1} + \frac{u_2^{(0)}}{A_2} \frac{\partial A_1}{\partial \alpha_2} + \frac{A_1 u_3^{(0)}}{R_1} \right],$$

$$S_{22}^{(0)} = \frac{1}{A_2} \left[\frac{\partial u_2^{(0)}}{\partial \alpha_2} + \frac{u_1^{(0)}}{A_1} \frac{\partial A_2}{\partial \alpha_1} + \frac{A_2 u_3^{(0)}}{R_2} \right],$$

$$2S_{12}^{(0)} = \frac{A_1}{A_2} \frac{\partial}{\partial \alpha_2} \left[\frac{u_1^{(0)}}{A_1} \right] + \frac{A_2}{A_1} \frac{\partial}{\partial \alpha_1} \left[\frac{u_2^{(0)}}{A_2} \right],$$

(5.3.5)

$$E_1^{(0)} = -\frac{1}{A_1} \frac{\partial \phi^{(0)}}{\partial \alpha_1}, \quad E_2^{(0)} = -\frac{1}{A_2} \frac{\partial \phi^{(0)}}{\partial \alpha_2}, \quad E_3^{(0)} = -\phi^{(1)},$$

$$E_1^{(1)} = -\frac{1}{A_1} \frac{\partial \phi^{(1)}}{\partial \alpha_1}, \quad E_2^{(1)} = -\frac{1}{A_2} \frac{\partial \phi^{(1)}}{\partial \alpha_2}, \quad E_3^{(1)} = 0.$$

(5.3.6)

With successive substitutions, Equations (5.3.1) and (5.3.2) can be written as five equations for $u_1^{(0)}$, $u_2^{(0)}$, $u_3^{(0)}$, $\phi^{(0)}$ and $\phi^{(1)}$. At the boundary of a shell with a unit exterior normal **n** and a unit tangent **s**, we may prescribe

$$N_{nn} \quad \text{or} \quad u_n^{(0)}, \quad N_{ns} \quad \text{or} \quad u_s^{(0)},$$

$$D_n^{(0)} \quad \text{or} \quad \phi^{(0)}, \quad D_n^{(1)} \quad \text{or} \quad \phi^{(1)}.$$

(5.3.7)

5.4 Vibration of Ceramic Shells

As examples for applications of the shell equations obtained, we analyze some simple vibrations of ceramic shells.

5.4.1 Radial vibration of a spherical shell

Consider a thin spherical ceramic shell of mean radius R and thickness $2h$ with $R \gg h$ (see Figure 5.4.1). The ceramic is poled in the thickness direction, with fully electroded inner and outer surfaces. The electrodes are shorted so that the lower order electric fields vanish. Consider the spherically symmetric radial vibration of the shell with only one displacement component $u_r^{(0)} = u_3^{(0)}(t)$.

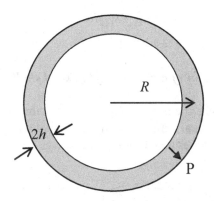

Figure 5.4.1. A spherical ceramic shell with radial poling.

For the motion we are considering, the membrane theory is sufficient. By symmetry we have

$$R_1 = R_2 = R, \quad S_{11}^{(0)} = S_{22}^{(0)} = \frac{u_3^{(0)}}{R}, \tag{5.4.1}$$

where the strain-displacement relation in Equation (5.3.5) has been used. Then, from the constitutive relation in Equation (5.3.3) and the material constants for polarized ceramics in Equations (2.3.35) and (2.3.36), we have

$$N_{11} = N_{22} = 2h(\gamma_{11}S_{11}^{(0)} + \gamma_{12}S_{22}^{(0)})$$

$$= 2h(\gamma_{11} + \gamma_{12})\frac{u_3^{(0)}}{R} = 2h(c_{11}^p + c_{12}^p)\frac{u_3^{(0)}}{R} \tag{5.4.2}$$

$$= 2h\frac{1}{s_{11}^E + s_{12}^E}\frac{u_3^{(0)}}{R}.$$

Substituting Equation (5.4.2) into Equation (5.3.1)$_3$, we obtain

$$-2\frac{1}{R}\frac{2h}{s_{11}^E + s_{12}^E}\frac{u_3^{(0)}}{R} = 2\rho h\ddot{u}_3^{(0)}. \tag{5.4.3}$$

For free vibrations, the resonant frequency is

$$\omega^2 = \frac{2}{\rho(s_{11}^E + s_{12}^E)R^2}, \tag{5.4.4}$$

which is the same as the result in [9].

5.4.2 Radial vibration of a circular cylindrical shell

Next we analyze the axi-symmetric radial vibration of an unbounded thin circular cylindrical ceramic shell with radial poling, electroded on its inner and outer surfaces (see Figure 5.4.2).

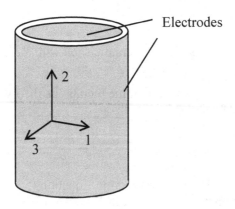

Figure 5.4.2. A thin circular cylindrical ceramic shell.

Let R be the mean radius, and $2h$ be the thickness of the shell. The electrodes are shorted so that the lower order electric fields vanish. The cylindrical coordinates (θ, z, r) correspond to $(\alpha_1, \alpha_2, \alpha_3)$. Consider cylindrically symmetric radial vibration of the shell with only one displacement component $u_r^{(0)} = u_3^{(0)}(t)$. For the motion we are considering, the membrane theory is sufficient. For a cylindrical shell we have

$$R_1 = R, \quad R_2 = \infty,$$
$$A_1 = R, \quad A_2 = 1. \tag{5.4.5}$$

The relevant strain and stress components are

$$S_{11}^{(0)} = \frac{u_3^{(0)}}{R}, \quad S_{22}^{(0)} = 0, \tag{5.4.6}$$

$$N_{11} = 2h(\gamma_{11}S_{11}^{(0)} + \gamma_{12}S_{22}^{(0)}) = 2h\gamma_{11}\frac{u_3^{(0)}}{R} = 2hc_{11}^p \frac{u_3^{(0)}}{R}$$
$$= 2h[c_{11}^E - \frac{(c_{13}^E)^2}{c_{33}^E}]\frac{u_3^{(0)}}{R} = 2h\frac{s_{11}^E}{(s_{11}^E)^2 - (s_{12}^E)^2}\frac{u_3^{(0)}}{R}. \tag{5.4.7}$$

Substituting Equation (5.4.7) into Equation (5.3.1)$_3$

$$-N_{11}\frac{1}{R} = 2\rho h\ddot{u}_3^{(0)}, \tag{5.4.8}$$

we obtain

$$-2h\frac{s_{11}^E}{(s_{11}^E)^2 - (s_{12}^E)^2}\frac{u_3^{(0)}}{R}\frac{1}{R} = 2\rho h\ddot{u}_3^{(0)}. \tag{5.4.9}$$

For time-harmonic motions, Equation (5.4.9) implies that

$$\omega^2 = \frac{1}{\rho R^2}\frac{s_{11}^E}{(s_{11}^E)^2 - (s_{12}^E)^2}, \tag{5.4.10}$$

which is the same as the result in [9].

5.5 A Shell on a Non-Thin Body

In some applications, we have a thin layer of one material on the surface of a body of another material. Such a structure can be modeled as a two-dimensional shell on a three-dimensional body with interface continuity conditions.

5.5.1 A piezoelectric shell on an elastic body

Consider static deformations of an elastic body coated with a thin piezoelectric film [57].

5.5.1.1 Governing equations

The piezoelectric shell has the following material matrices under the compact matrix notation:

$$
\begin{pmatrix}
c_{11} & c_{12} & c_{13} & 0 & 0 & 0 \\
c_{12} & c_{22} & c_{23} & 0 & 0 & 0 \\
c_{13} & c_{23} & c_{33} & 0 & 0 & 0 \\
0 & 0 & 0 & c_{44} & 0 & 0 \\
0 & 0 & 0 & 0 & c_{55} & 0 \\
0 & 0 & 0 & 0 & 0 & c_{66}
\end{pmatrix},
$$

$$
\begin{bmatrix}
e_{11} & e_{12} & e_{13} & e_{14} & e_{15} & e_{16} \\
e_{21} & e_{22} & e_{23} & e_{24} & e_{25} & e_{26} \\
e_{31} & e_{32} & e_{33} & e_{34} & e_{35} & e_{36}
\end{bmatrix},
\begin{pmatrix}
\varepsilon_{11} & \varepsilon_{12} & \varepsilon_{13} \\
\varepsilon_{12} & \varepsilon_{22} & \varepsilon_{23} \\
\varepsilon_{13} & \varepsilon_{23} & \varepsilon_{33}
\end{pmatrix}.
\tag{5.5.1}
$$

As special cases, the matrices in Equation (5.5.1) include piezoelectric ceramics poled in the 1, 2, or 3 directions. The shell is assumed to be very thin so that the membrane theory is sufficient. The membrane tensile and shear forces and electric displacements are given by the following constitutive relations:

$$N_{11} = N_1 = 2h(c_{11}^P S_1^{(0)} + c_{12}^P S_2^{(0)} - e_{k1}^P E_k^{(0)}),$$

$$N_{22} = N_2 = 2h(c_{12}^P S_1^{(0)} + c_{22}^P S_2^{(0)} - e_{k2}^P E_k^{(0)}),$$

$$N_{12} = N_6 = 2h(c_{66} S_6^{(0)} - e_{k6} E_k^{(0)}),$$

$$D_k^{(0)} = 2h(e_{k1}^P S_1^{(0)} + e_{k2}^P S_2^{(0)} + e_{k6} S_6^{(0)} - \varepsilon_{kl}^P E_l^{(0)}),$$

(5.5.2)

where

$$c_{11}^P = c_{11} - c_{13}^2 / c_{33}, \quad c_{12}^P = c_{12} - c_{13}c_{32} / c_{33},$$

$$c_{22}^P = c_{22} - c_{23}^2 / c_{33}, \quad e_{k1}^P = e_{k1} - c_{13}e_{k3} / c_{33}.$$

$$e_{k2}^P = e_{k2} - c_{23}e_{k3} / c_{33}, \quad \varepsilon_{kl}^P = \varepsilon_{kl} - e_{k3}e_{l3} / c_{33}.$$

(5.5.3)

The membrane equations of equilibrium and electrostatics take the form

$$\frac{\partial}{\partial \alpha_1}(A_2 N_1) + \frac{\partial}{\partial \alpha_2}(A_1 N_6)$$

$$+ N_6 \frac{\partial A_1}{\partial \alpha_2} - N_2 \frac{\partial A_2}{\partial \alpha_1} + A_1 A_2 (2\rho h f_1 + T_{31}\big|_{-h}^h) = 0,$$

$$\frac{\partial}{\partial \alpha_1}(A_2 N_6) + \frac{\partial}{\partial \alpha_2}(A_1 N_2)$$

$$+ N_6 \frac{\partial A_2}{\partial \alpha_1} - N_1 \frac{\partial A_1}{\partial \alpha_2} + A_1 A_2 (2\rho h f_2 + T_{32}\big|_{-h}^h) = 0, \qquad (5.5.4)$$

$$- N_1 \frac{1}{R_1} - N_2 \frac{1}{R_2} + 2\rho h f_3 + T_{33}\big|_{-h}^h = 0,$$

$$\frac{\partial}{\partial \alpha_1}(A_2 D_1^{(0)}) + \frac{\partial}{\partial \alpha_2}(A_1 D_2^{(0)})$$

$$+ (\frac{1}{R_1} + \frac{1}{R_2})A_1 A_2 D_3^{(0)} + A_1 A_2 D_3\big|_{-h}^h = 0.$$

With successive substitutions from Equations (5.5.2), (5.3.5) and (5.3.6), Equations (5.5.4)$_{1-3}$ can be written in the following compact form:

$$L_\beta(\mathbf{u}^{(0)}, \phi^{(0)}, \phi^{(1)}) + 2\rho h f_\beta + t_\beta^+ - t_\beta^- = 0, \qquad (5.5.5)$$

where L_β are linear differential operators, and $t_\beta^\pm = T_{3\beta}(\pm h)$ are the traction vectors at the major faces of the film. $\beta = 1, 2, 3$ is associated with the shell principle coordinates $(\alpha_1, \alpha_2, \alpha_3)$. At every point of the shell, there exists a set of transformation coefficients $\delta_{i\beta}$ between the local shell principal coordinates and the global Cartesian coordinates. Multiplying Equation (5.5.5) by the transformation coefficients, we have

$$L_i(\mathbf{u}^{(0)}, \phi^{(0)}, \phi^{(1)}) + 2\rho h f_i + t_i^+ - t_i^- = 0. \qquad (5.5.6)$$

The elastic body is governed by the equations of linear elasticity.

5.5.1.2 Boundary integral equation formulation

Since the film interacts with the body through the surface of the body, we use the boundary integral equation (BIE) formulation and its numerical solution technique – the boundary element method. The displacement $\mathbf{u}$ and surface traction $\mathbf{t}$ of the elastic body (in a domain Ω with boundary Γ) satisfy the following boundary integral equation:

$$\begin{aligned}
\mathbf{C}(P_0)\mathbf{u}(P_0) + \int_\Gamma \hat{\mathbf{T}}(P, P_0)\mathbf{u}(P)d\Gamma(P) \\
= \int_\Gamma \hat{\mathbf{U}}(P, P_0)\mathbf{t}(P)d\Gamma(P) + \int_\Omega \hat{\mathbf{U}}(P, P_0)\mathbf{b}(P)d\Omega(P),
\end{aligned} \qquad (5.5.7)$$

where $C_{ij} = \delta_{ij}/2$ for a smooth boundary, $\mathbf{b}$ is the body force vector of the elastic body, $\hat{\mathbf{U}}$ and $\hat{\mathbf{T}}$ are known second rank tensors and are related to the fundamental solution of the Navier operator in elasticity. They can be found in a book on boundary element method in elasticity. P is the source point and P_0 is the field point. The traction vector $\mathbf{t}$ on the surface of the elastic body is related to the traction on one of the faces of the film by $\mathbf{t} = -\mathbf{t}^-$. Therefore, from Equations (5.5.6) and (5.5.7), we obtain [57]

$$\mathbf{C}\mathbf{u} + \int_\Gamma \hat{\mathbf{T}}\mathbf{u}d\Gamma = -\int_\Gamma \hat{\mathbf{U}}(\mathbf{L} + 2\rho h\mathbf{f} + \mathbf{t}^+)d\Gamma + \int_\Omega \hat{\mathbf{U}}\mathbf{b}d\Omega, \qquad (5.5.8)$$

where the displacement continuity conditions between the body and the film have been used. Equation (5.5.8) is a system of boundary integral-differential equations because of the differential operator $\mathbf{L}$. If the body is only partially coated with a film, then BIE (5.5.7) applies to the non-coated portion of the surface.

5.5.1.3 An example

We examine the basic behavior of an elastic body with a piezoelectric film governed by Equation (5.5.8) in an example. Consider a two-dimensional plane strain problem of a circular elastic body of radius R shown in Figure 5.5.1.

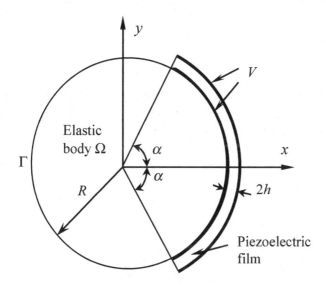

Figure 5.5.1. An elastic body coated with a piezoelectric film. The thick lines represent electrodes.

From $-\alpha$ to α, the body is coated with a ceramic film poled in the thickness direction. The film is electroded at its two major faces and can be used either as a sensor or an actuator. The voltage across the electrodes is denoted by V which implies that

$$\phi^{(0)} = 0, \quad \phi^{(1)} = V/2h. \tag{5.5.9}$$

The relevant membrane force, in the polar coordinate system defined by $x = r\cos\theta$ and $y = r\sin\theta$, is

$$N_\theta = 2hc_{11}^p(u_{\theta,\theta}^{(0)} + u_r^{(0)})/R + e_{31}^p V,$$

$$c_{11}^p = c_{11} - c_{13}^2/c_{33}, \quad e_{31}^p = e_{31} - c_{13}e_{33}/c_{33}. \tag{5.5.10}$$

From Equation (5.5.4), the tractions on the elastic body are

$$T_{rr} = -\frac{1}{R}N_\theta = -\frac{1}{R}(2hc_{11}^p \frac{u_{\theta,\theta}^{(0)} + u_r^{(0)}}{R} + e_{31}^p V) \cong -\frac{1}{R}e_{31}^p V,$$

$$T_{r\theta} = \frac{1}{R}\frac{\partial N_\theta}{\partial \theta} = \frac{1}{R}(2hc_{11}^p \frac{u_{\theta,\theta\theta}^{(0)} + u_{r,\theta}^{(0)}}{R} + e_{31}^p V_{,\theta}) \cong \frac{1}{R}e_{31}^p V_{,\theta}, \tag{5.5.11}$$

where the approximation is for the case when the film is relatively soft (small c_{11}^p). Since V is a piecewise constant function, its derivative leads to a delta function and the traction is effectively a normal distribution q and a pair of concentrated forces Q as shown in Figure 5.5.2 with

$$q = e_{31}^p V/R, \quad Q = e_{31}^p V. \tag{5.5.12}$$

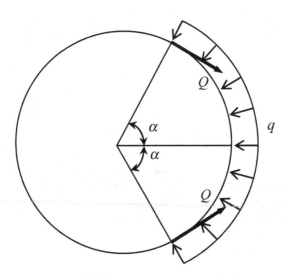

Figure 5.5.2. Actuating forces on the elastic body due to the film.

The presence of the concentrated Q can also be seen from the boundary condition of vanishing N_θ at the edge of the film and Equation (5.5.11). q is related to the curvature of the shell and does not exist for a flat film. Since the traction on the elastic body due to the film is now approximately known, the usual boundary element analysis of an elastic body can be performed. Otherwise boundary integral-differential equations will need to be solved as in the fourth section of the third chapter. As a numerical example we consider a PZT-7A piezoelectric film. For geometric parameters we choose $R = 20$ mm, $2h = 1$ mm. A voltage $V = 10$ volts is applied across the film thickness. For the elastic body we consider plastics with $E = 2.0 \times 10^9$ Pa and $v = 0.3$. Seventy-two quadratic boundary elements are used (see Figure 5.5.3). The three nodes at the (0, 20), (-20,0) and (0,-20) locations in the BEM model are fixed in the tangential direction.

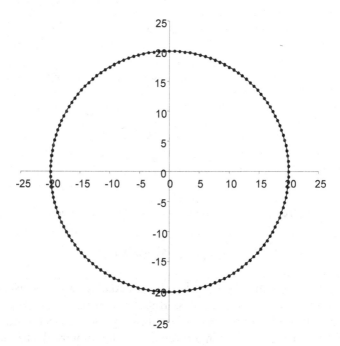

Figure 5.5.3. Discretization of the circular elastic body with 72 quadratic boundary elements (three nodes form one element).

The deformed shape of the body under the voltage is shown in Figure 5.5.4 for $\alpha = 30°$. The deformed shape is as expected under the applied loads and constraints. Conversely, if the body is deformed due to mechanical loads, a voltage will be produced across the electrodes of the film as a sensing signal.

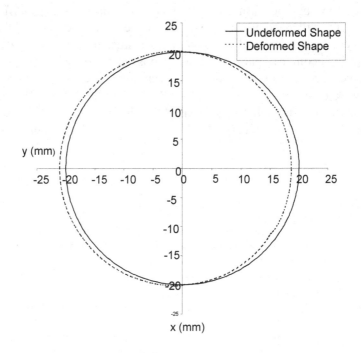

Figure 5.5.4. Deformed shape of the elastic body ($\alpha = 30°$).

5.5.2 An elastic shell on a piezoelectric body

Certain chemical and biological acoustic wave sensors detect a substance through the mass-frequency effect of the substance accumulated on the surface of a vibrating piezoelectric body. Next we analyze frequency shifts in a three-dimensional piezoelectric body due to a thin mass layer on its surface [58]. Consider a piezoelectric body with a thin elastic film of thickness $2h'$ and mass density ρ' on part of its surface (see Figure 5.5.5).

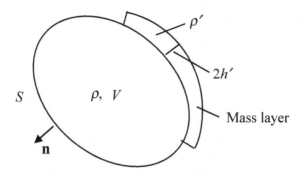

Figure 5.5.5. A piezoelectric body with a thin elastic layer on part of its surface.

5.5.2.1 Governing equations

The governing equations and boundary conditions of the body are

$$c_{jikl}u_{k,lj} + e_{kji}\phi_{,kj} = \rho\ddot{u}_i, \quad \text{in} \quad V,$$

$$-e_{ikl}u_{k,li} + \varepsilon_{ik}\phi_{,ki} = 0, \quad \text{in} \quad V,$$

$$u_i = 0, \quad \text{on} \quad S_u,$$

$$T_{ji}n_j = (c_{jikl}u_{k,l} + e_{kji}\phi_{,k})n_j = t_i, \quad \text{on} \quad S_T, \tag{5.5.13}$$

$$\phi = 0, \quad \text{on} \quad S_\phi,$$

$$D_i n_i = (e_{ikl}u_{k,l} - \varepsilon_{ik}\phi_{,k})n_i = 0, \quad \text{on} \quad S_D,$$

where t_i is due to the interaction between the body and the film. For the elastic layer, we use the two-dimensional equations of an elastic shell. For the lowest order effects of the mass and stiffness of the layer, the membrane theory is sufficient. The shell displacement vector is given by $u_\beta \cong u_\beta^{(0)}(\alpha_1, \alpha_2, t)$. $\beta = 1, 2,$ and 3 is associated with the shell principle coordinates $(\alpha_1, \alpha_2, \alpha_3)$. Then the shell membrane strains are

$$S_1^{(0)} = \frac{1}{A_1}\frac{\partial u_1^{(0)}}{\partial \alpha_1} + \frac{u_2^{(0)}}{A_1 A_2}\frac{\partial A_1}{\partial \alpha_2} + \frac{1}{R_1}u_3^{(0)},$$

$$S_2^{(0)} = \frac{1}{A_2}\frac{\partial u_2^{(0)}}{\partial \alpha_2} + \frac{u_1^{(0)}}{A_1 A_2}\frac{\partial A_2}{\partial \alpha_1} + \frac{1}{R_2}u_3^{(0)},\qquad(5.5.14)$$

$$S_6^{(0)} = \frac{A_2}{A_1}\frac{\partial}{\partial \alpha_1}(\frac{u_2^{(0)}}{A_2}) + \frac{A_1}{A_2}\frac{\partial}{\partial \alpha_2}(\frac{u_1^{(0)}}{A_1}).$$

The membrane stress resultants are given by the following constitutive relations:

$$N_1 = 2h'(\gamma_{11}S_1^{(0)} + \gamma_{12}S_2^{(0)}),$$

$$N_2 = 2h'(\gamma_{12}S_1^{(0)} + \gamma_{11}S_2^{(0)}),\qquad(5.5.15)$$

$$N_6 = 2h'\gamma_{66}S_6^{(0)}.$$

The membrane equations of motion are

$$\frac{\partial}{\partial \alpha_1}(A_2 N_1) + \frac{\partial}{\partial \alpha_2}(A_1 N_6)$$

$$+ N_6\frac{\partial A_1}{\partial \alpha_2} - N_2\frac{\partial A_2}{\partial \alpha_1} + A_1 A_2 f_1 = A_1 A_2 2\rho'h'\ddot{u}_1^{(0)},$$

$$\frac{\partial}{\partial \alpha_1}(A_2 N_6) + \frac{\partial}{\partial \alpha_2}(A_1 N_2)\qquad(5.5.16)$$

$$+ N_6\frac{\partial A_2}{\partial \alpha_1} - N_1\frac{\partial A_1}{\partial \alpha_2} + A_1 A_2 f_2 = A_1 A_2 2\rho'h'\ddot{u}_2^{(0)},$$

$$-\frac{1}{R_1}N_1 - \frac{2}{R_2}N_2 + f_3 = 2\rho'h'\ddot{u}_3^{(0)},$$

where f_i is the load per unit middle surface area of the shell. With successive substitutions, Equation (5.5.16) can be written in the following compact form

$$-L_\beta(\mathbf{u}^{(0)}) + f_\beta = 2\rho'h'\ddot{u}_\beta^{(0)},\qquad(5.5.17)$$

where L_β are linear differential operators. Multiplying Equation (5.5.17) by the transformation coefficients $\delta_{i\beta}$ between the local shell principal coordinates and the global Cartesian coordinates, we have

$$-L_i(\mathbf{u}^{(0)}) + f_i = \rho'h'\ddot{u}_i^{(0)}.\qquad(5.5.18)$$

We note that the t_i in Equation (5.5.13)$_4$ and the f_i in Equation (5.5.18) are actions and reactions, equal in magnitude and opposite in direction ($f_i = -t_i$). Substituting Equation (5.5.18) into Equation (5.5.13)$_4$, for time-harmonic motions with an exp($i\omega t$) time dependence, the eigenvalue problem for the free vibration of the body with the film can be written as

$$-c_{jikl}u_{k,lj} - e_{kji}\phi_{,kj} = \rho\lambda u_i, \quad \text{in} \quad V,$$

$$-e_{ikl}u_{k,li} + \varepsilon_{ik}\phi_{,ki} = 0, \quad \text{in} \quad V,$$

$$u_i = 0, \quad \text{on} \quad S_u,$$

$$T_{ji}n_j = (c_{jikl}u_{k,l} + e_{kji}\phi_{,k})n_j \qquad (5.5.19)$$

$$= \varepsilon[2\rho'h'\lambda u_i - L_i(\mathbf{u})], \quad \text{on} \quad S_T,$$

$$\phi = 0, \quad \text{on} \quad S_\phi,$$

$$D_i n_i = (e_{ikl}u_{k,l} - \varepsilon_{ik}\phi_{,k})n_i = 0, \quad \text{on} \quad S_D,$$

where we have denoted $\lambda = \omega^2$. ε is a small parameter introduced to formally show the smallness of the effect of the film. The real physical problem corresponds to $\varepsilon = 1$. Because of the continuity between the layer and the body, the displacement of the layer $\mathbf{u}^{(0)}$ is the same as the displacement $\mathbf{u}$ of the piezoelectric body at its surface S. Equation (5.5.19) can be written in a more compact form as

$$\mathbf{AU} = \lambda\mathbf{BU}, \quad \text{in} \quad V,$$

$$u_i = 0, \quad \text{on} \quad S_u,$$

$$T_{ji}(\mathbf{U})n_j = \varepsilon[2\rho'h'\lambda u_i - L_i(\mathbf{u})],, \quad \text{on} \quad S_T, \qquad (5.5.20)$$

$$\phi = 0, \quad \text{on} \quad S_\phi,$$

$$D_i(\mathbf{U})n_i = 0, \quad \text{on} \quad S_D,$$

where $\mathbf{U} = \{u_k, \phi\}$ is a 4-vector. The differential operators $\mathbf{A}$ and $\mathbf{B}$ are defined by

$$\mathbf{AU} = \{-c_{jikl}u_{k,lj} - e_{kji}\phi_{,kj}, -e_{ikl}u_{k,li} + \varepsilon_{ik}\phi_{,ki}\},$$

$$\mathbf{BU} = \{\rho u_i, 0\}. \qquad (5.5.21)$$

$T_{ji}(\mathbf{U})$ and $D_i(\mathbf{U})$ are the stress tensor and electric displacement vector in terms of the 4-vector.

5.5.2.2 Perturbation analysis

We look for a perturbation solution to Equation (5.5.20). Consider the following expansions:

$$\lambda \cong \lambda^{(0)} + \varepsilon \, \lambda^{(1)},$$

$$\mathbf{U} = \begin{Bmatrix} u_i \\ \phi \end{Bmatrix} \cong \begin{Bmatrix} u_i^{(0)} \\ \phi^{(0)} \end{Bmatrix} + \varepsilon \begin{Bmatrix} u_i^{(1)} \\ \phi^{(1)} \end{Bmatrix} = \mathbf{U}^{(0)} + \varepsilon \mathbf{U}^{(1)}, \qquad (5.5.22)$$

where $\omega^{(0)}$, $\mathbf{u}^{(0)}$ and $\phi^{(0)}$ are the frequency and modes when the elastic layer is not present. Beginning from Equation (5.5.22), for the rest of this section, the superscripts will be for perturbation orders, not for orders of shell displacements. Substituting Equation (5.5.22) into Equation (5.5.20), collecting terms of equal powers of ε, we obtain a series of perturbation problems of successive orders. We are interested in the lowest order effect of the layer. Therefore we collect coefficients of terms with powers of ε^0 and ε^1 only. The zero-order problem is

$$-c_{jikl}u_{k,lj}^{(0)} - e_{kji}\phi_{,kj}^{(0)} = \rho\lambda^{(0)}u_i^{(0)}, \quad \text{in} \quad V,$$

$$-e_{ikl}u_{k,li}^{(0)} + \varepsilon_{ik}\phi_{,ki}^{(0)} = 0, \quad \text{in} \quad V,$$

$$u_i^{(0)} = 0, \quad \text{on} \quad S_u,$$

$$(c_{jikl}u_{k,l}^{(0)} + e_{kji}\phi_{,k}^{(0)})n_j = 0, \quad \text{on} \quad S_T, \qquad (5.5.23)$$

$$\phi^{(0)} = 0, \quad \text{on} \quad S_\phi,$$

$$(e_{ikl}u_{k,l}^{(0)} - \varepsilon_{ik}\phi_{,k}^{(0)})n_i = 0, \quad \text{on} \quad S_D.$$

This represents free vibrations of the body without the surface mass layer. The solution to the zero-order problem, $\lambda^{(0)}$ and $\mathbf{U}^{(0)}$, is assumed known as usual in a perturbation analysis. The first-order problem below is to be solved:

$$-c_{jikl}u_{k,lj}^{(1)} - e_{kji}\phi_{,kj}^{(1)} = \rho\lambda^{(1)}u_i^{(0)} + \rho\lambda^{(0)}u_i^{(1)}, \quad \text{in} \quad V,$$

$$-e_{ikl}u_{k,li}^{(1)} + \varepsilon_{ik}\phi_{,ki}^{(1)} = 0, \quad \text{in} \quad V,$$

$$u_i^{(1)} = 0, \quad \text{on} \quad S_u,$$

$$(c_{jikl}u_{k,l}^{(1)} + e_{kji}\phi_{,k}^{(1)})n_j = [2\rho'h'\lambda^{(0)}u_i^{(0)} - L_i(\mathbf{u}^{(0)})], \quad \text{on} \quad S_T, \qquad (5.5.24)$$

$$\phi^{(1)} = 0, \quad \text{on} \quad S_\phi,$$

$$(e_{ikl}u_{k,l}^{(1)} - \varepsilon_{ik}\phi_{,k}^{(1)})n_i = 0, \quad \text{on} \quad S_D.$$

The differential equations for the first-order problem, Equations $(5.5.24)_{1,2}$, can be written as

$$\mathbf{AU}^{(1)} = \lambda^{(0)}\mathbf{BU}^{(1)} + \lambda^{(1)}\mathbf{BU}^{(0)}. \qquad (5.5.25)$$

Multiplying both sides of Equation (5.5.25) by $\mathbf{U}^{(0)}$ and integrating the resulting equation over V, we have

$$< \mathbf{AU}^{(1)}; \mathbf{U}^{(0)} >$$

$$= \int_V [(-c_{jikl}u_{k,lj}^{(1)} - e_{kji}\phi_{,kj}^{(1)})u_i^{(0)} + (-e_{ikl}u_{k,li}^{(1)} + \varepsilon_{ik}\phi_{,ki}^{(1)})\phi^{(0)}]dV \qquad (5.5.26)$$

$$= \lambda^{(0)} < \mathbf{BU}^{(1)}; \mathbf{U}^{(0)} > + \lambda^{(1)} < \mathbf{BU}^{(0)}; \mathbf{U}^{(0)} >,$$

where, for simplicity, we have used < ; > to represent the product of two 4-vectors and the integration over V. With integration by parts, it can be shown that

$$< \mathbf{AU}^{(0)}; \mathbf{U}^{(1)} >$$

$$= -\int_S [T_{ji}(\mathbf{U}^{(0)})n_j u_i^{(1)} + D_i(\mathbf{U}^{(0)})n_i\phi^{(1)}]dS$$

$$+ \int_S [T_{kl}(\mathbf{U}^{(1)})n_l u_k^{(0)} + D_k(\mathbf{U}^{(1)})n_k\phi^{(0)}]dS \qquad (5.5.27)$$

$$+ < \mathbf{U}^{(0)}; \mathbf{AU}^{(1)} >.$$

With the boundary conditions in Equations $(5.5.23)_{3\text{-}6}$ and $(5.5.24)_{3\text{-}6}$, Equation (5.5.27) becomes

$$< \mathbf{AU}^{(0)}; \mathbf{U}^{(1)} >$$

$$= \int_{S_T} [2\rho'h'\lambda^{(0)}u_k^{(0)} - L_k(\mathbf{u}^{(0)})]u_k^{(0)}dS + < \mathbf{U}^{(0)}; \mathbf{AU}^{(1)} >. \qquad (5.5.28)$$

Substituting Equation (5.5.28) into Equation (5.5.26),

$$
\begin{aligned}
&< \mathbf{A}\mathbf{U}^{(0)}; \mathbf{U}^{(1)} > - \int_{S_T} [2\rho' h' \lambda^{(0)} u_k^{(0)} - L_k(\mathbf{u}^{(0)})] u_k^{(0)} dS \\
&= \lambda^{(0)} < \mathbf{B}\mathbf{U}^{(1)}; \mathbf{U}^{(0)} > + \lambda^{(1)} < \mathbf{B}\mathbf{U}^{(0)}; \mathbf{U}^{(0)} >,
\end{aligned}
\tag{5.5.29}
$$

which can be further written as

$$
\begin{aligned}
&< \mathbf{A}\mathbf{U}^{(0)} - \lambda^{(0)} \mathbf{B}\mathbf{U}^{(0)}; \mathbf{U}^{(1)} > \\
&\quad - \int_{S_T} [2\rho' h' \lambda^{(0)} u_k^{(0)} - L_k(\mathbf{u}^{(0)})] u_k^{(0)} dS \\
&= \lambda^{(1)} < \mathbf{B}\mathbf{U}^{(0)}; \mathbf{U}^{(0)} >.
\end{aligned}
\tag{5.5.30}
$$

With Equations $(5.5.23)_{1,2}$, we obtain, from Equation (5.5.30)

$$
\begin{aligned}
\lambda^{(1)} &= - \frac{\int_{S_T} [2\rho' h' \lambda^{(0)} u_k^{(0)} - L_k(\mathbf{u}^{(0)})] u_k^{(0)} dS}{< \mathbf{B}\mathbf{U}^{(0)}; \mathbf{U}^{(0)} >} \\
&= - \frac{\int_{S_T} [2\rho' h' \lambda^{(0)} u_k^{(0)} - L_k(\mathbf{u}^{(0)})] u_k^{(0)} dS}{\int_V \rho u_i^{(0)} u_i^{(0)} dV}.
\end{aligned}
\tag{5.5.31}
$$

The above expression is for the eigenvalue $\lambda = \omega^2$. For ω we make the following expansion:

$$
\omega \cong \omega^{(0)} + \varepsilon \omega^{(1)}.
\tag{5.5.32}
$$

Then

$$
\begin{aligned}
\lambda = \omega^2 &\cong (\omega^{(0)} + \varepsilon \omega^{(1)})^2 \\
&\cong (\omega^{(0)})^2 + 2\varepsilon \omega^{(0)} \omega^{(1)} \cong \lambda^{(0)} + \varepsilon \lambda^{(1)}.
\end{aligned}
\tag{5.5.33}
$$

Hence

$$
\begin{aligned}
\frac{\varepsilon \omega^{(1)}}{\omega^{(0)}} &\cong \frac{1}{2(\omega^{(0)})^2} \varepsilon \lambda^{(1)} \\
&= - \frac{1}{2(\omega^{(0)})^2} \varepsilon \frac{\int_{S_T} [2\rho' h' \lambda^{(0)} u_k^{(0)} - L_k(\mathbf{u}^{(0)})] u_k^{(0)} dS}{\int_V \rho u_i^{(0)} u_i^{(0)} dV}.
\end{aligned}
\tag{5.5.34}
$$

Finally, setting $\varepsilon = 1$ in Equation (5.5.34), we obtain

$$\frac{\omega - \omega^{(0)}}{\omega^{(0)}} \cong \frac{1}{2(\omega^{(0)})^2} \times \frac{\int_{S_T} [L_k(\mathbf{u}^{(0)})u_k^{(0)} - 2\rho'h'(\omega^{(0)})^2 u_k^{(0)}u_k^{(0)}]dS}{\int_V \rho u_i^{(0)}u_i^{(0)}dV}.$$

(5.5.35)

The stiffness effect of the film is in L_k. When $L_k = 0$, Equation (5.5.35) reduces to

$$\frac{\omega - \omega^{(0)}}{\omega^{(0)}} \cong -\frac{\int_{S_T} \rho'h'u_k^{(0)}u_k^{(0)}dS}{\int_V \rho u_i^{(0)}u_i^{(0)}dV},$$

(5.5.36)

which is due to the inertial effect of the film mass only, and is the result of [59].

5.5.2.3 An example

As an example consider the radial vibration of a thin elastic ring (see Figure 5.5.6) of a mean radius R, thickness $2h$, Young's modulus E,

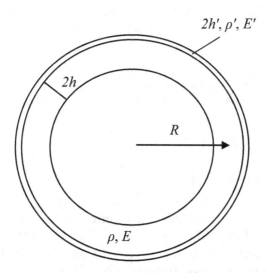

Figure 5.5.6. A thin elastic ring with a mass layer.

Poisson's ratio v, and mass density ρ. The mass layer has a thickness $2h'$, Young's modulus E' and mass density ρ'.

A ring can be considered as a circular cylindrical shell with the axial extensional force $N_z = 0$. When the mass layer is not present, in cylindrical coordinates, the lowest radial mode is given by ([9] or Equation (7.3.6) of the present book)

$$(\omega^{(0)})^2 = \frac{E}{R^2 \rho},$$

$$u_r^{(0)} = C, \quad u_\theta^{(0)} = 0, \quad u_z^{(0)} = -v\frac{C}{R}z \cong 0,$$

(5.5.37)

where C is an arbitrary constant. $u_z^{(0)}$ is due to Poisson's effect, which is relatively small and is neglected as an approximation. For the layer, the equation of motion in the radial direction is

$$-\frac{2h'E'}{R^2}u_r + f_r = 2\rho'h'\ddot{u}_r.$$

(5.5.38)

Hence

$$L_r(\mathbf{u}) = \frac{2h'E'}{R^2}u_r, \quad L_\theta(\mathbf{u}) \cong 0, \quad L_z(\mathbf{u}) \cong 0.$$

(5.5.39)

Substituting Equations (5.5.37) and (5.5.39) into the perturbation integral in Equation (5.5.35), we obtain the frequency shift due to the mass layer as

$$\frac{\omega - \omega^{(0)}}{\omega^{(0)}} \cong \frac{1}{2(\omega^{(0)})^2}\frac{1}{2\rho h}\left[\frac{2h'E'}{R^2} - 2\rho'h'(\omega^{(0)})^2\right]$$

$$= \frac{1}{2(\omega^{(0)})^2}\frac{1}{\rho h}\left(\frac{h'E'}{R^2} - \rho'h'\frac{E}{R^2\rho}\right)$$

$$= \frac{1}{2}\frac{h'}{h}\left(\frac{E'}{E} - \frac{\rho'}{\rho}\right).$$

(5.5.40)

Equation (5.5.40) shows that the inertial effect of the mass layer lowers the frequency, and the stiffness of the mass layer raises the frequency as expected. The result of Equation (5.5.40) and the result from [59] considering the inertial effect of mass only (which is the special case of

Equation (5.5.40) when $E' = 0$) are plotted in Figure 5.5.7 using $E = s_{11}$ and ρ from quartz. For E' and ρ' we consider two cases of gold and aluminum. It is seen that for a heavy mass layer of gold the inertial effect of the mass layer dominates. However, for a light mass layer of aluminum the stiffness effect is also important.

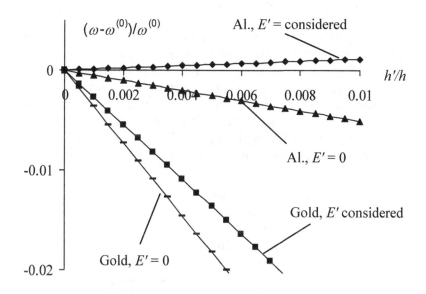

Figure 5.5.7. Effects of film mass and stiffness on frequency shifts.

From the one-dimensional equation for a composite ring, it can be shown that

$$\omega^2 = \frac{Eh + E'h'}{R^2(\rho h + \rho'h')}. \tag{5.5.41}$$

Compared to Equation (4.8.9) of [9] or Equation (7.3.6) of the present book, we see that Equation (5.5.41) represents the frequency of a ring with a mass density and a Young's modulus averaged from those of the two phases according to the volume fraction. It can be easily verified that for a thin mass layer, Equation (5.5.41) reduces to Equation (5.5.40).

Finally, we point out that the effect of a mass layer on frequency can also be studied from the variational (Rayleigh quotient) formulation of the eigenvalue problem for free vibrations of a piezoelectric body with a mass layer. The variational analysis is somewhat simpler than the above perturbation analysis using the differential operators of the shell equations.

5.5.3 An elastic shell in a viscous fluid

A crystal resonator when put in contact with a fluid changes its resonant frequencies. This effect has been used to make various fluid sensors for measuring fluid density or viscosity. Torsional modes of a circular cylindrical cylinder or shell have particles moving tangentially at the surface and are ideal for liquid sensing. We now study the propagation of torsional waves in a circular cylindrical elastic shell in contact with a viscous fluid [60] (see Figure 5.5.8). Two cases will be considered: a shell filled with a fluid and a shell immersed in a fluid.

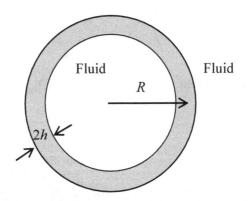

Figure 5.5.8. An elastic shell in a viscous fluid.

5.5.3.1 Governing equations

Consider the following velocity field of the azimuthal motion of a linear, viscous fluid in cylindrical coordinates:

$$v_r = v_z = 0, \quad v_\theta = v(r,z,t). \tag{5.5.42}$$

The nontrivial stress components are [61]

$$T_{r\theta} = \mu r \frac{\partial}{\partial r}\left(\frac{v}{r}\right), \quad T_{rz} = \mu \frac{\partial v}{\partial z}, \quad (5.5.43)$$

where μ is the viscosity of the fluid. The relevant equation of motion is [57]

$$\rho' \frac{\partial v}{\partial t} = \mu\left(\frac{\partial^2 v}{\partial r^2} + \frac{1}{r}\frac{\partial v}{\partial r} - \frac{v}{r^2}\right) + \mu \frac{\partial^2 v}{\partial z^2}, \quad (5.5.44)$$

where ρ' is the fluid density. We look for a wave solution in the form

$$v = V(r)\sin kz \exp(i\omega t), \quad (5.5.45)$$

where k is the wave number and ω is the frequency. Substitution of Equation (5.5.45) into Equation (5.5.44) results in the following equation for V:

$$\frac{\partial^2 V}{\partial \xi^2} + \frac{1}{\xi}\frac{\partial V}{\partial \xi} - \left(1 + \frac{1}{\xi^2}\right)V = 0, \quad (5.5.46)$$

where

$$\xi = \lambda r, \quad \lambda^2 = k^2 + i\omega\frac{\rho'}{\mu}. \quad (5.5.47)$$

Equation (5.5.46) is the modified Bessel equation of order one. Its general solution is a linear combination of $I_1(\lambda r)$ and $K_1(\lambda r)$, the first order modified Bessel functions of the first and second kind. Hence

$$v = [C_1 I_1(\lambda r) + C_2 K_1(\lambda r)]\sin kz \exp(i\omega t), \quad (5.5.48)$$

where C_1 and C_2 are undetermined constants.

Consider torsional motions of a circular cylindrical shell with thickness $2h$ and middle surface radius R. The membrane theory is sufficient. Let the middle surface displacement field be

$$\begin{aligned}
u_1^{(0)} &= u_\theta^{(0)} = u(z,t), \\
u_2^{(0)} &= u_z^{(0)} = 0, \\
u_3^{(0)} &= u_r^{(0)} = 0.
\end{aligned} \quad (5.5.49)$$

From Equation (5.3.1)$_1$ the equation that governs u is

$$\rho 2h\ddot{u} = G2h\frac{\partial^2 u}{\partial z^2} + T_{r\theta}(R^+) - T_{r\theta}(R^-), \qquad (5.5.50)$$

where $\tilde{n}$ is the mass density and G is the shear modulus of the shell. $T_{r\theta}(R^+)$ and $T_{r\theta}(R^-)$ are the shear stress components at the outer and inner surfaces of the shell. They are due to the interaction with the fluid. Substituting Equation (5.5.43) into Equation (5.5.50) we have

$$\rho 2h\ddot{u} = G2h\frac{\partial^2 u}{\partial z^2} + \left[\mu r\frac{\partial}{\partial r}\left(\frac{v}{r}\right)\right]_{R^+} - \left[\mu r\frac{\partial}{\partial r}\left(\frac{v}{r}\right)\right]_{R^-}. \qquad (5.5.51)$$

Differentiating Equation (5.5.51) with respect to time once, using the continuity of velocity (non-slip condition) between the shell and the fluid, we obtain

$$\rho 2h\ddot{v}|_R = G2h\frac{\partial^2 v}{\partial z^2}\bigg|_R + \left[\mu r\frac{\partial}{\partial r}\left(\frac{\dot{v}}{r}\right)\right]_{R^+} - \left[\mu r\frac{\partial}{\partial r}\left(\frac{\dot{v}}{r}\right)\right]_{R^-}. \qquad (5.5.52)$$

5.5.3.2 Interior problem

First consider a circular cylindrical shell filled with a fluid (interior problem). In this case, for the fluid region, since $K_1(0)$ is unbounded, we must have $C_2 = 0$ in Equation (5.5.48). Substituting the remaining C_1 term into Equation (5.5.52), we obtain

$$\omega^2 = c_T^2 k^2 + i\frac{\mu\omega}{\rho 2hR}\left[\frac{\lambda R I_0(\lambda R)}{I_1(\lambda R)} - 2\right], \qquad (5.5.53)$$

where $c_T^2 = G/\rho$ is the torsional wave speed of the elastic shell when the fluid is not present. Equation (5.5.53) shows the effect of the fluid on the wave behavior, and hence the potential of measuring fluid properties through the wave frequency or speed. In the special case of a light fluid with a low density, from Equation (5.5.47) we have

$$\lambda \cong k + \frac{i\omega\rho'}{2\mu k}. \qquad (5.5.54)$$

Then

$$I_0(\lambda R) \cong I_0(kR) + I_0'(kR)\frac{i\omega\rho' R}{2\mu k},$$

$$I_1(\lambda R) \cong I_1(kR) + I_1'(kR)\frac{i\omega\rho' R}{2\mu k}. \tag{5.5.55}$$

Substituting Equations (5.5.54) and (5.5.55) into Equation (5.5.53), to the lowest order effect of the fluid mass density, we obtain

$$\omega^2\left(1 + \frac{\beta}{k2h}\frac{\rho'}{\rho}\right) = c_T^2 k^2 + i\alpha\frac{\mu\omega}{\rho 2hR}, \tag{5.5.56}$$

where

$$\alpha = \frac{kRI_0(kR)}{I_1(kR)} - 2,$$

$$\beta = \frac{kR[I_1^2(kR) - I_0^2(kR)] + 2I_0(kR)I_1(kR)}{2I_1^2(kR)}. \tag{5.5.57}$$

In Equation (5.5.56) the effects of the fluid mass and viscosity become explicit. If we denote

$$\omega = \omega_0 + \Delta\omega, \quad \omega_0 = c_T k, \tag{5.5.58}$$

where ω_0 is the frequency when the fluid is not present, from Equation (5.5.56) we obtain

$$\frac{\Delta\omega}{\omega_0} \cong -\frac{\beta}{4kh}\frac{\rho'}{\rho} + i\frac{\alpha\mu}{4\omega_0\rho hR}. \tag{5.5.59}$$

Equation (5.5.59) shows that ρ' causes frequency shifts and μ causes dissipation, as expected.

5.5.3.3 Exterior problem

Next consider a circular cylindrical shell immersed in an unbounded fluid (exterior problem). In this case, since $I_1(\infty)$ is unbounded, we must have $C_1 = 0$ in Equation (5.5.48). Substituting the remaining C_2 term into Equation (5.5.52), we obtain

$$\omega^2 = c_T^2 k^2 + i \frac{\mu\omega}{\rho 2hR} \left[\frac{\lambda R K_0(\lambda R)}{K_1(\lambda R)} + 2 \right]. \tag{5.5.60}$$

Approximations similar to Equations (5.5.54) through (5.5.59) can also be made to Equation (5.5.60).

Chapter 6
Piezoelectric Beams

In this chapter equations for the extension and flexure with shear deformations of electroelastic beams with a rectangular cross section are derived from the three-dimensional equations by double power series expansions in the thickness and width directions. The theory for coupled flexure and thickness-shear of an elastic beam was developed by Timoshenko before the corresponding plate theory. The study of high frequency extensional vibrations of elastic beams began with [62,63]. Similar to the situation for extensional vibrations of plates [33,34], the dispersion relations of the equations in [62] lack a complex branch, and [63] removes this deficiency. A high frequency theory for torsion is given in [64], which takes into consideration the deformation of a cross section. Power series expansions were used to obtain one-dimensional equations for elastic [65] and piezoelectric beams [66,24]. The material in this chapter is from [24,67-69]. Beams under biasing fields can be found in [70].

6.1 Power Series Expansion

Consider a piezoelectric beam with a rectangular cross section as shown in Figure 6.1.1. It is assumed that the beam has a slender shape with $a \gg b, c$. The coordinate system consists of the centroidal principal axes.

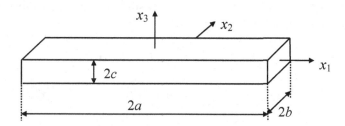

Figure 6.1.1. A piezoelectric beam with a rectangular cross section.

To develop a one-dimensional theory, we make the following expansions of the mechanical displacement vector and the electrostatic potential:

$$u_j = \sum_{m,n=0}^{\infty} x_2^m x_3^n u_j^{(m,n)}(x_1,t), \quad \phi = \sum_{m,n=0}^{\infty} x_2^m x_3^n \phi^{(m,n)}(x_1,t). \quad (6.1.1)$$

Then the strains and the electric field can be written as

$$S_{ij} = \sum_{m,n=0}^{\infty} x_2^m x_3^n S_{ij}^{(m,n)}, \quad E_i = \sum_{m,n=0}^{\infty} x_2^m x_3^n E_i^{(m,n)}, \quad (6.1.2)$$

where the strains and the electric field of various orders are as follows:

$$S_{ij}^{(m,n)} = \frac{1}{2}[u_{j,i}^{(m,n)} + u_{i,j}^{(m,n)} + (m+1)(\delta_{i2}u_j^{(m+1,n)} + \delta_{j2}u_i^{(m+1,n)})$$

$$+ (n+1)(\delta_{i3}u_j^{(m,n+1)} + \delta_{j3}u_i^{(m,n+1)})], \quad (6.1.3)$$

$$E_i^{(m,n)} = -\phi_{,i}^{(m,n)} - \delta_{i2}(m+1)\phi^{(m+1,n)} - \delta_{i3}(n+1)\phi^{(m,n+1)}.$$

The one-dimensional equations of motion and electrostatics are obtained from the variational principle in Equation (1.2.26)

$$T_{1j,1}^{(m,n)} - mT_{2j}^{(m-1,n)} - nT_{3j}^{(m,n-1)} + F_j^{(m,n)} = \rho \sum_{r,s=0}^{\infty} B_{mnrs}\ddot{u}_j^{(r,s)},$$

$$D_{1,1}^{(m,n)} - mD_2^{(m-1,n)} - nD_3^{(m,n-1)} + D^{(m,n)} = 0, \quad (6.1.4)$$

where the stress and the electric displacement of various orders are defined by

$$T_{ij}^{(m,n)} = \int_A T_{ij} x_2^m x_3^n dA,$$

$$D_i^{(m,n)} = \int_A D_i x_2^m x_3^n dA. \quad (6.1.5)$$

$A = 4bc$ is the cross sectional area. The surface traction, surface charge and body force of various orders are

$$F_j^{(m,n)} = b^m \int_{-c}^{c} [T_{2j}(b) - (-1)^m T_{2j}(-b)]x_3^n dx_3$$

$$+ c^n \int_{-b}^{b} [T_{3j}(c) - (-1)^n T_{3j}(-c)]x_2^m dx_2 + \int_A \rho f_j x_2^m x_3^n dA \; ,$$

$$D^{(m,n)} = b^m \int_{-c}^{c} [D_2(b) - (-1)^m D_2(-b)] x_3^n dx_3$$

$$+ c^n \int_{-b}^{b} [D_3(c) - (-1)^n D_3(-c)] x_2^m dx_2 . \tag{6.1.6}$$

B_{mnrs} is a geometric quantity given by

$$B_{mnrs} = \int_A x_2^{m+r} x_3^{n+s} dA$$

$$= \begin{cases} \dfrac{4b^{r+m+1} c^{s+n+1}}{(r+m+1)(s+n+1)}, & m+r, \quad n+s \quad \text{even,} \\ 0, & \text{else.} \end{cases} \tag{6.1.7}$$

The one-dimensional constitutive relations take the following form:

$$T_{ij}^{(m,n)} = \sum_{r,s=0}^{\infty} B_{mnrs} (c_{ijkl} S_{kl}^{(r,s)} - e_{kij} E_k^{(r,s)}),$$

$$D_i^{(m,n)} = \sum_{r,s=0}^{\infty} B_{mnrs} (e_{ikl} S_{kl}^{(r,s)} + \varepsilon_{ik} E_k^{(r,s)}). \tag{6.1.8}$$

Or, with the compressed matrix notation, for $p, q = 1-6$

$$T_p^{(m,n)} = \sum_{r,s=0}^{\infty} B_{mnrs} (c_{pq} S_q^{(r,s)} - e_{ip} E_i^{(r,s)}),$$

$$D_i^{(m,n)} = \sum_{r,s=0}^{\infty} B_{mnrs} (e_{ip} S_p^{(r,s)} + \varepsilon_{ik} E_k^{(r,s)}). \tag{6.1.9}$$

The above constitutive relations are not ready to be used. Important adjustments are necessary which will be performed later. Since piezoelectric materials are anisotropic, couplings among extension, flexure and torsion may occur due to anisotropy. For anisotropic beams, although various types of deformations may couple, usually there still exist modes which are essentially extensional or essentially flexural, which are the modes we are interested in. The essentially torsional modes will not be treated in this book.

6.2 Zero-Order Theory for Extension

First we extract a zero-order theory for extension from the general expansions.

6.2.1 Equations for zero-order theory

For a zero-order theory we make the truncation

$$u_j \cong u_j^{(0,0)}(x_1,t) + x_2 u_j^{(1,0)}(x_1,t) + x_3 u_j^{(0,1)}(x_1,t),$$

$$\phi \cong \phi^{(0,0)}(x_1,t) + x_2 \phi^{(1,0)}(x_1,t) + x_3 \phi^{(0,1)}(x_1,t), \tag{6.2.1}$$

where $u_2^{(1,0)}$ and $u_3^{(0,1)}$ are needed to describe Poisson's effects during extension. They cannot be directly set to zero, but will be eliminated later by stress relaxations. An equation for the extensional displacement $u_1^{(0,0)}$ will be obtained. From Equations (6.1.3) and (6.2.1), the zero-order strains and the electric fields to be considered are

$$S_{ij} \cong S_{ij}^{(0,0)},$$

$$E_k \cong E_k^{(0,0)} + x_2 E_k^{(1,0)} + x_3 E_k^{(0,1)}, \tag{6.2.2}$$

where

$$S_1^{(0,0)} = u_{1,1}^{(0,0)}, \quad S_{22}^{(0)} = u_2^{(1,0)}, \quad S_{33}^{(0)} = u_3^{(0,1)},$$

$$S_4^{(0,0)} = 2S_{23}^{(0,0)} = u_2^{(0,1)} + u_3^{(1,0)},$$

$$S_5^{(0,0)} = 2S_{31}^{(0,0)} = u_{3,1}^{(0,0)} + u_1^{(0,1)}, \tag{6.2.3}$$

$$S_6^{(0,0)} = 2S_{12}^{(0,0)} = u_{2,1}^{(0,0)} + u_1^{(1,0)},$$

$$E_1^{(0,0)} = -\phi_{,1}^{(0,0)}, \quad E_2^{(0,0)} = -\phi^{(1,0)}, \quad E_3^{(0,0)} = -\phi^{(0,1)},$$

$$E_1^{(1,0)} = -\phi_{,1}^{(1,0)}, \quad E_2^{(1,0)} = -2\phi^{(2,0)} = 0, \quad E_3^{(1,0)} = -\phi^{(1,1)} = 0, \tag{6.2.4}$$

$$E_1^{(0,1)} = -\phi_{,1}^{(0,1)}, \quad E_2^{(0,1)} = -\phi^{(1,1)} = 0, \quad E_3^{(0,1)} = -2\phi^{(0,2)} = 0.$$

From Equation (6.1.4) the equation for extension and electrostatics are

$$T_{11,1}^{(0,0)} + F_1^{(0,0)} = 4\rho b c \ddot{u}_1^{(0,0)}, \tag{6.2.5}$$

$$D_{1,1}^{(0,0)} + D^{(0,0)} = 0,$$

$$D_{1,1}^{(1,0)} - D_2^{(0,0)} + D^{(1,0)} = 0,$$

$$D_{1,1}^{(0,1)} - D_3^{(0,0)} + D^{(0,1)} = 0. \tag{6.2.6}$$

If there are electrodes present, some of the equations in Equation (6.2.6) need to be dropped or adjusted, as in the third section of the second chapter. The zero-order constitutive relations are obtained from Equation (6.1.9) by setting $m = n = 0$ and $r = s = 0$

$$T_p^{(0,0)} = 4bc(c_{pq}S_q^{(0,0)} - e_{ip}E_i^{(0,0)}),$$
$$D_i^{(0,0)} = 4bc(e_{iq}S_q^{(0,0)} + \varepsilon_{ik}^S E_k^{(0,0)}).$$
$$(6.2.7)$$

Equation (6.2.7) should be adjusted by setting the following zero-order stresses to zero

$$T_2^{(0,0)} = T_3^{(0,0)} = T_4^{(0,0)} = T_5^{(0,0)} = T_6^{(0,0)} = 0. \qquad (6.2.8)$$

In this case it is more convenient to re-derive the beam constitutive relations from the following three-dimensional constitutive relations:

$$S_p = s_{pq}T_q + d_{kp}E_k,$$
$$D_i = d_{iq}T_q + \varepsilon_{ij}^T E_j. \qquad (6.2.9)$$

Within the approximation we are interested in, from Equation (6.2.2)

$$S_p^{(0,0)} = s_{pq}T_q + d_{kp}(E_k^{(0,0)} + x_2 E_k^{(1,0)} + x_3 E_k^{(0,1)}),$$
$$D_i = d_{iq}T_q + \varepsilon_{ij}^T (E_j^{(0,0)} + x_2 E_j^{(1,0)} + x_3 E_j^{(0,1)}). \qquad (6.2.10)$$

Integrating Equation (6.2.10) over the cross section of the beam, we obtain

$$4bcS_p^{(0,0)} = s_{pq}T_q^{(0,0)} + 4bc\,d_{kp}E_k^{(0,0)}$$
$$= s_{p1}T_1^{(0,0)} + 4bc\,d_{kp}E_k^{(0,0)},$$
$$D_i^{(0,0)} = d_{iq}T_q^{(0,0)} + 4bc\varepsilon_{ij}^T E_j^{(0,0)}$$
$$= d_{i1}T_1^{(0,0)} + 4bc\varepsilon_{ij}^T E_j^{(0,0)}. \qquad (6.2.11)$$

From Equation (6.2.11)$_1$, for $p = 1$

$$4bcS_1^{(0,0)} = s_{11}T_1^{(0,0)} + 4bc\,d_{k1}E_k^{(0,0)}, \qquad (6.2.12)$$

which can be inverted to give the mechanical constitutive relation for extension

$$T_1^{(0,0)} = \frac{4bc}{s_{11}}(S_1^{(0,0)} - d_{k1}E_k^{(0,0)})$$
$$= 4bc(\tilde{c}_{11}S_1^{(0,0)} - \tilde{e}_{k1}E_k^{(0,0)}), \qquad (6.2.13)$$

where

$$\tilde{c}_{11} = \frac{1}{s_{11}}, \quad \tilde{e}_{k1} = \frac{d_{k1}}{s_{11}}. \tag{6.2.14}$$

Substitution of Equation (6.2.13) into Equation (6.2.11)$_2$ results in the following electric constitutive relations:

$$\begin{aligned}
D_i^{(0,0)} &= d_{i1}T_1^{(0,0)} + 4bc\varepsilon_{ij}^T E_j^{(0,0)} \\
&= d_{i1}\frac{4bc}{s_{11}}(S_1^{(0,0)} - d_{k1}E_k^{(0,0)}) + 4bc\varepsilon_{ij}^T E_j^{(0,0)} \\
&= 4bc\frac{d_{i1}}{s_{11}}S_1^{(0,0)} + 4bc(\varepsilon_{ik}^T - \frac{d_{i1}d_{k1}}{s_{11}})E_k^{(0,0)} \\
&= 4bc(\tilde{e}_{i1}S_1^{(0,0)} + \tilde{\varepsilon}_{ik}E_k^{(0,0)}),
\end{aligned} \tag{6.2.15}$$

where

$$\tilde{\varepsilon}_{ik} = \varepsilon_{ik}^T - \frac{d_{i1}d_{k1}}{s_{11}}. \tag{6.2.16}$$

The first-order constitutive relations needed are approximated by

$$D_i^{(1,0)} = \frac{4b^3c}{3}\varepsilon_{ik}^S E_k^{(1,0)}, \quad D_i^{(0,1)} = \frac{4bc^3}{3}\varepsilon_{ik}^S E_k^{(0,1)}. \tag{6.2.17}$$

In summary, we have

$$T_{11,1}^{(0,0)} + F_1^{(0,0)} = 4\rho bc\ddot{u}_1^{(0,0)}, \tag{6.2.18}$$

$$\begin{aligned}
D_{1,1}^{(0,0)} + D^{(0,0)} &= 0, \\
D_{1,1}^{(1,0)} - D_2^{(0,0)} + D^{(1,0)} &= 0, \\
D_{1,1}^{(0,1)} - D_3^{(0,0)} + D^{(0,1)} &= 0,
\end{aligned} \tag{6.2.19}$$

$$T_1^{(0,0)} = 4bc(\tilde{c}_{11}S_1^{(0,0)} - \tilde{e}_{k1}E_k^{(0,0)}), \tag{6.2.20}$$

$$D_i^{(0,0)} = 4bc(\tilde{e}_{i1}S_1^{(0,0)} + \tilde{\varepsilon}_{ik}E_k^{(0,0)}), \tag{6.2.21}$$

$$D_i^{(1,0)} = \frac{4b^3c}{3}\varepsilon_{ik}^S E_k^{(1,0)}, \quad D_i^{(0,1)} = \frac{4bc^3}{3}\varepsilon_{ik}^S E_k^{(0,1)}, \tag{6.2.22}$$

$$S_1^{(0,0)} = u_{1,1}^{(0,0)}, \tag{6.2.23}$$

$$E_1^{(0,0)} = -\phi_{,1}^{(0,0)}, \quad E_2^{(0,0)} = -\phi^{(1,0)}, \quad E_3^{(0,0)} = -\phi^{(0,1)},$$

$$E_1^{(1,0)} = -\phi_{,1}^{(1,0)}, \quad E_2^{(1,0)} = 0, \quad E_3^{(1,0)} = 0, \tag{6.2.24}$$

$$E_1^{(0,1)} = -\phi_{,1}^{(0,1)}, \quad E_2^{(0,1)} = 0, \quad E_3^{(0,1)} = 0,$$

$$\tilde{c}_{11} = \frac{1}{s_{11}}, \quad \tilde{e}_{k1} = \frac{d_{k1}}{s_{11}}, \quad \tilde{\varepsilon}_{ik} = \varepsilon_{ik}^T - \frac{d_{i1}d_{k1}}{s_{11}}. \tag{6.2.25}$$

With successive substitutions, Equations (6.2.18) and (6.2.19) can be written as four equations for $u_1^{(0,0)}$, $\phi^{(0,0)}$, $\phi^{(1,0)}$ and $\phi^{(0,1)}$. For end conditions we can prescribe

$$T_1^{(0,0)} \quad \text{or} \quad u_1^{(0,0)}, \quad \phi^{(0,0)} \quad \text{or} \quad D_1^{(0,0)},$$

$$\phi^{(1,0)} \quad \text{or} \quad D_1^{(1,0)}, \quad \phi^{(0,1)} \quad \text{or} \quad D_1^{(0,1)}. \tag{6.2.26}$$

6.2.2 Equations for ceramic beams

We discuss two cases of thickness poling and axial poling, respectively.

6.2.2.1 Thickness poling

Consider a ceramic beam poled in the x_3 direction (see Figure 6.2.1).

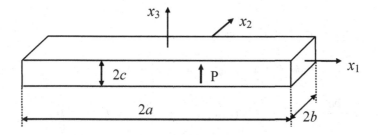

Figure 6.2.1. A ceramic beam poled in the x_3 direction.

The material matrices are given in Equation (2.4.87). Then from Equation (6.2.14) we have

$$\widetilde{c}_{11} = \frac{1}{s_{11}}, \quad \widetilde{e}_{11} = \widetilde{e}_{21} = 0, \quad \widetilde{e}_{31} = \frac{d_{31}}{s_{11}}, \tag{6.2.27}$$

$$[\widetilde{\varepsilon}_{ik}] = \begin{pmatrix} \widetilde{\varepsilon}_{11} & 0 & 0 \\ 0 & \widetilde{\varepsilon}_{11} & 0 \\ 0 & 0 & \widetilde{\varepsilon}_{33} \end{pmatrix}, \quad \begin{aligned} \widetilde{\varepsilon}_{11} &= \varepsilon_{11}^T, \\ \widetilde{\varepsilon}_{33} &= \varepsilon_{33}^T - \frac{d_{31}d_{31}}{s_{11}}. \end{aligned} \tag{6.2.28}$$

The constitutive relations needed are

$$T_1^{(0,0)} = 4bc(\widetilde{c}_{11}u_{1,1}^{(0,0)} + \widetilde{e}_{31}\phi^{(0,1)}), \tag{6.2.29}$$

$$D_1^{(0,0)} = -4bc\widetilde{\varepsilon}_{11}\phi_{,1}^{(0,0)},$$
$$D_2^{(0,0)} = -4bc\widetilde{\varepsilon}_{11}\phi^{(1,0)}, \tag{6.2.30}$$
$$D_3^{(0,0)} = 4bc(\widetilde{e}_{31}u_{1,1}^{(0,0)} - \widetilde{\varepsilon}_{33}\phi^{(0,1)}),$$

$$D_1^{(1,0)} = -\frac{4b^3c}{3}\varepsilon_{11}^S\phi_{,1}^{(1,0)}, \quad D_2^{(1,0)} = D_3^{(1,0)} = 0,$$
$$D_1^{(0,1)} = -\frac{4bc^3}{3}\varepsilon_{11}^S\phi_{,1}^{(0,1)}, \quad D_2^{(0,1)} = D_3^{(0,1)} = 0. \tag{6.2.31}$$

The equations of motion and charge are

$$4bc(\widetilde{c}_{11}u_{1,11}^{(0,0)} + \widetilde{e}_{31}\phi_{,1}^{(0,1)}) + F_1^{(0,0)} = 4\rho bc\ddot{u}_1^{(0,0)}, \tag{6.2.32}$$

$$4bc\widetilde{\varepsilon}_{11}\phi_{,11}^{(0,0)} + D^{(0,0)} = 0,$$
$$-\frac{4b^3c}{3}\varepsilon_{11}^S\phi_{,11}^{(1,0)} + 4bc\widetilde{\varepsilon}_{11}\phi^{(1,0)} + D^{(1,0)} = 0, \tag{6.2.33}$$
$$-\frac{4bc^3}{3}\varepsilon_{11}^S\phi_{,11}^{(0,1)} - 4bc(\widetilde{e}_{31}u_{1,1}^{(0,0)} - \widetilde{\varepsilon}_{33}\phi^{(0,1)}) + D^{(0,1)} = 0.$$

6.2.2.2 Axial poling

Next consider a ceramic beam poled in the x_1 direction (see Figure 6.2.2).

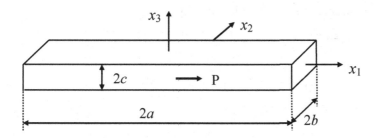

Figure 6.2.2. A ceramic beam poled in the x_1 direction.

The material matrices are given by Equation (2.4.97). From Equation (6.2.14) we have

$$\tilde{c}_{11} = \frac{1}{s_{33}}, \quad \tilde{e}_{11} = \frac{d_{33}}{s_{33}}, \quad \tilde{e}_{21} = \tilde{e}_{31} = 0, \tag{6.2.34}$$

$$[\tilde{\varepsilon}_{ik}] = \begin{pmatrix} \tilde{\varepsilon}_{11} & 0 & 0 \\ 0 & \tilde{\varepsilon}_{22} & 0 \\ 0 & 0 & \tilde{\varepsilon}_{33} \end{pmatrix}, \quad \begin{aligned} \tilde{\varepsilon}_{11} &= \varepsilon_{33}^T - \frac{d_{33}d_{33}}{s_{33}}, \\ \tilde{\varepsilon}_{22} &= \varepsilon_{11}^T, \quad \tilde{\varepsilon}_{33} = \varepsilon_{11}^T. \end{aligned} \tag{6.2.35}$$

The constitutive relations are

$$T_1^{(0,0)} = 4bc(\tilde{c}_{11}u_{1,1}^{(0,0)} + \tilde{e}_{11}\phi_{,1}^{(0,0)}), \tag{6.2.36}$$

$$\begin{aligned} D_1^{(0,0)} &= 4bc(\tilde{e}_{11}u_{1,1}^{(0,0)} - \tilde{\varepsilon}_{11}\phi_{,1}^{(0,0)}), \\ D_2^{(0,0)} &= -4bc\tilde{\varepsilon}_{22}\phi^{(1,0)}, \\ D_3^{(0,0)} &= -4bc\tilde{\varepsilon}_{22}\phi^{(0,1)}, \end{aligned} \tag{6.2.37}$$

$$D_1^{(1,0)} = -\frac{4b^3c}{3}\varepsilon_{33}^S\phi_{,1}^{(1,0)}, \quad D_2^{(1,0)} = D_3^{(1,0)} = 0,$$

$$D_1^{(0,1)} = -\frac{4bc^3}{3}\varepsilon_{33}^S\phi_{,1}^{(0,1)}, \quad D_2^{(1,0)} = D_3^{(1,0)} = 0. \tag{6.2.38}$$

The equations of motion and charge are

$$4bc(\tilde{c}_{11}u_{1,11}^{(0,0)} + \tilde{e}_{11}\phi_{,11}^{(0,0)}) + F_1^{(0,0)} = 4\rho bc\ddot{u}_1^{(0,0)}, \tag{6.2.39}$$

$$4bc(\tilde{e}_{11}u_{1,11}^{(0,0)} - \tilde{\varepsilon}_{11}\phi_{,11}^{(0,0)}) + D^{(0,0)} = 0,$$

$$-\frac{4b^3c}{3}\varepsilon_{33}^S\phi_{,11}^{(1,0)} + 4bc\tilde{\varepsilon}_{22}\phi^{(1,0)} + D^{(1,0)} = 0, \tag{6.2.40}$$

$$-\frac{4bc^3}{3}\varepsilon_{33}^S\phi_{,11}^{(0,1)} + 4bc\tilde{\varepsilon}_{22}\phi^{(0,1)} + D^{(0,1)} = 0.$$

6.2.3 Extensional vibration of a ceramic beam

Consider a ceramic beam poled in the x_3 direction as shown in Figure 6.2.3.

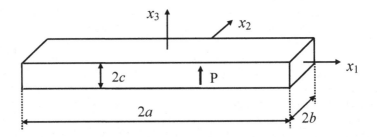

Figure 6.2.3. A ceramic beam poled in the x_3 direction.

We assume that $a \gg b \gg c$. The two ends of the beam and the lateral surfaces are traction-free. If the surfaces of the areas at $x_3 = \pm c$ are fully electroded with a driving voltage $\pm V/2$ across the electrodes, the electric potentials are

$$\phi^{(0,0)} = 0, \quad \phi^{(0,1)} = V/2c. \tag{6.2.41}$$

From Equations (6.2.32) and (6.2.29) we have the following boundary value problem:

$$\tilde{c}_{11}u_{1,11}^{(0,0)} = \rho\ddot{u}_1^{(0,0)}, \quad -a < x_1 < a,$$

$$T_1^{(0,0)} = 4bc(\tilde{c}_{11}u_{1,1}^{(0,0)} + \tilde{e}_{31}\phi^{(0,1)}) = 0, \quad x_1 = \pm a. \tag{6.2.42}$$

Equation (6.2.42) shows that the applied voltage effectively acts like two extensional end forces on the beam. For free vibrations, $V = 0$ and the electrodes are shorted. We look for free vibration solutions in the form

$$u_1^{(0,0)}(x_1,t) = u_1(x_1)\exp(i\omega t). \tag{6.2.43}$$

Then the eigenvalue problem is

$$u_{1,11} + \rho\omega^2 s_{11}u_1 = 0, \quad -a < x_1 < a,$$

$$u_{1,1} = 0, \quad x_1 = \pm a. \tag{6.2.44}$$

The solution of $\omega = 0$ and $u = $ constant represents a rigid body mode. For the rest of the modes we try $u = \sin kx_1$. Then, from Equation (6.2.44)$_1$, $k = \omega\sqrt{\rho s_{11}}$. To satisfy Equation (6.2.44)$_2$ we must have

$$\cos ka = 0, \quad \Rightarrow \quad k_{(n)}a = \frac{n\pi}{2}, \quad n = 1,3,5,\cdots, \tag{6.2.45}$$

or

$$\omega_{(n)}\sqrt{\rho s_{11}}\,a = \frac{n\pi}{2}, \quad \omega_{(n)} = \frac{n\pi}{2a\sqrt{\rho s_{11}}}, \quad n = 1,3,5,\cdots. \tag{6.2.46}$$

Similarly, by considering $u = \cos kx_1$, the following frequencies can be determined:

$$\omega_{(n)} = \frac{n\pi}{2a\sqrt{\rho s_{11}}}, \quad n = 2,4,6,\cdots. \tag{6.2.47}$$

The frequencies in Equations (6.2.46) and (6.2.47) are integral multiples of $\omega_{(1)}$ and are called harmonics. $\omega_{(1)}$ is called the fundamental and the rest are called the overtones. These results are the same as those in [9].

6.3 First-Order Theory

Next we truncate the general expansions to obtain a first-order theory for coupled extension, flexure, thickness-shear and width-shear [24].

6.3.1 Coupled extension, flexure and shear

For a first-order theory, to include all first-order strains, we make the following truncations:

$$
u_j \cong u_j^{(0,0)} + x_2 u_j^{(1,0)} + x_3 u_j^{(0,1)} + x_2^2 u_j^{(2,0)} + x_2 x_3 u_j^{(1,1)} + x_3^2 u_j^{(0,2)},
$$
$$
\phi \cong \phi^{(0,0)} + x_2 \phi^{(1,0)} + x_3 \phi^{(0,1)},
$$

(6.3.1)

where $u_2^{(1,0)}$, $u_2^{(2,0)}$, $u_3^{(0,1)}$ and $u_3^{(0,2)}$ are needed to describe Poisson's effects accompanying extension and flexure. They cannot be directly set to zero, but will be eliminated later by stress relaxations. Equations for the extensional displacement $u_1^{(0,0)}$, flexural displacements $u_2^{(0,0)}$ and $u_3^{(0,0)}$, and thickness-shear and width-shear displacements $u_1^{(1,0)}$ and $u_1^{(0,1)}$ will be obtained. From Equations (6.1.3) and (6.3.1), the strains and electric fields are

$$
S_{ij} \cong S_{ij}^{(0,0)} + x_2 S_{ij}^{(1,0)} + x_3 S_{ij}^{(0,1)},
$$
$$
E_k \cong E_k^{(0,0)} + x_2 E_k^{(1,0)} + x_3 E_k^{(0,1)},
$$

(6.3.2)

where the zero-order strains are given by

$$
S_1^{(0,0)} = u_{1,1}^{(0,0)}, \quad S_{22}^{(0)} = u_2^{(1,0)}, \quad S_{33}^{(0)} = u_3^{(0,1)},
$$
$$
S_4^{(0,0)} = 2S_{23}^{(0,0)} = u_2^{(0,1)} + u_3^{(1,0)},
$$
$$
S_5^{(0,0)} = 2S_{31}^{(0,0)} = u_{3,1}^{(0,0)} + u_1^{(0,1)},
$$
$$
S_6^{(0,0)} = 2S_{12}^{(0,0)} = u_{2,1}^{(0,0)} + u_1^{(1,0)}.
$$

(6.3.3)

For the first-order strains, only the following expressions are needed:

$$
S_1^{(1,0)} = u_{1,1}^{(1,0)},
$$

(6.3.4)

$$
S_1^{(0,1)} = u_{1,1}^{(0,1)}.
$$

(6.3.5)

The rest of the first-order strains will be eliminated by stress relaxations. The relevant electric fields are

$$E_1^{(0,0)} = -\phi_{,1}^{(0,0)}, \quad E_2^{(0,0)} = -\phi^{(1,0)}, \quad E_3^{(0,0)} = -\phi^{(0,1)},$$

$$E_1^{(1,0)} = -\phi_{,1}^{(1,0)}, \quad E_2^{(1,0)} = -2\phi^{(2,0)} = 0, \quad E_3^{(1,0)} = -\phi^{(1,1)} = 0, \quad (6.3.6)$$

$$E_1^{(0,1)} = -\phi_{,1}^{(0,1)}, \quad E_2^{(0,1)} = -\phi^{(1,1)} = 0, \quad E_3^{(0,1)} = -2\phi^{(0,2)} = 0.$$

We keep the equations of motion corresponding to the extension $u_1^{(0,0)}$, flexure $u_2^{(0,0)}$ and $u_3^{(0,0)}$, and shear deformations $u_1^{(1,0)}$ and $u_1^{(0,1)}$:

$$T_{11,1}^{(0,0)} + F_1^{(0,0)} = \rho A \ddot{u}_1^{(0,0)},$$

$$T_{12,1}^{(0,0)} + F_2^{(0,0)} = \rho A \ddot{u}_2^{(0,0)},$$

$$T_{13,1}^{(0,0)} + F_3^{(0,0)} = \rho A \ddot{u}_3^{(0,0)}, \quad (6.3.7)$$

$$T_{11,1}^{(1,0)} - T_{21}^{(0,0)} + F_1^{(1,0)} = \rho A r_3^2 \ddot{u}_1^{(1,0)},$$

$$T_{11,1}^{(0,1)} - T_{31}^{(0,0)} + F_1^{(0,1)} = \rho A r_2^2 \ddot{u}_1^{(0,1)},$$

where

$$A = 4bc, \quad r_2^2 = c^2/3, \quad r_3^2 = b^2/3. \quad (6.3.8)$$

r_2 and r_3 are the radii of gyration of the cross section about the x_2 and x_3 axes. For electrostatics we keep the equations for $\phi^{(0,0)}$, $\phi^{(1,0)}$, $\phi^{(0,1)}$

$$D_{1,1}^{(0,0)} + D^{(0,0)} = 0,$$

$$D_{1,1}^{(1,0)} - D_2^{(0,0)} + D^{(1,0)} = 0, \quad (6.3.9)$$

$$D_{1,1}^{(0,1)} - D_3^{(0,0)} + D^{(0,1)} = 0.$$

For the zero-order mechanical constitutive relations we have, from Equation $(6.1.8)_1$, for $m = n = 0$,

$$T_p^{(0,0)} = 4bc(c_{pq}S_q^{(0,0)} - e_{ip}E_i^{(0,0)}). \quad (6.3.10)$$

Equation (6.3.10) should be adjusted by setting the following zero-order contour stresses to zero [65]

$$T_2^{(0,0)} = T_3^{(0,0)} = T_4^{(0,0)} = 0, \quad (6.3.11)$$

which permits the free development of the corresponding zero-order contour strains. Equation (6.3.11) will be used to eliminate $S_2^{(0,0)}$, $S_3^{(0,0)}$ and $S_4^{(0,0)}$ from Equation (6.3.10). In view of the conditions in Equation (6.3.11), we now introduce an index convention which will be of considerable use to us in the sequel. The convention is that we let the subscripts α, β take the values 1, 5, 6 and λ, μ, ν take the remaining values 2, 3, 4. Then Equation (6.3.10) can be written as

$$T_\alpha^{(0,0)} = 4bc(c_{\alpha\beta}S_\beta^{(0,0)} + c_{\alpha\mu}S_\mu^{(0,0)} - e_{i\alpha}E_i^{(0,0)}),$$
$$T_\lambda^{(0,0)} = 4bc(c_{\lambda\beta}S_\beta^{(0,0)} + c_{\lambda\mu}S_\mu^{(0,0)} - e_{i\lambda}E_i^{(0,0)}).$$

(6.3.12)

With this convention Equation (6.3.11) can be written as

$$T_\lambda^{(0,0)} = 0,$$

(6.3.13)

which implies, with Equation (6.3.12)$_2$, that

$$4bcS_\lambda^{(0,0)} = -4bcc_{\lambda\mu}^{-1}c_{\mu\beta}S_\beta^{(0,0)} + 4bcc_{\lambda\mu}^{-1}e_{i\mu}E_i^{(0,0)},$$

(6.3.14)

where $c_{\lambda\mu}^{-1}$ is the matrix inverse of $c_{\lambda\mu}$:

$$c_{\lambda\mu}^{-1}c_{\mu\nu} = \delta_{\lambda\nu}.$$

(6.3.15)

Substitution of Equation (6.3.14) into Equation (6.3.12)$_1$ yields the relaxed zero-order mechanical constitutive relations

$$T_\alpha^{(0,0)} = 4bc(\widehat{c}_{\alpha\beta}S_\beta^{(0,0)} - \widehat{e}_{i\alpha}E_i^{(0,0)}),$$

(6.3.16)

where

$$\widehat{c}_{\alpha\beta} = c_{\alpha\beta} - c_{\alpha\lambda}c_{\lambda\mu}^{-1}c_{\mu\beta}, \quad \widehat{e}_{i\alpha} = e_{i\alpha} - e_{i\mu}c_{\mu\lambda}^{-1}c_{\lambda\alpha}.$$

(6.3.17)

For the zero-order electric constitutive relations we begin with Equation (6.1.8)$_2$ for $m = n = 0$

$$D_i^{(0,0)} = 4bc(e_{iq}S_q^{(0,0)} + \varepsilon_{ik}E_k^{(0,0)})$$
$$= 4bc(e_{i\alpha}S_\alpha^{(0,0)} + e_{i\lambda}S_\lambda^{(0,0)} + \varepsilon_{ik}E_k^{(0,0)}).$$

(6.3.18)

Substituting Equation (6.3.14) into Equation (6.3.18), we obtain

$$D_i^{(0,0)} = 4bc(\widehat{e}_{i\beta}S_\beta^{(0,0)} + \widehat{\varepsilon}_{ik}E_k^{(0,0)}),$$

(6.3.19)

where

$$\widehat{\varepsilon}_{ik} = \varepsilon_{ik} + e_{i\lambda}c_{\lambda\mu}^{-1}e_{k\mu}.$$

(6.3.20)

If two shear correction factors κ_2 and κ_3 are introduced by the replacement of the following thickness-shear and width-shear strains in the strain energy:

$$S_{12}^{(0,0)} \to \kappa_2 S_{12}^{(0,0)}, \quad S_{13}^{(0,0)} \to \kappa_3 S_{13}^{(0,0)}, \tag{6.3.21}$$

the zero-order constitutive relations will be modified into

$$\begin{aligned}
T_\alpha^{(0,0)} &= 4bc(\widehat{c}_{\alpha\beta}' S_\beta^{(0,0)} - \widehat{e}_{i\alpha}' E_i^{(0,0)}), \\
D_i^{(0,0)} &= 4bc(\widehat{e}_{i\beta}' S_\beta^{(0,0)} + \widehat{\varepsilon}_{ik} E_k^{(0,0)}).
\end{aligned} \tag{6.3.22}$$

Equation (6.3.22) depends on the strains $S_1^{(0,0)}$, $S_5^{(0,0)}$ and $S_6^{(0,0)}$, which are related to the extensional, flexural and shear displacements.

For the first-order mechanical constitutive relations from Equation $(6.1.8)_1$ we obtain, by setting $m = 1$ and $n = 0$,

$$T_p^{(1,0)} = \frac{4b^3 c}{3}(c_{pq} S_q^{(1,0)} - e_{ip} E_i^{(1,0)}). \tag{6.3.23}$$

Equation (6.3.23) should be adjusted by setting the following first-order beam stresses to zero:

$$T_2^{(1,0)} = T_3^{(1,0)} = T_4^{(1,0)} = T_5^{(1,0)} = T_6^{(1,0)} = 0, \tag{6.3.24}$$

which permits the free development of the corresponding first-order strains. Equation (6.3.24) can be used to eliminate $S_2^{(1,0)}$, $S_3^{(1,0)}$ $S_4^{(1,0)}$, $S_5^{(1,0)}$ and $S_6^{(1,0)}$ from Equation (6.3.23). In this case we re-derive the beam constitutive relations from the following three-dimensional constitutive relations:

$$\begin{aligned}
S_p &= s_{pq} T_q + d_{kp} E_k, \\
D_i &= d_{iq} T_q + \varepsilon_{ij}^T E_j.
\end{aligned} \tag{6.3.25}$$

Within the approximation we are interested in, from Equation (6.3.2)

$$\begin{aligned}
S_p^{(0,0)} + x_2 S_p^{(1,0)} &+ x_3 S_p^{(0,1)} \\
&= s_{pq} T_q + d_{kp}(E_k^{(0,0)} + x_2 E_k^{(1,0)} + x_3 E_k^{(0,1)}), \\
D_i &= d_{iq} T_q + \varepsilon_{ij}^T (E_j^{(0,0)} + x_2 E_j^{(1,0)} + x_3 E_j^{(0,1)}).
\end{aligned} \tag{6.3.26}$$

Integrating the product of Equation (6.3.26) with x_2 over the cross section of the beam, we obtain

$$\frac{4b^3c}{3} S_p^{(1,0)} = s_{pq} T_q^{(1,0)} + \frac{4b^3c}{3} d_{kp} E_k^{(1,0)}$$

$$= s_{p1} T_1^{(1,0)} + \frac{4b^3c}{3} d_{kp} E_k^{(1,0)},$$

$$D_i^{(1,0)} = d_{iq} T_q^{(1,0)} + \frac{4b^3c}{3} \varepsilon_{ij}^T E_j^{(1,0)}$$

$$= d_{i1} T_1^{(1,0)} + \frac{4b^3c}{3} \varepsilon_{ij}^T E_j^{(1,0)}.$$

(6.3.27)

From Equation (6.3.27)$_1$, for $p = 1$

$$\frac{4b^3c}{3} S_1^{(1,0)} = s_{11} T_1^{(1,0)} + \frac{4b^3c}{3} d_{k1} E_k^{(1,0)},$$ (6.3.28)

which can be inverted to give

$$T_1^{(1,0)} = \frac{1}{s_{11}} \frac{4b^3c}{3} (S_1^{(1,0)} - d_{k1} E_k^{(1,0)})$$

$$= \frac{4b^3c}{3} (\tilde{c}_{11} S_1^{(1,0)} - \tilde{e}_{k1} E_k^{(1,0)}).$$

(6.3.29)

Substitution of Equation (6.3.29) into Equation (6.3.27)$_2$ results in the electric constitutive relation

$$D_i^{(1,0)} = d_{i1} T_1^{(1,0)} + \frac{4b^3c}{3} \varepsilon_{ij}^T E_j^{(1,0)}$$

$$= d_{i1} \frac{1}{s_{11}} \frac{4b^3c}{3} (S_1^{(1,0)} - d_{k1} E_k^{(1,0)}) + \frac{4b^3c}{3} \varepsilon_{ij}^T E_j^{(1,0)}$$

$$= \frac{4b^3c}{3} \frac{d_{i1}}{s_{11}} S_1^{(1,0)} + \frac{4b^3c}{3} (\varepsilon_{ik}^T - \frac{d_{i1} d_{k1}}{s_{11}}) E_k^{(1,0)}$$

$$= \frac{4b^3c}{3} (\tilde{e}_{i1} S_1^{(1,0)} + \tilde{\varepsilon}_{ik} E_k^{(1,0)}).$$

(6.3.30)

The above first-order constitutive relation is for flexure in the x_2 direction. Similarly, for flexure in the x_3 direction,

$$T_1^{(0,1)} = \frac{4bc^3}{3}(\widetilde{c}_{11}S_1^{(0,1)} - \widetilde{e}_{i1}E_i^{(0,1)}),$$

$$D_i^{(0,1)} = \frac{4bc^3}{3}(\widetilde{e}_{i1}S_1^{(0,1)} + \widetilde{\varepsilon}_{ik}E_k^{(0,1)}).$$

(6.3.31)

In Equations (6.3.29) through (6.3.31) only the shear strains $S_1^{(1,0)}$ and $S_1^{(0,1)}$ appear, which depend on the shear displacements. In summary, we have obtained

$$T_{11,1}^{(0,0)} + F_1^{(0,0)} = 4\rho bc\ddot{u}_1^{(0,0)},$$

$$T_{12,1}^{(0,0)} + F_2^{(0,0)} = 4\rho bc\ddot{u}_2^{(0,0)},$$

$$T_{13,1}^{(0,0)} + F_3^{(0,0)} = 4\rho bc\ddot{u}_3^{(0,0)},$$

(6.3.32)

$$T_{11,1}^{(1,0)} - T_{21}^{(0,0)} + F_1^{(1,0)} = \frac{4}{3}\rho b^3 c\ddot{u}_1^{(1,0)},$$

$$T_{11,1}^{(0,1)} - T_{31}^{(0,0)} + F_1^{(0,1)} = \frac{4}{3}\rho bc^3\ddot{u}_1^{(0,1)},$$

$$D_{1,1}^{(0,0)} + D^{(0,0)} = 0,$$

$$D_{1,1}^{(1,0)} - D_2^{(0,0)} + D^{(1,0)} = 0,$$

(6.3.33)

$$D_{1,1}^{(0,1)} - D_3^{(0,0)} + D^{(0,1)} = 0,$$

$$T_\alpha^{(0,0)} = 4bc(\widehat{c}'_{\alpha\beta}S_\beta^{(0,0)} - \widehat{e}'_{i\alpha}E_i^{(0,0)}), \quad \alpha, \beta = 1,5,6,$$

$$D_i^{(0,0)} = 4bc(\widehat{e}'_{i\beta}S_\beta^{(0,0)} + \widehat{\varepsilon}_{ik}E_k^{(0,0)}),$$

(6.3.34)

$$T_1^{(1,0)} = \frac{4b^3c}{3}(\widetilde{c}_{11}S_1^{(1,0)} - \widetilde{e}_{k1}E_k^{(1,0)}),$$

$$D_i^{(1,0)} = \frac{4b^3c}{3}(\widetilde{e}_{i1}S_1^{(1,0)} + \widetilde{\varepsilon}_{ik}E_k^{(1,0)}),$$

(6.3.35)

$$T_1^{(0,1)} = \frac{4bc^3}{3}(\tilde{c}_{11}S_1^{(0,1)} - \tilde{e}_{i1}E_i^{(0,1)}),$$

$$D_i^{(0,1)} = \frac{4bc^3}{3}(\tilde{e}_{i1}S_1^{(0,1)} + \tilde{\varepsilon}_{ik}E_k^{(0,1)}),$$
(6.3.36)

$$S_1^{(0,0)} = u_{1,1}^{(0,0)},$$

$$S_5^{(0,0)} = u_{3,1}^{(0,0)} + u_1^{(0,1)}, \quad S_6^{(0,0)} = u_{2,1}^{(0,0)} + u_1^{(1,0)},$$
(6.3.37)

$$S_1^{(1,0)} = u_{1,1}^{(1,0)}, \quad S_1^{(0,1)} = u_{1,1}^{(0,1)},$$
(6.3.38)

$$E_1^{(0,0)} = -\phi_{,1}^{(0,0)}, \quad E_2^{(0,0)} = -\phi^{(1,0)}, \quad E_3^{(0,0)} = -\phi^{(0,1)},$$

$$E_1^{(1,0)} = -\phi_{,1}^{(1,0)}, \quad E_2^{(1,0)} = 0, \quad E_3^{(1,0)} = 0,$$
(6.3.39)

$$E_1^{(0,1)} = -\phi_{,1}^{(0,1)}, \quad E_2^{(0,1)} = 0, \quad E_3^{(0,1)} = 0,$$

$$\hat{c}_{\alpha\beta} = c_{\alpha\beta} - c_{\alpha\lambda}c_{\lambda\mu}^{-1}c_{\mu\beta}, \quad \hat{e}_{i\alpha} = e_{i\alpha} - e_{i\mu}c_{\mu\lambda}^{-1}c_{\lambda\alpha},$$

$$\hat{\varepsilon}_{ik} = \varepsilon_{ik} + e_{i\lambda}c_{\lambda\mu}^{-1}e_{k\mu}, \quad \alpha,\beta = 1,5,6, \quad \lambda,\mu,\nu = 2,3,4,$$
(6.3.40)

$$\tilde{c}_{11} = \frac{1}{s_{11}}, \quad \tilde{e}_{k1} = \frac{d_{k1}}{s_{11}}, \quad \tilde{\varepsilon}_{ik} = \varepsilon_{ik}^T - \frac{d_{i1}d_{k1}}{s_{11}}.$$

With successive substitutions, Equations (6.3.32) and (6.3.33) can be written as eight equations for $u_1^{(0,0)}$, $u_2^{(0,0)}$, $u_3^{(0,0)}$, $u_1^{(1,0)}$, $u_1^{(0,1)}$, $\phi^{(0,0)}$, $\phi^{(1,0)}$ and $\phi^{(0,1)}$. For end conditions we can prescribe

$$T_1^{(0,0)} \quad \text{or} \quad u_1^{(0,0)}, \quad T_6^{(0,0)} \quad \text{or} \quad u_2^{(0,0)}, \quad T_5^{(0,0)} \quad \text{or} \quad u_3^{(0,0)},$$

$$T_1^{(1,0)} \quad \text{or} \quad u_1^{(1,0)}, \quad T_1^{(0,1)} \quad \text{or} \quad u_1^{(0,1)},$$
(6.3.41)

$$\phi^{(0,0)} \quad \text{or} \quad D_1^{(0,0)}, \quad \phi^{(1,0)} \quad \text{or} \quad D_1^{(1,0)}, \quad \phi^{(0,1)} \quad \text{or} \quad D_1^{(0,1)}.$$

6.3.2 Reduction to classical flexure

For the theory of elementary flexure without shear deformations, we set the rotatory inertia terms in Equations $(6.3.32)_{4,5}$ to zero:

$$T_{11,1}^{(1,0)} - T_{21}^{(0,0)} + F_1^{(1,0)} = 0,$$
$$T_{11,1}^{(0,1)} - T_{31}^{(0,0)} + F_1^{(0,1)} = 0,$$

(6.3.42)

which can be used to solve for $T_{21}^{(0,0)}$ and $T_{31}^{(0,0)}$. Substitute the resulting expressions into Equations $(6.3.32)_{2,3}$

$$T_{11,11}^{(1,0)} + F_{1,1}^{(1,0)} + F_2^{(0,0)} = \rho A \ddot{u}_2^{(0,0)},$$
$$T_{11,11}^{(0,1)} + F_{1,1}^{(0,1)} + F_3^{(0,0)} = \rho A \ddot{u}_3^{(0,0)},$$

(6.3.43)

which are the equations for elementary flexure. We also need to set the zero-order shear strains to zero

$$S_6^{(0,0)} = u_{2,1}^{(0,0)} + u_1^{(1,0)} = 0,$$
$$S_5^{(0,0)} = u_{3,1}^{(0,0)} + u_1^{(0,1)} = 0,$$

(6.3.44)

so that the first-order flexural strains can be represented by

$$S_1^{(1,0)} = -u_{2,11}^{(0,0)}, \quad S_1^{(0,1)} = -u_{3,11}^{(0,0)}.$$

(6.3.45)

The equations for the classical theory are

$$T_{11,1}^{(0,0)} + F_1^{(0,0)} = 4\rho bc \ddot{u}_1^{(0,0)},$$
$$T_{12,1}^{(0,0)} + F_2^{(0,0)} = 4\rho bc \ddot{u}_2^{(0,0)},$$
$$T_{13,1}^{(0,0)} + F_3^{(0,0)} = 4\rho bc \ddot{u}_3^{(0,0)},$$

(6.3.46)

$$D_{1,1}^{(0,0)} + D^{(0,0)} = 0,$$
$$D_{1,1}^{(1,0)} - D_2^{(0,0)} + D^{(1,0)} = 0,$$
$$D_{1,1}^{(0,1)} - D_3^{(0,0)} + D^{(0,1)} = 0,$$

(6.3.47)

$$T_1^{(0,0)} = 4bc(\hat{c}_{11}' S_1^{(0,0)} - \hat{e}_{i1}' E_i^{(0,0)}),$$
$$D_i^{(0,0)} = 4bc(\hat{e}_{i1}' S_1^{(0,0)} + \hat{\varepsilon}_{ik}' E_k^{(0,0)}),$$

(6.3.48)

$$T_{21}^{(0,0)} = T_{11,1}^{(1,0)} + F_1^{(1,0)},$$
$$T_{31}^{(0,0)} = T_{11,1}^{(0,1)} + F_1^{(0,1)},$$

(6.3.49)

$$T_1^{(1,0)} = \frac{4b^3c}{3}(\tilde{c}_{11}S_1^{(1,0)} - \tilde{e}_{k1}E_k^{(1,0)}),$$

$$D_i^{(1,0)} = \frac{4b^3c}{3}(\tilde{e}_{i1}S_1^{(1,0)} + \tilde{\varepsilon}_{ik}E_k^{(1,0)}),$$

(6.3.50)

$$T_1^{(0,1)} = \frac{4bc^3}{3}(\tilde{c}_{11}S_1^{(0,1)} - \tilde{e}_{i1}E_i^{(0,1)}),$$

$$D_i^{(0,1)} = \frac{4bc^3}{3}(\tilde{e}_{i1}S_1^{(0,1)} + \tilde{\varepsilon}_{ik}E_k^{(0,1)}),$$

(6.3.51)

$$S_1^{(0,0)} = u_{1,1}^{(0,0)}, \quad S_1^{(1,0)} = -u_{2,11}^{(0,0)}, \quad S_1^{(0,1)} = -u_{3,11}^{(0,0)},$$ (6.3.52)

$$E_1^{(0,0)} = -\phi_{,1}^{(0,0)}, \quad E_2^{(0,0)} = -\phi^{(1,0)}, \quad E_3^{(0,0)} = -\phi^{(0,1)},$$

$$E_1^{(1,0)} = -\phi_{,1}^{(1,0)}, \quad E_2^{(1,0)} = 0, \quad E_3^{(1,0)} = 0,$$ (6.3.53)

$$E_1^{(0,1)} = -\phi_{,1}^{(0,1)}, \quad E_2^{(0,1)} = 0, \quad E_3^{(0,1)} = 0.$$

With successive substitutions, Equations (6.3.46) and (6.3.47) can be written as six equations for $u_1^{(0,0)}$, $u_2^{(0,0)}$, $u_3^{(0,0)}$, $\phi^{(0,0)}$, $\phi^{(1,0)}$ and $\phi^{(0,1)}$. For end conditions we can prescribe

$$T_1^{(0,0)} \quad \text{or} \quad u_1^{(0,0)}, \quad T_6^{(0,0)} \quad \text{or} \quad u_2^{(0,0)}, \quad T_5^{(0,0)} \quad \text{or} \quad u_3^{(0,0)},$$

$$T_1^{(1,0)} \quad \text{or} \quad u_{2,1}^{(0,0)}, \quad T_1^{(0,1)} \quad \text{or} \quad u_{3,1}^{(0,0)},$$ (6.3.54)

$$\phi^{(0,0)} \quad \text{or} \quad D_1^{(0,0)}, \quad \phi^{(1,0)} \quad \text{or} \quad D_1^{(1,0)}, \quad \phi^{(0,1)} \quad \text{or} \quad D_1^{(0,1)}.$$

6.3.3 Thickness-shear approximation

Consider the beam in Figure 6.3.1. The material of the beam is of general anisotropy.

6.3.3.1 Equations for coupled flexure and thickness-shear

We study motions dominated by coupled thickness-shear $u_1^{(0,1)}$ and flexure $u_3^{(0,0)}$. The major mechanical resultants are the shear force $T_{13}^{(0,0)}$

and bending moment $T_{11}^{(0,1)}$. Coupling to extension is neglected. We also assume $b \gg c$. The beam may be electroded at $x_3 = \pm c$ or at both ends, but not at $x_2 = \pm b$. We perform the thickness-shear approximation in a way that is sufficient for later applications to a piezoelectric transformer [67] in this section, not in the most general manner.

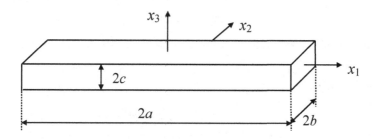

Figure 6.3.1. A piezoelectric beam.

First we summarize the equations for coupled thickness-shear and flexure without extension. In the absence of body and surface mechanical loads, from Equations (6.3.7) and (6.3.9), the relevant equations of motion and the lowest order charge equation are

$$T_{13,1}^{(0,0)} = \rho 4bc \ddot{u}_3^{(0,0)},$$

$$T_{11,1}^{(0,1)} - T_{31}^{(0,0)} = \rho \frac{4bc^3}{3} \ddot{u}_1^{(0,1)}, \tag{6.3.55}$$

$$D_{1,1}^{(0,0)} + D^{(0,0)} = 0.$$

When there is no extension we re-derive constitutive relations from

$$S_{ij} = s_{ijkl}T_{kl} + d_{kij}E_k,$$

$$D_i = d_{ijk}T_{jk} + \varepsilon_{ij}^T E_j. \tag{6.3.56}$$

The zero-order constitutive relations can be obtained by substituting Equation (6.3.2) into Equation (6.3.56) and integrating the resulting equation over the cross section of the beam

$$4bcS_p^{(0,0)} = s_{pq}T_q^{(0,0)} + 4bcd_{kp}E_k^{(0,0)},$$

$$D_i^{(0,0)} = d_{iq}T_q^{(0,0)} + 4bc\varepsilon_{ij}^T E_j^{(0,0)}. \tag{6.3.57}$$

We keep the dominating shear force $T_5^{(0,0)}$ and make the following stress relaxation:

$$T_q^{(0,0)} = 0, \quad q = 1,2,3,4,6 .\tag{6.3.58}$$

Then, for $p = 5$, Equation (6.3.57) becomes

$$
\begin{aligned}
4bcS_5^{(0,0)} &= s_{55}T_5^{(0,0)} + 4bcd_{k5}E_k^{(0,0)}, \\
D_i^{(0,0)} &= d_{i5}T_5^{(0,0)} + 4bc\varepsilon_{ij}^T E_j^{(0,0)}.
\end{aligned}
\tag{6.3.59}
$$

Equation (6.3.59) can be inverted as

$$
\begin{aligned}
T_5^{(0,0)} &= \frac{4bc}{s_{55}}(\kappa^2 S_5^{(0,0)} - \kappa d_{k5}E_k^{(0,0)}), \\
D_i^{(0,0)} &= 4bc(\frac{d_{i5}}{s_{55}}\kappa S_5^{(0,0)} + \widetilde{\varepsilon}_{ij} E_j^{(0,0)}),
\end{aligned}
\tag{6.3.60}
$$

where $\kappa(=\kappa_1)$ is a shear correction factor, and

$$\widetilde{\varepsilon}_{ij} = \varepsilon_{ij}^T - \frac{d_{i5}d_{j5}}{s_{55}} .\tag{6.3.61}$$

Integrating the product of Equation $(6.3.56)_1$ with x_3 over the cross section, we obtain

$$\frac{4bc^3}{3}S_p^{(0,1)} = s_{pq}T_q^{(0,1)} + \frac{4bc^3}{3}d_{kp}E_k^{(0,1)} .\tag{6.3.62}$$

For flexure, $T_1^{(0,1)}$ is the dominating resultant. We make the following stress relaxation:

$$T_q^{(0,1)} = 0, \quad q = 2,3,4,5,6 .\tag{6.3.63}$$

Then, for $p = 1$, Equation (6.3.62) becomes

$$\frac{4bc^3}{3}S_1^{(0,1)} = s_{11}T_1^{(0,1)} + \frac{4bc^3}{3}d_{k1}E_k^{(0,1)},\tag{6.3.64}$$

which can be inverted as

$$T_{11}^{(0,1)} = \frac{4bc^3}{3s_{11}}(S_1^{(0,1)} - d_{11}E_1^{(0,1)}) .\tag{6.3.65}$$

The relevant strains and electric fields are

$$S_5^{(0,0)} = u_{3,1}^{(0,0)} + u_1^{(0,1)}, \quad S_1^{(0,1)} = u_{1,1}^{(0,1)},$$

$$E_1^{(0,0)} = -\phi_{,1}^{(0,0)}, \quad E_3^{(0,0)} = -\phi^{(0,1)}, \quad E_1^{(0,1)} = -\phi_{,1}^{(0,1)}. \tag{6.3.66}$$

With successive substitutions, Equation (6.3.55) can be written as three equations for $u_3^{(0,0)}$ and $u_1^{(0,1)}$ as well as the electrostatic potential $\phi^{(0,0)}$

$$\frac{1}{s_{55}}[\kappa^2 (u_{3,11}^{(0,0)} + u_{1,1}^{(0,1)}) + \kappa d_{15}\phi_{,11}^{(0,0)} + \kappa d_{35}\phi_{,1}^{(0,1)}] = \rho \ddot{u}_3^{(0,0)},$$

$$\frac{1}{s_{11}}(u_{1,11}^{(0,1)} + d_{11}\phi_{,11}^{(0,1)}) - \frac{3}{c^2 s_{55}}[\kappa^2 (u_{3,1}^{(0,0)} + u_1^{(0,1)}) \tag{6.3.67}$$

$$+ \kappa d_{15}\phi_{,1}^{(0,0)} + \kappa d_{35}\phi^{(0,1)}] = \rho \ddot{u}_1^{(0,1)},$$

$$\frac{d_{15}}{s_{55}}\kappa(u_{3,11}^{(0,0)} + u_{1,1}^{(0,1)}) - \tilde{\varepsilon}_{11}\phi_{,11}^{(0,0)} - \tilde{\varepsilon}_{13}\phi_{,1}^{(0,1)} + \frac{1}{4bc}D^{(0,0)} = 0,$$

where $E_2^{(0,0)}$ has been dropped in view of the ceramic piezoelectric transformer we are going to analyze later using these equations. The expression for $D_3^{(0,0)}$ in Equation (6.3.60)$_3$ is not used in Equation (6.3.67), but it is useful for determining the electric charge on an electrode when a beam is electroded at $x_3 = \pm c$. For a beam electroded at $x_3 = \pm c$, $\phi^{(0,0)}$ is no more than a function of time and Equation (6.3.67)$_3$ is not needed. For an unelectroded beam, Equation (6.3.67)$_3$ is needed to determine $\phi^{(0,0)}$.

6.3.3.2 Thickness-shear approximation

Next we study the thickness-shear approximation. When the beam is vibrating essentially at thickness-shear modes, the coupling to flexural motion is weak and can be eliminated by the thickness-shear approximation. For this simplification we proceed as follows. Consider the following simple wave solution:

$$\{u_3^{(0,0)}, \quad u_1^{(0,1)}, \quad \phi^{(0,0)}, \quad \phi^{(0,1)}\}$$

$$= \{A, \quad B, \quad C, \quad D\}\exp[i(\xi x_1 + \omega t)]. \tag{6.3.68}$$

Substituting Equation (6.3.68) into (6.3.67)$_1$ yields the following relation:

$$\frac{1}{s_{55}}[\kappa^2(-\xi^2 A + i\xi B) - \kappa d_{15}\xi^2 C + \kappa d_{35}i\xi D] = -\rho\omega^2 A. \qquad (6.3.69)$$

For long waves, the wave number ξ is small. Hence we drop the quadratic terms of ξ in Equation (6.3.69). Furthermore, since we consider vibrations at frequencies very close to the lowest thickness-shear frequency ω_∞ of an infinite beam, we set $\omega^2 \approx \omega_\infty^2$ in Equation (6.3.69). We then obtain the following approximation of Equation (6.3.69):

$$A = -\frac{1}{\rho\omega_\infty^2 s_{55}}(\kappa^2 i\xi B + \kappa d_{35}i\xi D), \qquad (6.3.70)$$

which is equivalent to the following differential relation:

$$u_3^{(0,0)} = -\frac{1}{\rho\omega_\infty^2 s_{55}}(\kappa^2 u_{1,1}^{(0,1)} + \kappa d_{35}\phi_{,1}^{(0,1)}). \qquad (6.3.71)$$

Substituting Equation (6.3.71) into Equation (6.3.67)$_2$, we obtain the following equation for long thickness-shear waves:

$$\begin{aligned}
&\frac{1}{s_{11}^*}u_{1,11}^{(0,1)} - \frac{3\kappa^2}{c^2 s_{55}}u_1^{(0,1)} \\
&+ \frac{d_{11}^*}{s_{11}^*}\phi_{,11}^{(0,1)} - \frac{3\kappa d_{35}}{c^2 s_{55}}\phi^{(0,1)} - \frac{3\kappa d_{15}}{c^2 s_{55}}\phi_{,1}^{(0,0)} = \rho\ddot{u}_1^{(0,1)},
\end{aligned} \qquad (6.3.72)$$

where

$$\frac{1}{s_{11}^*} = \frac{1}{s_{11}} + \frac{3\kappa^4}{\rho\omega_\infty^2 c^2 s_{55}^2}, \frac{d_{11}^*}{s_{11}^*} = \frac{d_{11}}{s_{11}} + \frac{3\kappa^3 d_{35}}{\rho\omega_\infty^2 c^2 s_{55}^2}. \qquad (6.3.73)$$

For the electric potential we drop the second derivative term of $u_3^{(0,0)}$ in Equation (6.3.67)$_3$, which is small for long waves, and obtain

$$\frac{d_{15}\kappa}{s_{55}}u_{1,1}^{(0,1)} - \tilde{\varepsilon}_{11}\phi_{,11}^{(0,0)} - \tilde{\varepsilon}_{13}\phi_{,1}^{(0,1)} + \frac{1}{4bc}D^{(0,0)} = 0. \qquad (6.3.74)$$

Thus we have eliminated the flexural displacement $u_3^{(0,0)}$ and obtained two equations, Equations (6.3.72) and (6.3.74), for the shear displacement $u_3^{(0,1)}$ and the electric potential $\phi^{(0,0)}$. Under the above thickness-shear approximation, the resultants are approximated by dropping second derivative terms

$$T_{13}^{(0,0)} = \frac{4bc}{s_{55}}(\kappa^2 u_1^{(0,1)} + \kappa d_{15}\phi_{,1}^{(0,0)} + \kappa d_{35}\phi^{(0,1)}),$$

$$T_{11}^{(0,1)} = \frac{4bc^3}{3s_{11}}(u_{1,1}^{(0,1)} + d_{11}\phi_{,1}^{(0,1)}),$$

$$\text{(6.3.75)}$$

$$D_1^{(0,0)} = 4bc(\frac{d_{15}}{s_{55}}\kappa u_1^{(0,1)} - \varepsilon_{11}^*\phi_{,1}^{(0,0)} - \varepsilon_{13}^*\phi^{(0,1)}),$$

$$D_3^{(0,0)} = 4bc(\frac{d_{35}}{s_{55}}\kappa u_1^{(0,1)} - \varepsilon_{31}^*\phi_{,1}^{(0,0)} - \varepsilon_{33}^*\phi^{(0,1)}).$$

6.3.4 Equations for ceramic bimorphs

Since a uniform electric field produces strains in a piezoelectric beam but not curvatures, a two-layered beam (bimorph) is usually used to generate bending [68,69]. We study two cases of thickness and axial poling below.

6.3.4.1 Thickness poling

Consider the ceramic bimorph in Figure 6.3.2. When the polarization is reversed, the elastic and dielectric constants remain the same but the piezoelectric constants change their signs.

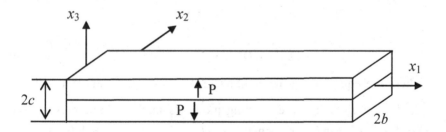

Figure 6.3.2. A ceramic bimorph with thickness poling.

We want to obtain one-dimensional equations for the elementary (classical) flexural motion of the beam bimorph. The major components of the mechanical displacement and electric potential are

$$u_1(x_1,x_2,x_3,t) \cong -x_2 u_{2,1}^{(0,0)}(x_1,t) - x_3 u_{3,1}^{(0,0)}(x_1,t),$$

$$u_2(x_1,x_2,x_3,t) \cong u_2^{(0,0)}(x_1,t),$$

$$u_3(x_1,x_2,x_3,t) \cong u_3^{(0,0)}(x_1,t),$$ (6.3.76)

$$\phi(x_1,x_2,x_3,t) \cong \phi^{(0,0)}(x_1,t) + x_2\phi^{(1,0)}(x_1,t) + x_3\phi^{(0,1)}(x_1,t),$$

where the thickness- and width-stretch displacements $u_2^{(1,0)}$ and $u_3^{(0,1)}$ are not explicitly given, but they are not zero. $\phi^{(0,0)}$ is responsible for the axial electric field in the x_1 direction, $\phi^{(1,0)}$ and $\phi^{(0,1)}$ are related to the lateral electric fields in the x_2 and x_3 directions. Equation (6.3.76) implies $S_1 \cong x_2 S_1^{(1,0)} + x_3 S_1^{(0,1)}$ and the following beam strains and electric fields:

$$S_1^{(1,0)} = -u_{2,11}^{(0,0)}, \quad S_1^{(0,1)} = -u_{3,11}^{(0,0)}, \tag{6.3.77}$$

$$E_1^{(0,0)} = -\phi_{,1}^{(0,0)}, \quad E_2^{(0,0)} = -\phi^{(1,0)}, \quad E_3^{(0,0)} = -\phi^{(0,1)}. \tag{6.3.78}$$

$S_1^{(1,0)}$ and $S_1^{(0,1)}$ represent bending curvatures in the x_2 and x_3 directions. We are only interested in the zero-order electric field. The equations of flexure and electrostatics are

$$T_{11,11}^{(1,0)} + F_{1,1}^{(1,0)} + F_2^{(0,0)} = 4\rho bc\ddot{u}_2^{(0,0)},$$

$$T_{11,11}^{(0,1)} + F_{1,1}^{(0,1)} + F_3^{(0,0)} = 4\rho bc\ddot{u}_3^{(0,0)}, \tag{6.3.79}$$

$$D_{1,1}^{(0,0)} + D^{(0,0)} = 0, \tag{6.3.80}$$

where the equations corresponding to $\phi^{(1,0)}$ and $\phi^{(0,1)}$ are not included. $T_{11}^{(1,0)}$ and $T_{11}^{(0,1)}$ are the bending moments in the x_2 and x_3 directions. For long and thin beams, since $T_{22} = T_{23} = 0$ at $x_2 = \pm b$ and $T_{33} = T_{32} = 0$ at $x_3 = \pm c$, these stress components are also very small inside the beam and are approximately zero everywhere. When the beam is not in pure bending, there are shear stresses T_{12} and T_{13} related to bending which are responsible for the shear resultants $T_{12}^{(0,0)}$ and $T_{13}^{(0,0)}$. For ceramics, shear and extension are not coupled. Therefore, in calculating the bending strains, the only stress component that needs to be considered is T_{11}.

With the compact notation, $T_1 = s_{11}^{-1}(S_1 - d_{k1}E_k)$ and the bending moment in the x_2 direction is

$$
\begin{aligned}
T_1^{(1,0)} &= \int_A T_{11}x_2 dA = \int_A \frac{1}{s_{11}}(S_1 - d_{k1}E_k)x_2 dA \\
&= S_1^{(1,0)} \int_A \frac{x_2^2}{s_{11}} dA - E_k^{(0,0)} \int_A x_2 \frac{d_{k1}}{s_{11}} dA = \frac{4b^3 c}{3s_{11}} S_1^{(1,0)},
\end{aligned}
\tag{6.3.81}
$$

which is not coupled to the electric field considered as expected. Similarly, the bending moment in the x_3 direction can be written as

$$
\begin{aligned}
T_1^{(0,1)} &= \int_A T_{11}x_3 dA = \int_A \frac{1}{s_{11}}(S_1 - d_{k1}E_k)x_3 dA \\
&= S_1^{(0,1)} \int_A \frac{x_3^2}{s_{11}} dA - E_k^{(0,0)} \int_A x_3 \frac{d_{k1}}{s_{11}} dA \\
&= \frac{4bc^3}{3s_{11}} S_1^{(0,1)} - \frac{2c^2 b d_{31}}{s_{11}} E_3^{(0,0)},
\end{aligned}
\tag{6.3.82}
$$

which is electrically coupled to $\phi^{(0,1)}$ as expected. The electric constitutive relation of interest is found to be

$$
\begin{aligned}
D_3^{(0,0)} &= \int_A D_3 dA = \int_A (d_{3p}T_p + \varepsilon_{3j}E_j)dA \\
&= \int_A (d_{31}T_1 + \varepsilon_{3j}E_j)dA \\
&= \int_A [d_{31}s_{11}^{-1}(S_1 - d_{k1}E_k) + \varepsilon_{3j}E_j]dA \\
&= \frac{2c^2 b d_{31}}{s_{11}} S_1^{(0,1)} + 4bc\widetilde{\varepsilon}_{ik}E_k^{(0,0)},
\end{aligned}
\tag{6.3.83}
$$

where $S_1 \cong x_2 S_1^{(1,0)} + x_3 S_1^{(0,1)}$ has been used and

$$
\widetilde{\varepsilon}_{ik} = \varepsilon_{ik}^T - d_{i1}d_{k1}/s_{11}.
\tag{6.3.84}
$$

Within the classical theory, the transverse shearing forces in the beam are related to bending moments by

$$
\begin{aligned}
T_{12}^{(0,0)} &= T_{11,1}^{(1,0)} + F_1^{(1,0)}, \\
T_{13}^{(0,0)} &= T_{11,1}^{(0,1)} + F_1^{(0,1)}.
\end{aligned}
\tag{6.3.85}
$$

6.3.4.2 Axial poling

Next consider the case when the two layers of a bimorph have opposite axial poling directions (see Figure 6.3.3).

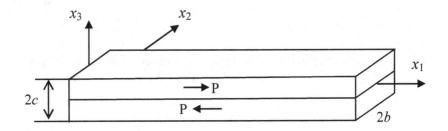

Figure 6.3.3. A ceramic bimorph with axial poling.

In this case the beam constitutive relations take the form

$$T_1^{(1,0)} = \frac{4b^3c}{3s_{33}} S_1^{(1,0)},$$

$$T_1^{(0,1)} = \frac{4bc^3}{3s_{33}} S_1^{(0,1)} - \frac{2c^2bd_{33}}{s_{33}} E_1^{(0,0)}, \qquad (6.3.86)$$

$$D_1^{(0,0)} = \frac{2c^2bd_{33}}{s_{33}} S_1^{(0,1)} + 4bc(\varepsilon_{33}^T - d_{33}^2 / s_{33})E_1^{(0,0)}. \qquad (6.3.87)$$

Equation (6.3.86) shows that $T_1^{(0,1)}$ is coupled to $E_1^{(0,0)}$ as expected.

6.3.5 A transformer — free vibration analysis

As an example, we use the beams equations developed to analyze a piezoelectric transformer [67] (see Figure 6.3.4).

It is assumed that $a,\overline{a} >> b,\overline{b} >> c,\overline{c}$. The driving portion $-\overline{a} < x_1 < 0$ is electroded at $x_3 = \pm\overline{c}$, with electrodes in the areas bounded by the thick lines. In the receiving portion $0 < x_1 < a$, the beam is electroded at the end $x_1 = a$. The driving portion and the receiving portion may have slightly different thickness $2\overline{c}$ and $2c$, and width $2\overline{b}$ and $2b$. V_1 is the input voltage and V_2 is the output voltage. The transformer is assumed to be made of an arbitrary piezoelectric material. When it is made of polarized ceramics, the polarizations in the driving and receiving portions are as shown in the figure.

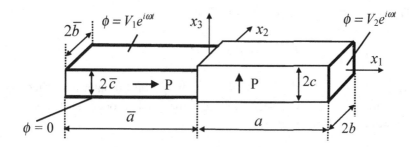

Figure 6.3.4. A thickness-shear piezoelectric transformer.

6.3.5.1 Governing equations

For the driving portion $-\bar{a} < x_1 < 0$, the electric potential is the known driving potential

$$\phi = \frac{x_3 + \bar{c}}{2\bar{c}} V_1 \exp(i\omega t),$$

$$\phi^{(0,0)} = \frac{1}{2} V_1 \exp(i\omega t), \quad \phi^{(0,1)} = \frac{1}{2\bar{c}} V_1 \exp(i\omega t). \tag{6.3.88}$$

The equation of motion under the thickness-shear approximation is from Equation (6.3.72)

$$\frac{1}{\bar{s}_{11}^*} u_{1,11}^{(0,1)} - \frac{3\bar{\kappa}^2}{\bar{c}^2 \bar{s}_{55}} u_1^{(0,1)} = \bar{\rho} \ddot{u}_1^{(0,1)}, \quad -\bar{a} < x_1 < 0, \tag{6.3.89}$$

where we have denoted all the geometric and material parameters of the driving portion by an over bar. The boundary condition of vanishing shear force is, from Equation (6.3.75),

$$T_{13}^{(0,0)} = \frac{4\bar{b}\bar{c}}{\bar{s}_{55}} (\bar{\kappa}^2 u_1^{(0,1)} + \bar{\kappa}\bar{d}_{35} \phi^{(0,1)}) = 0, \quad x_1 = -\bar{a}. \tag{6.3.90}$$

For the receiving portion $0 < x_1 < a$, $\phi^{(0,0)}$ is the major part of the electric potential. The equation of motion is from Equation (6.3.72):

$$\frac{1}{s_{11}^*}u_{1,11}^{(0,1)} - \frac{3\kappa^2}{c^2 s_{55}}u_1^{(0,1)} - \frac{3\kappa d_{15}}{c^2 s_{55}}\phi_{,1}^{(0,0)} = \rho\ddot{u}_1^{(0,1)}, \quad 0 < x_1 < a. \quad (6.3.91)$$

The electrostatic equation is from Equation (6.3.74):

$$\frac{d_{15}\kappa}{s_{55}}u_{1,1}^{(0,1)} - \tilde{\varepsilon}_{11}\phi_{,11}^{(0,0)} = 0, \quad 0 < x_1 < a, \quad (6.3.92)$$

where the term $D^{(0,0)}$ for the electric charge on the lateral surfaces $x_2 = \pm b$ and $x_3 = \pm c$ has been dropped for unelectroded surfaces. From Equation (6.3.92) we obtain

$$\phi_{,1}^{(0,0)} = \frac{d_{15}\kappa}{\tilde{\varepsilon}_{11}s_{55}}u_1^{(0,1)} + C_1, \quad (6.3.93)$$

where C_1 is an integration constant. Physically C_1 is related to the charge and hence the current on the electrode at $x_1 = a$. Substituting Equation (6.3.93) into Equation (6.3.91), we obtain

$$\frac{1}{s_{11}^*}u_{1,11}^{(0,1)} - \frac{3\kappa^2}{c^2 s_{55}^*}u_1^{(0,1)} - \frac{3\kappa d_{15}}{c^2 s_{55}}C_1 = \rho\ddot{u}_1^{(0,1)}, \quad 0 < x_1 < a,$$

$$\frac{1}{s_{55}^*} = \frac{1}{s_{55}}(1 + \frac{d_{15}^2}{\tilde{\varepsilon}_{11}s_{55}}). \quad (6.3.94)$$

For boundary conditions we need

$$T_{13}^{(0,0)} = \frac{4bc}{s_{55}}(\kappa^2 u_1^{(0,1)} + \kappa d_{15}\phi_{,1}^{(0,0)})$$

$$= \frac{4bc}{s_{55}^*}(\kappa^2 u_1^{(0,1)} + \frac{s_{55}^*\kappa d_{15}}{s_{55}}C_1), \quad (6.3.95)$$

$$D_1^{(0,0)} = 4bc(\frac{d_{15}\kappa}{s_{55}}u_1^{(0,1)} - \tilde{\varepsilon}_{11}\phi_{,1}^{(0,0)}) = -\tilde{\varepsilon}_{11}4bcC_1.$$

Then the boundary conditions take the following form:

$$T_{13}^{(0,0)} = \frac{4bc}{s_{55}^*}(\kappa^2 u_1^{(0,1)} + \frac{s_{55}^* \kappa d_{15}}{s_{55}} C_1) = 0, \quad x_1 = a,$$

$$\phi^{(0,0)} = V_2 \exp(i\omega t), \quad x_1 = a, \qquad (6.3.96)$$

$$\phi^{(0,0)}(0^+) = \frac{1}{2}V_1 \exp(i\omega t).$$

Since V_2 is unknown, some circuit condition across the receiving electrodes is needed to determine its value. For continuity conditions we impose

$$u_1^{(0,1)}(0^-) = u_1^{(0,1)}(0^+),$$
$$u_{1,1}^{(0,1)}(0^-) = u_{1,1}^{(0,1)}(0^+), \qquad (6.3.97)$$

in which the second equation is an approximation.

6.3.5.2 Free vibration analysis

Consider free vibrations for which the driving voltage V_1 or $\phi^{(0,0)}$ and $\phi^{(0,1)}$ vanish in the driving portion, which means that the driving electrodes are shorted. The receiving electrodes are open with no current or charge on the electrodes ($D_1^{(0,0)} = 0$). Hence C_1 is zero. This is the simplest circuit condition for the receiving electrodes. Then all the equations and boundary conditions become homogeneous. We need to solve an eigenvalue problem for ω^2

$$-u_{1,11}^{(0,1)} + \frac{3\bar{\kappa}^2 \bar{s}_{11}^*}{\bar{c}^2 \bar{s}_{55}} u_1^{(0,1)} = \bar{\rho}\bar{s}_{11}^* \omega^2 u_1^{(0,1)}, -\bar{a} < x_1 < 0,$$

$$-u_{1,11}^{(0,1)} + \frac{3\kappa^2 s_{11}^*}{c^2 s_{55}^*} u_1^{(0,1)} = \rho s_{11}^* \omega^2 u_1^{(0,1)}, 0 < x_1 < a, \qquad (6.3.98)$$

$$u_1^{(0,1)}(-\bar{a}) = 0, u_1^{(0,1)}(a) = 0,$$

$$u_1^{(0,1)}(0^-) = u_1^{(0,1)}(0^+), u_{1,1}^{(0,1)}(0^-) = u_{1,1}^{(0,1)}(0^+).$$

We try the following solution which already satisfies the two boundary conditions in (6.3.98)$_{3,4}$ at the two ends of the transformer:

$$u_1^{(0,1)} = \begin{cases} A\sin \bar{k}(x_1 + \bar{a}), -\bar{a} < x_1 < 0, \\ B\sin k(a - x_1), 0 < x_1 < a, \end{cases} \qquad (6.3.99)$$

where A , B , $\bar{k}$ and k are undetermined constants. Substitution of Equation (6.3.99) into Equation (6.3.98)$_{1,2}$ yields

$$\bar{k}^2 = \bar{s}_{11}^*(\bar{\rho}\omega^2 - \frac{3\bar{\kappa}^2}{\bar{c}^2\bar{s}_{55}}), k^2 = s_{11}^*(\rho\omega^2 - \frac{3\kappa^2}{c^2 s_{55}^*}), \qquad (6.3.100)$$

which shows that the solution of $u_1^{(0,1)}$ may have exponential or sinusoidal behaviors depending on the signs of $\bar{k}^2$ and k^2. This is related to the energy-trapping phenomenon of thickness-shear modes. For the operating mode of a transformer, sinusoidal behavior in both portions is desired. Hence we consider the case when both $\bar{k}^2$ and k^2 are positive:

$$\bar{\rho}\omega^2 - \frac{3\bar{\kappa}^2}{\bar{c}^2\bar{s}_{55}} > 0, \quad \rho\omega^2 - \frac{3\kappa^2}{c^2 s_{55}^*} > 0. \qquad (6.3.101)$$

Substituting Equation (6.3.99) into the continuity conditions in Equation (6.3.98)$_{5,6}$ we have

$$A\sin(\bar{k}\bar{a}) = B\sin(ka),$$
$$\bar{k}A\cos(\bar{k}\bar{a}) = -Bk\cos(ka), \qquad (6.3.102)$$

which, for nontrivial solutions of A and B, yields the frequency equation

$$\frac{\tan(\bar{k}\bar{a})}{\tan(ka)} = -\frac{\bar{k}}{k}. \qquad (6.3.103)$$

The corresponding mode shape function is

$$u_1^{(0,1)} = \begin{cases} \dfrac{\sin \bar{k}(x_1 + \bar{a})}{\sin(\bar{k}\bar{a})}, & -\bar{a} < x_1 < 0, \\ \dfrac{\sin k(a - x_1)}{\sin(ka)}, & 0 < x_1 < a. \end{cases} \qquad (6.3.104)$$

Then the electric potential can be found as

$$\phi^{(0,0)} = \begin{cases} 0, & -\bar{a} < x_1 < 0, \\ \cos k(a - x_1) - \cos(ka), & 0 < x_1 < a. \end{cases} \qquad (6.3.105)$$

For a transformer it is also desired that the wave number k of the operating mode in the receiving portion satisfies

$$k^2 = s_{11}^*(\rho\omega^2 - \frac{3\kappa^2}{c^2 s_{55}^*}) \le (\frac{\pi}{a})^2, \qquad (6.3.106)$$

so that the receiving portion is not longer than one-half of the wave length in the x_1 direction of the thickness-shear mode. Then the shear deformation does not change its sign and the voltage generated by the shear deformation accumulates spatially without cancellation. Solutions satisfying this condition will be obtained for a ceramic transformer below.

6.3.5.3 Ceramic transformers

Consider a ceramic transformer of constant width $b = \overline{b}$. The driving portion is polarized in the x_1 direction and the receiving portion is polarized in the x_3 direction. For ceramics, usually

$$| d_{15} | > | d_{31} |, \quad | d_{15} | > | d_{33} |. \tag{6.3.107}$$

For example,

$$
\begin{array}{cccc}
 & d_{15} & d_{31} & d_{33} \\
PZT-2 & 440 & -60 & 152 \\
PZT-5H & 741 & -274 & 593 \\
PZT-7A & 362 & -60 & 150
\end{array}
\tag{6.3.108}
$$

This suggests that the thickness-shear transformer discussed here which operates with d_{15} may be more effective in transforming than the conventional Rosen extensional transformers which use d_{31} and d_{33} [9]. For a ceramic transformer, the equations derived above take specific forms. We have

$$
\begin{aligned}
\overline{\omega}_\infty^2 &= \frac{\pi^2}{4\overline{c}^2 \rho s_{44}}, \quad -\overline{a} < x_1 < 0, \\
\omega_\infty^2 &= \frac{\pi^2}{4c^2 \rho s_{44}(1-k_{15}^2)}, \quad 0 < x_1 < a,
\end{aligned}
\tag{6.3.109}
$$

because for free vibrations the electrodes in the driving portion are shorted and the electrodes for the receiving portion are open. In addition,

$$
\begin{aligned}
\frac{1}{\overline{s}_{11}^*} &= \frac{1}{s_{33}} + \frac{3\overline{\kappa}^4}{\rho\overline{\omega}_\infty^2 \overline{c}^2 s_{44}^2} = \frac{1}{s_{33}} + \frac{\pi^2}{12}\frac{1}{s_{44}}, \\
\frac{1}{s_{11}^*} &= \frac{1}{s_{11}} + \frac{3\kappa^4}{\rho\omega_\infty^2 c^2 s_{44}^2} = \frac{1}{s_{11}} + \frac{\pi^2}{12}\frac{1-k_{15}^2}{s_{44}},
\end{aligned}
\tag{6.3.110}
$$

$$\widetilde{\varepsilon}_{11} = \varepsilon_{11}^{T} - \frac{d_{15}d_{15}}{s_{44}} = \varepsilon_{11}(1 - k_{15}^2),$$

$$\frac{1}{s_{55}^*} = \frac{1}{s_{44}}(1 + \frac{d_{15}^2}{\widetilde{\varepsilon}_{11}s_{44}}) = \frac{1}{s_{44}}\frac{1}{1 - k_{15}^2},$$

(6.3.111)

$$\overline{k}^2 = \overline{s}_{11}^*(\rho\omega^2 - \frac{3\overline{\kappa}^2}{\overline{c}^2 s_{44}}) = \overline{s}_{11}^*\rho(\omega^2 - \overline{\omega}_\infty^2),$$

$$k^2 = s_{11}^*(\rho\omega^2 - \frac{3\kappa^2}{c^2 s_{55}^*}) = s_{11}^*\rho(\omega^2 - \omega_\infty^2).$$

(6.3.112)

First consider the case of a transformer with constant thickness $\overline{c} = c$ and equal length of the driving and receiving portions $\overline{a} = a$. We have the inequality

$$\overline{\omega}_\infty^2 = \frac{\pi^2}{4c^2\rho s_{44}} < \omega_\infty^2 = \frac{\pi^2}{4c^2\rho s_{44}(1 - k_{15}^2)}.$$

(6.3.113)

Before we start solving the frequency equation, we note that the transformer is assumed to be long and, for voltage accumulation, the entire receiving portion of the transformer should be vibrating in phase. Therefore, the transformer should be vibrating at a frequency very close to and slightly higher than the infinite beam thickness-shear frequency of the receiving portion, with a very small wave number k. Hence we should approximately have

$$\omega^2 \approx \omega_\infty^2 = \frac{\pi^2}{4c^2\rho s_{44}(1 - k_{15}^2)}.$$

(6.3.114)

Then, at this frequency, for the driving portion, a finite wave number can be approximately determined by

$$\overline{k}^2 = s_{11}^*\rho(\omega^2 - \overline{\omega}_\infty^2) \approx \overline{s}_{11}^*\rho(\omega_\infty^2 - \overline{\omega}_\infty^2).$$

(6.3.115)

For PZT-5H, by trial and error, it can be quickly found that, when a = 20 cm and c = 1 cm, the first root of the frequency equation is 59.66 kHz. It is indeed very close to, and slightly higher than the infinite beam frequency ω_∞ of the receiving portion, which is found to be 59.31 kHz. We then have $\overline{k}a = 5.4\pi$ and $ka = 0.86\pi$. The corresponding shear and potential distributions are shown in Figure 6.3.5. The electric potential rises in the receiving portion and a voltage is generated between $x_1 = 0$

and $x_1 = a$. Hence this mode can be used as an operating mode of the transformer. While in the receiving portion the shear deformation does not change sign along the beam and voltage is accumulated, in the driving portion the shear changes sign a few times. This is fine for a transformer, because this mode can be excited by a few pairs of electrodes in the driving portion with alternating signs of driving voltages among pairs of driving electrodes. Transformers with a short driving portion can also be designed, with fewer pairs of driving electrodes.

$$u_1^{(1)}, \phi^{(0,0)}$$

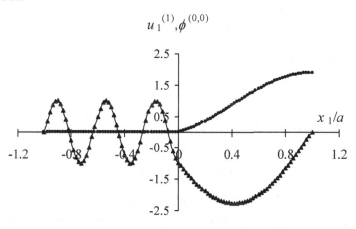

Figure 6.3.5. Shear and potential distributions of the operating mode ($\bar{c} = c$). $u_1^{(1)}$ is marked by triangles, and $\phi^{(0,0)}$ by circles.

The reason that the transformer needs several pairs of driving electrodes is the change of sign of the shear deformation in the driving portion. This is a result of the fact that the infinite beam frequency of the driving portion is lower than that of the receiving portion. The infinite beam frequencies of the driving and receiving portions can be made close by adjusting $\bar{c}$ and c. This can lead to a reduction of the wave number in the driving portion and hence fewer pairs of driving electrodes. To lower the infinite beam shear frequency of the receiving portion, we increase the beam thickness c of the receiving portion such that $\bar{c} < c$, where $\bar{c}$ and c will be properly adjusted to suit our need as follows. We consider the case of equal length of the driving and receiving portions

($\bar{a} = a$). It can be verified that the following is a limit solution to the frequency equation

$$\bar{k}a = \frac{\pi^+}{2}, \quad \tan(\bar{k}a) = -\infty,$$

$$ka = \frac{\pi^-}{2}, \quad \tan(ka) = \infty. \tag{6.3.116}$$

Then

$$\bar{k}^2 a^2 = a^2 \bar{s}_{11}^* \rho(\omega^2 - \bar{\omega}_\infty^2) = \pi^2 / 4,$$

$$k^2 a^2 = a^2 s_{11}^* \rho(\omega^2 - \omega_\infty^2) = \pi^2 / 4, \tag{6.3.117}$$

which leads to the frequency

$$\omega^2 = \bar{\omega}_\infty^2 + \frac{\pi^2}{4a^2 \bar{s}_{11}^* \rho} = \omega_\infty^2 + \frac{\pi^2}{4a^2 s_{11}^* \rho}. \tag{6.3.118}$$

From Equations (6.3.118) and (6.3.109) we have the condition

$$\frac{1}{\bar{c}^2 s_{44}} + \frac{1}{a^2 \bar{s}_{11}^*} = \frac{1}{c^2 s_{44}(1 - k_{15}^2)} + \frac{1}{a^2 s_{11}^*}, \tag{6.3.119}$$

which can be satisfied by adjusting the geometrical parameters $\bar{c}$ and c. The mode shapes of $u_1^{(1)}$ and $\phi^{(0,0)}$ are shown in Figure 6.3.6, which can be driven by one pair of electrodes.

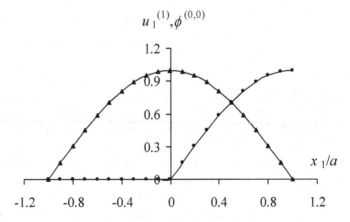

Figure 6.3.6. Shear and potential distributions of the operating mode ($\bar{c} < c$). $u_1^{(1)}$ is marked by triangles, and $\phi^{(0,0)}$ by circles.

Similarly, it can be verified that the following is also a limit solution to the frequency equation

$$\bar{k}a = \pi^+, \quad \tan(\bar{k}a) = 0^+,$$
$$ka = \pi^-, \quad \tan(ka) = 0^-, \tag{6.3.120}$$

$$\bar{k}^2 a^2 = a^2 \bar{s}_{11}^* \rho(\omega^2 - \bar{\omega}_\infty^2) = \pi^2,$$
$$k^2 a^2 = a^2 s_{11}^* \rho(\omega^2 - \omega_\infty^2) = \pi^2, \tag{6.3.121}$$

$$\omega^2 = \bar{\omega}_\infty^2 + \frac{\pi^2}{a^2 \bar{s}_{11}^* \rho} = \omega_\infty^2 + \frac{\pi^2}{a^2 s_{11}^* \rho}, \tag{6.3.122}$$

$$\frac{1}{4\bar{c}^2 s_{44}} + \frac{1}{a^2 \bar{s}_{11}^*} = \frac{1}{4c^2 s_{44}(1 - k_{15}^2)} + \frac{1}{a^2 s_{11}^*}. \tag{6.3.123}$$

The mode shapes for the shear deformation and electric potential in this case is shown in Figure 6.3.7, which can also be driven by one pair of electrodes.

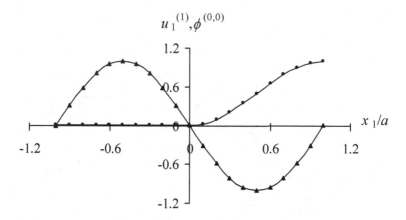

Figure 6.3.7. Shear and potential distributions of the operating mode ($\bar{c} < c$). $u_1^{(1)}$ is marked by triangles, and $\phi^{(0,0)}$ by circles.

Chapter 7
Piezoelectric Rings

In this chapter we consider motions of a piezoelectric ring. Although the ring is in a plane, its motion can be three-dimensional. We are interested in coupled extensional and flexural motions with shear deformations, but not torsion. Most of the equations for rings can be obtained from the shell equations, but the constitutive relations should be from the beams equations.

7.1 First-Order Theory

Consider a differential element of a ring in the x_3-x_1 plane as shown in Figure 7.1.1. For a ring, cylindrical coordinates (θ, z, r) corresponding to $(\alpha_1, \alpha_2, \alpha_3)$ are sufficient. Specifically, $x_3 = r\cos\theta$, $x_1 = r\sin\theta$, and $x_2 = z$.

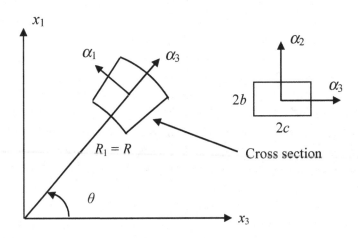

Figure 7.1.1. An element of a ring.

For a first-order shear deformation theory we make the following expansions of the displacement and electric potential:

$$u_j \cong u_j^{(0,0)}(\alpha_1,t) + \alpha_2 u_j^{(1,0)}(\alpha_1,t) + \alpha_3 u_j^{(0,1)}(\alpha_1,t)$$

$$+ \alpha_2^2 u_j^{(2,0)}(\alpha_1,t) + \alpha_2\alpha_3 u_j^{(1,1)}(\alpha_1,t) + \alpha_3^2 u_j^{(0,2)}(\alpha_1,t), \qquad (7.1.1)$$

$$\phi \cong \phi^{(0,0)}(\alpha_1,t) + \alpha_2\phi^{(1,0)}(\alpha_1,t) + \alpha_3\phi^{(0,1)}(\alpha_1,t),$$

where the thickness-stretch displacements $u_2^{(1,0)}$ and $u_3^{(0,1)}$ will be eliminated by stress relaxations. The geometry of a ring can be reduced from that of a shell with

$$A_1 = R, \quad A_2 = 1,$$
$$R_1 = R, \quad R_2 = \infty. \qquad (7.1.2)$$

Then the strains and electric field can be written as:

$$S_{kl} = S_{kl}^{(0,0)} + \alpha_2 S_{kl}^{(1,0)} + \alpha_3 S_{kl}^{(0,1)},$$
$$E_k = E_k^{(0,0)} + \alpha_2 E_k^{(1,0)} + \alpha_3 E_k^{(0,1)}. \qquad (7.1.3)$$

The zero-order strains are

$$S_{11}^{(0,0)} = \frac{1}{A_1}\frac{\partial u_1^{(0,0)}}{\partial \alpha_1} + \frac{u_3^{(0,0)}}{R}, \quad S_{22}^{(0,0)} = u_2^{(1,0)},$$

$$S_{33}^{(0,0)} = u_3^{(0,1)}, \quad 2S_{23}^{(0,0)} = u_2^{(0,1)} + u_3^{(1,0)},$$

$$2S_{31}^{(0,0)} = u_1^{(0,1)} - \frac{u_1^{(0,0)}}{R} + \frac{1}{A_1}\frac{\partial u_3^{(0,0)}}{\partial \alpha_1}, \qquad (7.1.4)$$

$$2S_{12}^{(0,0)} = u_1^{(1,0)} + \frac{1}{A_1}\frac{\partial u_2^{(0,0)}}{\partial \alpha_1}.$$

$S_{22}^{(0)}$, $S_{33}^{(0)}$ and $S_{23}^{(0)}$ will be eliminated by stress relaxations. For the first-order strains, we only need the following expressions:

$$S_{11}^{(1,0)} = \frac{1}{A_1}\frac{\partial u_1^{(1,0)}}{\partial \alpha_1} + \frac{u_3^{(1,0)}}{R} \cong \frac{1}{A_1}\frac{\partial u_1^{(1,0)}}{\partial \alpha_1},$$

$$S_{11}^{(0,1)} = \frac{1}{A_1}\frac{\partial u_1^{(0,1)}}{\partial \alpha_1} + \frac{u_3^{(0,1)}}{R} \cong \frac{1}{A_1}\frac{\partial u_1^{(0,1)}}{\partial \alpha_1}. \qquad (7.1.5)$$

The rest of the first-order strains will be eliminated by stress relaxations. The relevant electric fields are

$$E_1^{(0,0)} = -\frac{1}{A_1}\frac{\partial \phi^{(0)}}{\partial \alpha_1}, \quad E_2^{(0,0)} = -\phi^{(1,0)}, \quad E_3^{(0,0)} = -\phi^{(0,1)},$$

$$E_1^{(1,0)} = -\frac{1}{A_1}\frac{\partial \phi^{(1,0)}}{\partial \alpha_1}, \quad E_2^{(1,0)} = 0, \quad E_3^{(1,0)} = 0, \qquad (7.1.6)$$

$$E_1^{(0,1)} = -\frac{1}{A_1}\frac{\partial \phi^{(0,1)}}{\partial \alpha_1}, \quad E_2^{(0,1)} = 0, \quad E_3^{(0,1)} = 0.$$

The equations of motion and the charge equations are

$$\frac{1}{A_1}\frac{\partial N_{11}}{\partial \alpha_1} + Q_{13}\frac{1}{R} + F_1^{(0,0)} = 4\rho bc\ddot{u}_1^{(0,0)},$$

$$\frac{1}{A_1}\frac{\partial Q_{12}}{\partial \alpha_1} + F_2^{(0,0)} = 4\rho bc\ddot{u}_2^{(0,0)}, \qquad (7.1.7)$$

$$\frac{1}{A_1}\frac{\partial Q_{13}}{\partial \alpha_1} - N_{11}\frac{1}{R} + F_3^{(0,0)} = 4\rho bc\ddot{u}_3^{(0,0)},$$

$$\frac{1}{A_1}\frac{\partial M_{11}^{(1,0)}}{\partial \alpha_1} - Q_{12} + F_1^{(1,0)} = \frac{4\rho b^3 c}{3}\ddot{u}_1^{(1,0)},$$

$$\frac{1}{A_1}\frac{\partial M_{11}^{(0,1)}}{\partial \alpha_1} - Q_{13} + F_1^{(0,1)} = \frac{4\rho bc^3}{3}\ddot{u}_1^{(0,1)}, \qquad (7.1.8)$$

and

$$\frac{1}{A_1}\frac{\partial D_1^{(0,0)}}{\partial \alpha_1} + \frac{1}{R}D_3^{(0,0)} + D^{(0,0)} = 0,$$

$$\frac{1}{A_1}\frac{\partial D_1^{(1,0)}}{\partial \alpha_1} + \frac{1}{R}D_3^{(1,0)} - D_2^{(0,0)} + D^{(1,0)} = 0, \qquad (7.1.9)$$

$$\frac{1}{A_1}\frac{\partial D_1^{(0,1)}}{\partial \alpha_1} + \frac{1}{R}D_3^{(0,1)} - D_3^{(0,0)} + D^{(0,1)} = 0,$$

where the resultants are defined by

$$\{N_{11}, Q_{12}, Q_{13}\} = \{T_{11}^{(0)}, T_{12}^{(0)}, T_{13}^{(0)}\} = \int_A T_{ij}\, dA,$$

$$M_{11}^{(m,n)} = T_{11}^{(m,n)} = \int_A T_{ab}\alpha_2^m \alpha_3^n\, dA, \tag{7.1.10}$$

$$D_i^{(m,n)} = \int_A D_i \alpha_2^m \alpha_3^n\, dA.$$

$A = 4bc$ is the cross sectional area. The surface and body loads are

$$F_j^{(m,n)} = b^m \int_{-c}^{c} [T_{2j}(b) - (-1)^m T_{2j}(-b)]\alpha_3^n\, d\alpha_3$$

$$+ c^n \int_{-b}^{b} [T_{3j}(c) - (-1)^n T_{3j}(-c)]\alpha_2^m\, d\alpha_2$$

$$+ \int_A \rho f_j \alpha_2^m \alpha_3^n\, dA, \tag{7.1.11}$$

$$D^{(m,n)} = b^m \int_{-c}^{c} [D_2(b) - (-1)^m D_2(-b)]\alpha_3^n\, d\alpha_3$$

$$+ c^n \int_{-b}^{b} [D_3(c) - (-1)^n D_3(-c)]\alpha_2^m\, d\alpha_2.$$

The constitutive relations are the same as those in the first-order theory of beams

$$T_\alpha^{(0,0)} = 4bc(\hat{c}_{\alpha\beta}' S_\beta^{(0,0)} - \hat{e}_{i\alpha}' E_i^{(0,0)}), \quad \alpha, \beta = 1, 5, 6,$$
$$D_i^{(0,0)} = 4bc(\hat{e}_{i\beta}' S_\beta^{(0,0)} + \hat{\varepsilon}_{ik}' E_k^{(0,0)}), \tag{7.1.12}$$

$$T_1^{(1,0)} = \frac{4b^3 c}{3}(\tilde{c}_{11} S_1^{(1,0)} - \tilde{e}_{k1} E_k^{(1,0)}),$$
$$D_i^{(1,0)} = \frac{4b^3 c}{3}(\tilde{e}_{i1} S_1^{(1,0)} + \tilde{\varepsilon}_{ik} E_k^{(1,0)}), \tag{7.1.13}$$

$$T_1^{(0,1)} = \frac{4bc^3}{3}(\tilde{c}_{11} S_1^{(0,1)} - \tilde{e}_{i1} E_i^{(0,1)}),$$
$$D_i^{(0,1)} = \frac{4bc^3}{3}(\tilde{e}_{i1} S_1^{(0,1)} + \tilde{\varepsilon}_{ik} E_k^{(0,1)}), \tag{7.1.14}$$

$$\hat{c}_{\alpha\beta} = c_{\alpha\beta} - c_{\alpha\lambda}c_{\lambda\mu}^{-1}c_{\mu\beta}, \quad \hat{e}_{i\alpha} = e_{i\alpha} - e_{i\mu}c_{\mu\lambda}^{-1}c_{\lambda\alpha},$$

$$\hat{\varepsilon}_{ik} = \varepsilon_{ik} + e_{i\lambda}c_{\lambda\mu}^{-1}e_{k\mu}, \quad \alpha,\beta = 1,5,6, \quad \lambda,\mu,\nu = 2,3,4, \tag{7.1.15}$$

$$\tilde{c}_{11} = \frac{1}{s_{11}}, \quad \tilde{e}_{k1} = \frac{d_{k1}}{s_{11}}, \quad \tilde{\varepsilon}_{ik} = \varepsilon_{ik}^T - \frac{d_{i1}d_{k1}}{s_{11}}.$$

With successive substitutions, Equations (7.1.7) through (7.1.9) can be written as eight equations for $u_1^{(0,0)}$, $u_2^{(0,0)}$, $u_3^{(0,0)}$, $u_1^{(1,0)}$, $u_1^{(0,1)}$, $\phi^{(0,0)}$, $\phi^{(1,0)}$ and $\phi^{(0,1)}$. For end conditions we can prescribe

$$T_1^{(0,0)} \quad \text{or} \quad u_1^{(0,0)}, \quad T_6^{(0,0)} \quad \text{or} \quad u_2^{(0,0)}, \quad T_5^{(0,0)} \quad \text{or} \quad u_3^{(0,0)},$$

$$T_1^{(1,0)} \quad \text{or} \quad u_1^{(1,0)}, \quad T_1^{(0,1)} \quad \text{or} \quad u_1^{(0,1)}, \tag{7.1.16}$$

$$\phi^{(0,0)} \quad \text{or} \quad D_1^{(0,0)}, \quad \phi^{(1,0)} \quad \text{or} \quad D_1^{(1,0)}, \quad \phi^{(0,1)} \quad \text{or} \quad D_1^{(0,1)}.$$

7.2 Classical Theory

The classical theory for flexure can be obtained from the first-order theory by two approximations, i.e., dropping the rotatory inertia and making the shear strains vanish. The resulting equations are

$$\frac{1}{A_1}\frac{\partial N_{11}}{\partial \alpha_1} + Q_{13}\frac{1}{R} + F_1^{(0,0)} = 4\rho bc\ddot{u}_1^{(0,0)},$$

$$\frac{1}{A_1}\frac{\partial Q_{12}}{\partial \alpha_1} + F_2^{(0,0)} = 4\rho bc\ddot{u}_2^{(0,0)}, \tag{7.2.1}$$

$$\frac{1}{A_1}\frac{\partial Q_{13}}{\partial \alpha_1} - N_{11}\frac{1}{R} + F_3^{(0,0)} = 4\rho bc\ddot{u}_3^{(0,0)},$$

$$\frac{1}{A_1}\frac{\partial D_1^{(0,0)}}{\partial \alpha_1} + \frac{1}{R}D_3^{(0,0)} + D^{(0,0)} = 0,$$

$$\frac{1}{A_1}\frac{\partial D_1^{(1,0)}}{\partial \alpha_1} + \frac{1}{R}D_3^{(1,0)} - D_2^{(0,0)} + D^{(1,0)} = 0, \tag{7.2.2}$$

$$\frac{1}{A_1}\frac{\partial D_1^{(0,1)}}{\partial \alpha_1} + \frac{1}{R}D_3^{(0,1)} - D_3^{(0,0)} + D^{(0,1)} = 0,$$

$$T_1^{(0,0)} = 4bc(\hat{c}'_{11} S_1^{(0,0)} - \hat{e}'_{i1} E_i^{(0,0)}),$$
$$D_i^{(0,0)} = 4bc(\hat{e}'_{i1} S_1^{(0,0)} + \hat{\varepsilon}_{ik} E_k^{(0,0)}), \qquad (7.2.3)$$

$$T_1^{(1,0)} = \frac{4b^3 c}{3}(\tilde{c}_{11} S_1^{(1,0)} - \tilde{e}_{k1} E_k^{(1,0)}),$$
$$D_i^{(1,0)} = \frac{4b^3 c}{3}(\tilde{e}_{i1} S_1^{(1,0)} + \tilde{\varepsilon}_{ik} E_k^{(1,0)}), \qquad (7.2.4)$$

$$T_1^{(0,1)} = \frac{4bc^3}{3}(\tilde{c}_{11} S_1^{(0,1)} - \tilde{e}_{i1} E_i^{(0,1)}),$$
$$D_i^{(0,1)} = \frac{4bc^3}{3}(\tilde{e}_{i1} S_1^{(0,1)} + \tilde{\varepsilon}_{ik} E_k^{(0,1)}), \qquad (7.2.5)$$

$$Q_{12} = \frac{1}{A_1}\frac{\partial M_{11}^{(1,0)}}{\partial \alpha_1} + F_1^{(1,0)}, \quad Q_{13} = \frac{1}{A_1}\frac{\partial M_{11}^{(0,1)}}{\partial \alpha_1} + F_1^{(0,1)}, \qquad (7.2.6)$$

$$S_{11}^{(0,0)} = \frac{1}{A_1}\frac{\partial u_1^{(0,0)}}{\partial \alpha_1} + \frac{u_3^{(0,0)}}{R},$$
$$S_{11}^{(1,0)} \cong \frac{1}{A_1}\frac{\partial u_1^{(1,0)}}{\partial \alpha_1}, \quad S_{11}^{(0,1)} \cong \frac{1}{A_1}\frac{\partial u_1^{(0,1)}}{\partial \alpha_1}, \qquad (7.2.7)$$

$$u_1^{(1,0)} = -\frac{1}{A_1}\frac{\partial u_2^{(0,0)}}{\partial \alpha_1}, \quad u_1^{(0,1)} = -\frac{1}{A_1}\frac{\partial u_3^{(0,0)}}{\partial \alpha_1} + \frac{u_1^{(0,0)}}{R}. \qquad (7.2.8)$$

With successive substitutions, Equations (7.2.1) and (7.2.2) can be written as six equations for $u_1^{(0,0)}$, $u_2^{(0,0)}$, $u_3^{(0,0)}$, $\phi^{(0,0)}$, $\phi^{(1,0)}$ and $\phi^{(0,1)}$. For end conditions we can prescribe

$$T_1^{(0,0)} \quad \text{or} \quad u_1^{(0,0)}, \quad T_{12}^{(0,0)} \quad \text{or} \quad u_2^{(0,0)}, \quad T_{13}^{(0,0)} \quad \text{or} \quad u_3^{(0,0)},$$

$$T_1^{(1,0)} \quad \text{or} \quad \frac{\partial u_2^{(0,0)}}{\partial \alpha_1}, \quad T_1^{(0,1)} \quad \text{or} \quad \frac{\partial u_3^{(0,0)}}{\partial \alpha_1}, \qquad (7.2.8)$$

$$\phi^{(0,0)} \quad \text{or} \quad D_1^{(0,0)}, \quad \phi^{(1,0)} \quad \text{or} \quad D_1^{(1,0)}, \quad \phi^{(0,1)} \quad \text{or} \quad D_1^{(0,1)}.$$

A more physical development of the classical theory of a piezoelectric ring can be found in [71].

7.3 Radial Vibration of a Ceramic Ring

As an application of the equations obtained, consider axi-symmetric radial vibrations of a thin ceramic ring with radial poling, electroded on its inner and outer surfaces (see Figure 7.3.1).

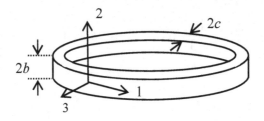

Figure 7.3.1. A ceramic ring with radial poling.

Let R be the mean radius. We assume that $R \gg 2b \gg 2c$. The electrodes are shorted so that the lowest order electric fields vanish. From Equation (7.2.1)$_3$ we have

$$-N_{11}\frac{1}{R} = 4\rho bc\ddot{u}_3^{(0,0)}, \tag{7.3.1}$$

where

$$N_{11} = T_1^{(0,0)} = 4bc\hat{c}'_{11}S_1^{(0,0)}, \tag{7.3.2}$$

$$S_1^{(0,0)} = \frac{u_3^{(0,0)}}{R}, \tag{7.3.3}$$

$$\hat{c}'_{11} = \frac{1}{s_{11}}. \tag{7.3.4}$$

With successive substitutions, Equation (7.3.1) becomes

$$-\frac{1}{s_{11}}\frac{u_3^{(0,0)}}{R^2} = \rho\ddot{u}_3^{(0,0)}. \tag{7.3.5}$$

For time-harmonic motions, the frequency is given by

$$\omega^2 = \frac{1}{\rho s_{11} R^2},$$

(7.3.6)

which is the same as the result given in [9].

Chapter 8
Piezoelectric Parallelepipeds

In this chapter, zero-dimensional equations for motions of a piezoelectric parallelepiped are derived from the three-dimensional equations for linear piezoelectricity by triple power series expansions of the mechanical displacement and electric potential in the length, width and thickness directions. The equations obtained are convenient to use when modeling the motion of a piezoelectric parallelepiped in a particular vibration mode. Many piezoelectric devices are in essentially single-mode vibrations. For these devices the zero-dimensional theory can yield results useful for many purposes. The lowest order equations obtained can describe motions with homogeneous deformations and uniform electric fields. The material in this chapter is mainly from [72].

8.1 Power Series Expansion

Consider a piezoelectric rectangular parallelepiped as shown in Figure 8.1.1. The coordinate system is formed by the centroidal principal axes.

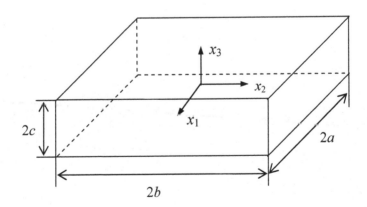

Figure 8.1.1. A rectangular piezoelectric parallelepiped.

To develop a zero-dimensional theory, we make the following expansions of the mechanical displacement vector and the electrostatic potential:

$$u_j(x_1, x_2, x_3, t) = \sum_{l,m,n=0}^{\infty} x_1^l x_2^m x_3^n u_j^{(l,m,n)}(t),$$

$$\phi(x_1, x_2, x_3, t) = \sum_{l,m,n=0}^{\infty} x_1^l x_2^m x_3^n \phi^{(l,m,n)}(t).$$

(8.1.1)

Then the strains and the electric field can be written as

$$S_{ij} = \sum_{l,m,n=0}^{\infty} x_1^l x_2^m x_3^n S_{ij}^{(l,m,n)},$$

$$E_i = \sum_{l,m,n=0}^{\infty} x_1^l x_2^m x_3^n E_i^{(l,m,n)},$$

(8.1.2)

where

$$S_{ij}^{(l,m,n)} = \frac{1}{2}[(l+1)(\delta_{i1}u_j^{(l+1,m,n)} + \delta_{j1}u_i^{(l+1,m,n)})$$
$$+ (m+1)(\delta_{i2}u_j^{(l,m+1,n)} + \delta_{j2}u_i^{(l,m+1,n)})$$
$$+ (n+1)(\delta_{i3}u_j^{(l,m,n+1)} + \delta_{j3}u_i^{(l,m,n+1)})],$$

(8.1.3)

$$E_i^{(l,m,n)} = -\delta_{i1}(l+1)\phi^{(l+1,m,n)}$$
$$- \delta_{i2}(m+1)\phi^{(l,m+1,n)} - \delta_{i3}(n+1)\phi^{(l,m,n+1)}.$$

(8.1.4)

Substituting Equation (8.1.1) into the variational formulation of linear piezoelectricity in Equation (1.2.26), with integration by parts, for independent variations of $\delta u_j^{(l,m,n)}$ and $\delta\phi^{(l,m,n)}$, we obtain the following zero-dimensional equations of motion and electrostatics:

$$-lT_{1j}^{(l-1,m,n)} - mT_{2j}^{(l,m-1,n)} - nT_{3j}^{(l,m,n-1)}$$
$$+ F_j^{(l,m,n)} = \rho \sum_{p,q,r=0}^{\infty} B_{l+p,m+q,n+r} \ddot{u}_j^{(p,q,r)},$$

(8.1.5)

$$-lD_1^{(l-1,m,n)} - mD_2^{(l,m-1,n)} - nD_3^{(l,m,n-1)} + D^{(l,m,n)} = 0,$$

where the stress and the electric displacement of various orders are defined by

$$T_{ij}^{(l,m,n)} = \int_V T_{ij} x_1^l x_2^m x_3^n dV,$$

$$D_i^{(l,m,n)} = \int_V D_i x_1^l x_2^m x_3^n dV. \qquad (8.1.6)$$

$V = 8abc$ is the volume of the parallelepiped. Surface traction, surface charge and body force of various orders have the following expressions:

$$F_j^{(l,m,n)} = T_j^{(l,m,n)} + \rho f_j^{(l,m,n)},$$

$$T_j^{(l,m,n)} = \int_{-b}^b dx_2 \int_{-c}^c dx_3 [T_{1j}(x_1 = a) - (-1)^l T_{1j}(x_1 = -a)] a^l x_2^m x_3^n$$

$$+ \int_{-a}^a dx_1 \int_{-c}^c dx_3 [T_{2j}(x_2 = b) - (-1)^m T_{2j}(x_2 = -b)] b^m x_1^l x_3^n \qquad (8.1.7)$$

$$+ \int_{-a}^a dx_1 \int_{-b}^b dx_2 [T_{3j}(x_3 = c) - (-1)^n T_{3j}(x_3 = -c)] c^n x_1^l x_2^m,$$

$$f_j^{(l,m,n)} = \int_V \rho f_j x_1^l x_2^m x_3^n dV,$$

$$D^{(l,m,n)} = \int_{-b}^b dx_2 \int_{-c}^c dx_3 [D_1(x_1 = a) - (-1)^l D_1(x_1 = -a)] a^l x_2^m x_3^n$$

$$+ \int_{-a}^a dx_1 \int_{-c}^c dx_3 [D_2(x_2 = b) - (-1)^m D_2(x_2 = -b)] b^m x_1^l x_3^n \qquad (8.1.8)$$

$$+ \int_{-a}^a dx_1 \int_{-b}^b dx_2 [D_3(x_3 = c) - (-1)^n D_3(x_3 = -c)] c^n x_1^l x_2^m.$$

$B_{l+p,m+q,n+r}$ is a geometric quantity defined by

$$B_{l+p,m+q,n+r} = \int_V x_1^{l+p} x_2^{m+q} x_3^{n+r} dV$$

$$= \begin{cases} \dfrac{8a^{l+p+1} b^{m+q+1} c^{n+r+1}}{(l+p+1)(m+q+1)(n+r+1)}, & l+p, m+q, n+r \text{ even,} \\ 0, & \text{else.} \end{cases} \qquad (8.1.9)$$

The zero-dimensional constitutive relations are

$$T_{ij}^{(l,m,n)} = \sum_{p,q,r=0}^{\infty} B_{l+p,m+q,n+r} (c_{ijkl} S_{kl}^{(p,q,r)} - e_{kij} E_k^{(p,q,r)}),$$

$$D_i^{(l,m,n)} = \sum_{p,q,r=0}^{\infty} B_{l+p,m+q,n+r} (e_{ikl} S_{kl}^{(p,q,r)} + \varepsilon_{ik} E_k^{(p,q,r)}),$$

(8.1.10)

or, with the compact matrix notation, for $M, N = 1, 2, \ldots, 6$

$$T_M^{(l,m,n)} = \sum_{p,q,r=0}^{\infty} B_{l+p,m+q,n+r} (c_{MN} S_N^{(p,q,r)} - e_{kM} E_k^{(p,q,r)}),$$

$$D_i^{(l,m,n)} = \sum_{p,q,r=0}^{\infty} B_{l+p,m+q,n+r} (e_{iN} S_N^{(p,q,r)} + \varepsilon_{ik} E_k^{(p,q,r)}).$$

(8.1.11)

With successive substitutions, these equations can be written as ordinary differential equations for the mechanical displacements and electric potentials of various orders.

8.2 Zero-Order Equations

We keep $u_j^{(0,0,0)}$, $u_j^{(1,0,0)}$, $u_j^{(0,1,0)}$, $u_j^{(0,0,1)}$, $\phi^{(1,0,0)}$, $\phi^{(0,1,0)}$, and $\phi^{(0,0,1)}$, and neglect all higher order displacements and potentials. For $(l,m,n) = (0,0,0)$, $(1,0,0)$, $(0,1,0)$ and $(0,0,1)$, we have the following equations of motion and electrostatics:

$$F_j^{(0,0,0)} = \rho 8abc \ddot{u}_j^{(0,0,0)},$$

$$D^{(0,0,0)} = 0,$$

(8.2.1)

$$-T_{1j}^{(0,0,0)} + F_j^{(1,0,0)} = \rho \frac{8a^3 bc}{3} \ddot{u}_j^{(1,0,0)},$$

$$-D_1^{(0,0,0)} + D^{(1,0,0)} = 0,$$

(8.2.2)

$$-T_{2j}^{(0,0,0)} + F_j^{(0,1,0)} = \rho \frac{8ab^3 c}{3} \ddot{u}_j^{(0,1,0)},$$

$$-D_2^{(0,0,0)} + D^{(0,1,0)} = 0,$$

(8.2.3)

$$-T_{3j}^{(0,0,0)} + F_j^{(0,0,1)} = \rho \frac{8abc^3}{3} \ddot{u}_j^{(0,0,1)},$$

$$-D_3^{(0,0,0)} + D^{(0,0,1)} = 0.$$

(8.2.4)

The constitutive relations for the above equations are

$$T_M^{(0,0,0)} = 8abc(c_{MN}S_N^{(0,0,0)} - e_{kM}E_k^{(0,0,0)}),$$

$$D_i^{(0,0,0)} = 8abc(e_{iN}S_N^{(0,0,0)} + \varepsilon_{ik}E_k^{(0,0,0)}),$$

(8.2.5)

in which the zero-order strains and electric fields are given by

$$S_1^{(0,0,0)} = u_1^{(1,0,0)}, \quad S_4^{(0,0,0)} = u_2^{(0,0,1)} + u_3^{(0,1,0)},$$

$$S_2^{(0,0,0)} = u_2^{(0,1,0)}, \quad S_5^{(0,0,0)} = u_3^{(1,0,0)} + u_1^{(0,0,1)},$$

$$S_3^{(0,0,0)} = u_3^{(0,0,1)}, \quad S_6^{(0,0,0)} = u_1^{(0,1,0)} + u_2^{(1,0,0)},$$

(8.2.6)

$$E_1^{(0,0,0)} = -\phi^{(1,0,0)}, \quad E_2^{(0,0,0)} = -\phi^{(0,1,0)}, \quad E_3^{(0,0,0)} = -\phi^{(0,0,1)}. \quad (8.2.7)$$

These equations can be written as equations for $u_j^{(0,0,0)}$, $u_j^{(1,0,0)}$, $u_j^{(0,1,0)}$, $u_j^{(0,0,1)}$, $\phi^{(1,0,0)}$, $\phi^{(0,1,0)}$ and $\phi^{(0,0,1)}$, which govern the rigid body motion, homogeneous stretching and shearing deformations of the parallelepiped and uniform electric fields.

8.3 A Piezoelectric Gyroscope

As an application of the zero-dimensional theory developed above, we revisit the thickness-shear ceramic plate piezoelectric gyroscope analyzed in the fourth section of the second chapter. Consider the rectangular ceramic plate shown in Figure 8.3.1.

The plate is poled in the thickness direction. A driving voltage V_1 is applied across the lateral electrodes at $x_1 = \pm a$ to excite the plate into thickness-shear motion u_1 in the x_1 direction. When the plate is rotating about the x_3 axis, the Coriolis force F_2 causes a thickness-shear motion u_2 in the x_2 direction. This shear in the x_2 direction generates a voltage V_2 between $x_2 = \pm b$, which can be used to detect the angular rate Ω.

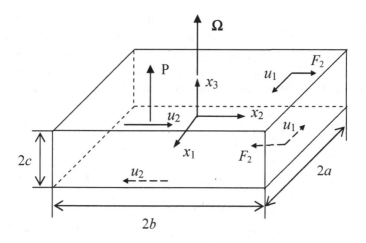

Figure 8.3.1. A thickness-shear piezoelectric gyroscope.

8.3.1 Governing equations

The lowest order zero-dimensional equations in the previous section are for a parallelepiped. In the derivation, there are no assumptions made on the relative magnitude of the length a, width b, and thickness c. The equations can be used to analyze the rectangular plate gyroscope in Figure 8.3.1 as long as the plate is in almost homogeneous thickness-shear vibration, which is the operating mode of the gyroscope. The equations for shear motions $u_1^{(0,0,1)}$ and $u_2^{(0,0,1)}$ are

$$-T_{31}^{(0,0,0)} = \rho \frac{8abc^3}{3}(\ddot{u}_1^{(0,0,1)} - 2\Omega\dot{u}_2^{(0,0,1)} - \Omega^2 u_1^{(0,0,1)}),$$

$$-T_{32}^{(0,0,0)} = \rho \frac{8abc^3}{3}(\ddot{u}_2^{(0,0,1)} + 2\Omega\dot{u}_1^{(0,0,1)} - \Omega^2 u_2^{(0,0,1)}),$$

(8.3.1)

where we have included Coriolis and centripetal accelerations and omitted the surface and body force terms which are not present in our gyroscope problem. The coordinate system is assumed to be rotating with the gyroscope. For piezoelectric gyroscopes, the Coriolis acceleration is responsible for the sensing mechanism. The centripetal acceleration can be neglected for most purposes. This is because piezoelectric gyroscopes are small sensors operating near resonance with high frequencies. Usually Ω is much smaller than the operating frequency. Therefore the centripetal acceleration, the third terms on the right hand sides of Equation $(8.3.1)_{1,2}$, which are quadratic in Ω, are small compared with the Corioils acceleration, the second terms of the equations. The relevant constitutive relations take the following form:

$$T_{32}^{(0,0,0)} = 8abc(c_{44}\kappa^2 S_4^{(0,0,0)} - e_{15}\kappa E_2^{(0,0,0)}),$$
$$T_{31}^{(0,0,0)} = 8abc(c_{44}\kappa^2 S_5^{(0,0,0)} - e_{15}\kappa E_1^{(0,0,0)}),$$
(8.3.2)

$$D_1^{(0,0,0)} = 8abc(e_{15}\kappa S_5^{(0,0,0)} + \varepsilon_{11}E_1^{(0,0,0)}),$$
$$D_2^{(0,0,0)} = 8abc(e_{15}\kappa S_4^{(0,0,0)} + \varepsilon_{11}E_2^{(0,0,0)}).$$
(8.3.3)

In Equations (8.3.2) and (8.3.3) we have introduced a thickness-shear correction factor κ to compensate for the error caused by truncating the series. For a ceramic plate we have $\kappa^2 = \pi^2/12$ when the plate is poled in the thickness direction. For convenience we introduce the notation

$$u_1^{(0,0,1)} = U_1, \quad u_2^{(0,0,1)} = U_2,$$
$$\phi^{(1,0,0)} = V_1/2a, \quad \phi^{(0,1,0)} = V_2/2b.$$
(8.3.4)

Then

$$S_4^{(0,0,0)} = U_2, \quad S_5^{(0,0,0)} = U_1,$$
$$E_1^{(0,0,0)} = -V_1/2a, \quad E_2^{(0,0,0)} = -V_2/2b.$$
(8.3.5)

With successive substitutions, we obtain

$$-c_{44}\kappa^2 U_1 - e_{15}\kappa V_1/2a = \rho\frac{c^2}{3}(\ddot{U}_1 - 2\Omega\dot{U}_2 - \Omega^2 U_1),$$
$$-c_{44}\kappa^2 U_2 - e_{15}\kappa V_2/2b = \rho\frac{c^2}{3}(\ddot{U}_2 + 2\Omega\dot{U}_1 - \Omega^2 U_2),$$
(8.3.6)

$$D_1^{(0,0,0)} = 8abc(e_{15}\kappa U_1 - \varepsilon_{11}V_1/2a),$$
$$D_2^{(0,0,0)} = 8abc(e_{15}\kappa U_2 - \varepsilon_{11}V_2/2b).$$
(8.3.7)

The total electric charge on the electrodes at $x_1 = a$ or $x_2 = b$ and the electric currents flowing out of them are given by

$$Q_1 = -D_1^{(0,0,0)}/(2a), \quad Q_2 = -D_2^{(0,0,0)}/(2b),$$
$$I_1 = -\dot{Q}_1, \quad I_2 = -\dot{Q}_2.$$
(8.3.8)

The driving voltage V_1 is usually considered known and is time-harmonic. The sensing electrodes at $x_2 = \pm b$ are usually connected by an output circuit with impedance Z for harmonic motions. In the special cases when $Z = 0$ or ∞, we have short or open output circuit with $V_2 = 0$ or $I_2 = 0$. In general, neither V_2 nor I_2 is known and a circuit equation is needed. Let the known time-harmonic driving voltage be $V_1 = i\overline{V}_1 \exp(i\omega t)$. Introducing the complex notation

$$
\begin{aligned}
U_1 &= i\overline{U}_1 \exp(i\omega t), \quad U_2 = \overline{U}_2 \exp(i\omega t), \\
I_2 &= \overline{I}_2 \exp(i\omega t), \quad V_2 = \overline{V}_2 \exp(i\omega t),
\end{aligned}
\tag{8.3.9}
$$

we can write the circuit condition as

$$
\overline{I}_2 = \overline{V}_2 / Z .
\tag{8.3.10}
$$

From Equations $(8.3.7)_2$, $(8.3.8)_{2,4}$ and $(8.3.10)$, we obtain

$$
4aci\omega(e_{15}\kappa\overline{U}_2 - \varepsilon_{11}\overline{V}_2 / 2b) = \overline{V}_2 / Z .
\tag{8.3.11}
$$

Then we have the following equations:

$$
\begin{aligned}
&-c_{44}\kappa^2\overline{U}_1 - e_{15}\kappa\overline{V}_1 / 2a \\
&= \rho\frac{c^2}{3}(-\omega^2\overline{U}_1 - 2\Omega\omega\overline{U}_2 - \Omega^2\overline{U}_1), \\
&-c_{44}\kappa^2\overline{U}_2 - e_{15}\kappa\overline{V}_2 / 2b \\
&= \rho\frac{c^2}{3}(-\omega^2\overline{U}_2 - 2\Omega\omega\overline{U}_1 - \Omega^2\overline{U}_2), \\
&4ac\omega(e_{15}\kappa\overline{U}_2 - \varepsilon_{11}\overline{V}_2 / 2b) = \overline{V}_2 / iZ(\omega).
\end{aligned}
\tag{8.3.12}
$$

Equation (8.3.12) represent three linear algebraic equations for $\overline{U}_1$, $\overline{U}_2$ and $\overline{V}_2$ in which $\overline{V}_1$ is the inhomogeneous driving term. In general Z is a function of ω. The specific form of this function depends on the structure of the load circuit.

8.3.2 Free vibration

For free vibrations we set $\overline{V}_1 = 0$, which physically means that the driving electrodes are shorted. Then Equation (8.3.12) reduces to a system of homogeneous equations for $\overline{U}_1$, $\overline{U}_2$ and $\overline{V}_2$. For nontrivial solutions, the determinant of the coefficient matrix must vanish, which leads to the following frequency equation:

$$\Delta(\omega) = (\omega^2 + \Omega^2 - \omega_\infty^2)^2$$
$$- 4\Omega^2 \omega^2 - \lambda(\omega)\omega_\infty^2(\omega^2 + \Omega^2 - \omega_\infty^2) = 0, \tag{8.3.13}$$

where we have denoted

$$\omega_\infty^2 = \frac{3c_{44}\kappa^2}{\rho c^2}, \quad \lambda(\omega) = \frac{k_{15}^2}{1 + Z_2(\omega)/Z(\omega)},$$
$$k_{15}^2 = \frac{e_{15}^2}{\varepsilon_{11}c_{44}}, \quad Z_2 = \frac{1}{i\omega C_2}, \quad C_2 = \frac{\varepsilon_{11}4ac}{2b}. \tag{8.3.14}$$

ω_∞ is the lowest thickness-shear resonant frequency of an elastic plate with shear constant c_{44}, mass density ρ, and thickness $2c$. The correction factor κ makes this frequency to be the same as that obtained from the three-dimensional equations. k_{15}^2 is an electro-mechanical coupling factor that represents the strength of this coupling of the material. C_2 is a static capacitance. Equation (8.3.13) can be written in the following form:

$$\omega^2 = (1 + \frac{1}{2}\lambda)\omega_\infty^2 + \Omega^2 \pm \frac{1}{2}\lambda\omega_\infty^2\sqrt{1 + \frac{16 + 8\lambda}{\lambda^2}\frac{\Omega^2}{\omega_\infty^2}}. \tag{8.3.15}$$

Strictly speaking, Equation (8.3.15) is not a frequency solution to Equation (8.3.13) because, in general, λ is a function of ω. In the special case of shorted output electrodes we have $Z = 0$, $\lambda = 0$. Then Equation (8.3.15) reduces to

$$\omega = \omega_\infty \pm \Omega, \tag{8.3.16}$$

which is a frequency solution. When the output electrodes are open we have $Z = \infty$, $\lambda = k_{15}^2$. In this case Equation (8.3.15) also represents a frequency solution. Piezoelectric gyroscopes usually operate under the condition that $\Omega \ll \omega_\infty$. From Equation (8.3.15) we can see that, for small Ω and open output electrodes, the effect of Ω on ω is quadratic. This is different from the special case of shorted output electrodes as shown by Equation (8.3.16). If the load circuit is essentially capacitive with a capacitance C, we have $Z = 1/(i\omega C)$ and $Z_2/Z = C/C_2$ which is independent of ω. Then Equation (8.3.15) represents a frequency solution.

As an example, consider PZT-5H. The relation of ω versus Ω for $Z/Z_2 = 0$, 0.2, and ∞ is plotted in Figure 8.3.2. For each value of Z, there are two frequencies that represent the two lowest modes of thickness-shear vibration. For the case of shorted output electrodes ($Z = 0$), there are no electric fields because the driving electrodes are also shorted. In this case the piezoelectric stiffening effect due to electric fields does not exist. Therefore the two resonant frequencies are the same when there is no rotation, because ceramics are transversely isotropic in the x_1-x_2 plane. However, rotation will cause these two frequencies to split. When the receiving electrodes are not shorted, there is an electric field in the x_2 direction that causes stiffening of the material and hence higher resonant frequencies. In this case, even if the plate is not rotating, the two shear frequencies are different.

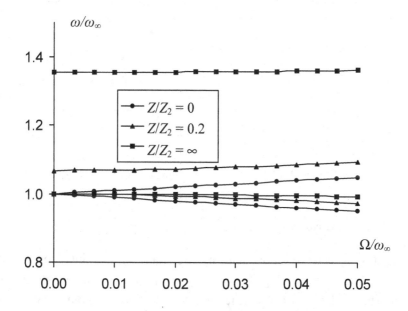

Figure 8.3.2. Resonant frequency versus the rotation rate Ω.

ω versus Z for the case of a capacitive output circuit is plotted in Figure 8.3.3 for fixed $\Omega/\omega_\infty = 0$, 0.025, and 0.05. The figure shows that the resonant frequencies vary according to the load. Together with the dependence of frequency on the rotation rate shown in Figure 8.3.2, the load dependence of frequency further complicates the design of these gyroscopes because the resonant frequencies have to be predicted and controlled accurately for the gyroscope to operate in resonant conditions with high sensitivity.

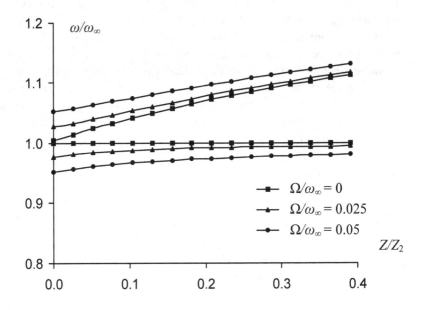

Figure 8.3.3. Resonant frequency versus the load Z.

8.3.3 Forced vibration

For forced vibration analysis, we want to take into consideration some effect of damping. We use $c_{44}^* = c_{44}(1 + iQ^{-1})$ for the shear elastic constant, where c_{44} and Q (the quality factor) are real and the value of Q for ceramics is usually on the order of 10^2 to 10^3. In the following, we fix the value of Q as 10^2 in our calculations. From Equation (8.3.12), we obtain the forced vibration solution for the output voltage sensitivity as

$$\frac{\overline{V}_2}{\overline{V}_1} = -(k_{15}^*)^2 \frac{b}{a} \frac{2\Omega\omega(\omega_\infty^*)^2}{(1 + Z_2 / Z)\Delta(\omega)}, \qquad (8.3.17)$$

where

$$(k_{15}^*)^2 = \frac{e_{15}^2}{\varepsilon_{11} c_{44}^*}, \quad (\omega_\infty^*)^2 = \frac{3 c_{44}^* \kappa^2}{\rho c^2}. \tag{8.3.18}$$

Voltage sensitivity as a function of the driving frequency is plotted in Figure 8.3.4 for a fixed Ω and two values of Z. It is seen that near the two resonant frequencies the sensitivity assumes maxima. The distance between the two resonant frequencies depends on Z, as also suggested by Figure 8.3.3. Numerical tests also show that if smaller values of Q are used in the calculation, the peaks become narrower and higher. Theoretically, higher peaks imply higher sensitivity. However, in reality, narrower peaks require better control in tuning the device into resonant conditions.

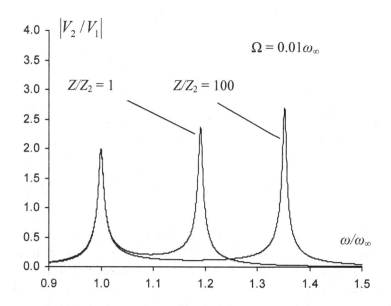

Figure 8.3.4. Sensitivity versus the driving frequency ω.

The dependence of the voltage sensitivity on the rotation rate Ω is shown in Figure 8.3.5 for a fixed driving frequency near resonance and for different values of the load Z. When Ω is much smaller than ω_0, which is true in most applications of piezoelectric gyroscopes, the relation between the sensitivity and Ω is essentially linear. Therefore, in the analysis of piezoelectric gyroscopes, very often the centrifugal force which represents higher order effects of Ω is neglected and the contribution to sensitivity is totally from the Coriolis force which is linear in Ω.

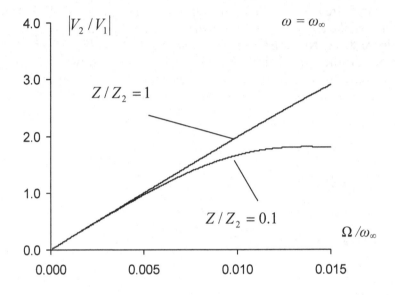

Figure 8.3.5. Sensitivity versus the rotation rate Ω.

The variation of sensitivity according to the load Z is also of interest in practice and is given in Figure 8.3.6 for a fixed driving frequency near resonance and for different values of Ω. For small loads the sensing electrodes are almost shorted and the voltage sensitivity is small although the output current may be large. As the load increases, the sensitivity increases and exhibits an almost linear range. When the load is large enough the output electrodes are essentially open with the output voltage saturated and a very small output current.

Finally, we note that in the case of open output electrodes ($Z = \infty$) and without material damping, if the effects of piezoelectric coupling and rotation on resonant frequencies are neglected, Equation (8.3.17) reduces to

$$\frac{\bar{V}_2}{\bar{V}_1} \cong -k_{15}^2 \frac{b}{a} \frac{2\Omega\omega\omega_\infty^2}{(\omega^2 - \omega_\infty^2)^2} \cong -k_{15}^2 \frac{b}{2a} \frac{\Omega\omega}{(\omega - \omega_\infty)^2}, \qquad (8.3.19)$$

which shows the most basic behavior of the gyroscope (compare to Equation (2.4.136)).

Figures 8.3.4 through 8.3.6 are qualitatively similar to those of the mass-rod gyroscope in [9].

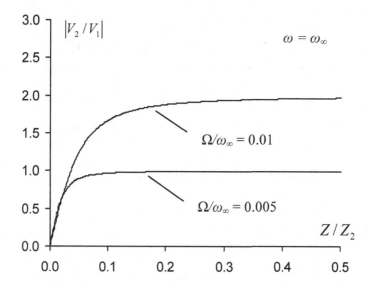

Figure 8.3.6. Sensitivity versus the load Z.

8.4 A Transformer — Forced Vibration Analysis

A free vibration analysis of a thickness-shear piezoelectric transformer was performed in the third section of the sixth chapter using one-dimensional equations. Frequency equation and modes were obtained, showing the mechanism of the transformer. For a complete analysis of a piezoelectric transformer, a forced vibration analysis is necessary. Although the thickness-shear transformer analyzed in the third section of the sixth chapter is a thin rod (see Figure 8.4.1), its operating modes shown in Figure 6.3.6 or Figure 6.3.7 are slowly varying in the driving and receiving portions when the transformer is long, and therefore can be approximated by the zero-dimensional equations. In this section we perform a forced vibration analysis of the transformer using zero-dimensional equations.

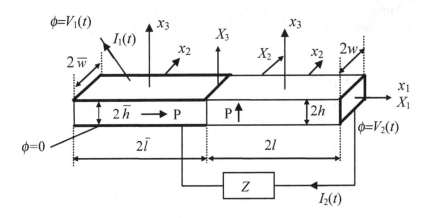

Figure 8.4.1. A thickness-shear piezoelectric transformer.

8.4.1 Governing equations

The driving portion is determined by $-\bar{l} < x_1 < 0$. The receiving portion is $0 < x_1 < l$. V_1 is the input voltage and V_2 is the output voltage. For each portion of the transformer there exists a local Cartesian coordinate system x_i with its origin at the center of the portion and directions along the global system X_i. The thickness-shear motion we are considering can be approximately represented by

$$u_1 \cong x_3 u_1^{(0,0,1)}, \quad u_2 \cong 0, \quad u_3 \cong 0 . \tag{8.4.1}$$

In the driving portion the electric potential is the known driving potential with

$$\phi \cong \phi^{(0,0,0)} + x_3 \phi^{(0,0,1)},$$
$$\phi^{(0,0,0)} = V_1 / 2, \quad \phi^{(0,0,1)} = V_1 /(2\bar{h}). \tag{8.4.2}$$

From Equation $(8.2.4)_1$, setting $j = 1$, we obtain the equation of motion as

$$-T_{31}^{(0,0,0)} + F_1^{(0,0,1)} = \rho \frac{8\overline{l}\overline{w}\overline{h}^3}{3} \ddot{u}_1^{(0,0,1)}, \qquad (8.4.3)$$

where

$$F_1^{(0,0,1)} = \int_{-\overline{w}}^{\overline{w}} \int_{-\overline{h}}^{\overline{h}} T_{11}(X_1 = 0^-)X_3 dX_2 dX_3. \qquad (8.4.4)$$

Relevant constitutive relations are taken from Equation (8.2.5)

$$T_{31}^{(0,0,0)} = 8\overline{l}\overline{w}\overline{h}(c_{44}\kappa^2 S_5^{(0,0,0)} - e_{15}\kappa E_3^{(0,0,0)}),$$
$$D_3^{(0,0,0)} = 8\overline{l}\overline{w}\overline{h}(e_{15}\kappa S_5^{(0,0,0)} + \varepsilon_{11}E_3^{(0,0,0)}). \qquad (8.4.5)$$

In Equation (8.4.5) we have introduced a thickness-shear correction factor κ. For a ceramic parallelepiped we can use $\kappa^2 = \pi^2/12$ as an approximation. For the charge and current on the electrode at $X_3 = \overline{h}$, we have

$$Q_1 = -D_3^{(0,0,0)}/(2\overline{h}), \quad I_1 = -\dot{Q}_1. \qquad (8.4.6)$$

In the receiving portion, the electric potential can be written as

$$\phi \cong \phi^{(0,0,0)} + x_1\phi^{(1,0,0)},$$
$$\phi^{(0,0,0)} = V_2/2, \quad \phi^{(1,0,0)} = V_2/(2l), \qquad (8.4.7)$$

where V_2 is unknown. The equation of motion is

$$-T_{31}^{(0,0,0)} + F_1^{(0,0,1)} = \rho \frac{8lwh^3}{3} \ddot{u}_1^{(0,0,1)}, \qquad (8.4.8)$$

where

$$F_1^{(0,0,1)} = \int_{-\overline{w}}^{\overline{w}} \int_{-\overline{h}}^{\overline{h}} -T_{11}(X_1 = 0^+)X_3 dX_2 dX_3. \qquad (8.4.9)$$

Relevant constitutive relations are

$$T_{31}^{(0,0,0)} = 8lwh(c_{44}\kappa^2 S_5^{(0,0,0)} - e_{15}\kappa E_1^{(0,0,0)}),$$
$$D_1^{(0,0,0)} = 8lwh(e_{15}\kappa S_5^{(0,0,0)} + \varepsilon_{11}E_1^{(0,0,0)}). \qquad (8.4.10)$$

The electric charge on the electrode at $X_1 = 2l$ and the electric current flows out of the electrode are given by

$$Q_2 = -D_1^{(0,0,0)}/(2l), \quad I_2 = -\dot{Q}_2. \qquad (8.4.11)$$

In deriving the above we have assumed that there is no body force and made use of the traction-free boundary conditions at $X_1 = 2l$, $X_2 = \pm w$, and $X_3 = \pm h$.

Substituting Equation (8.4.10)$_1$ into Equation (8.4.8), and Equation (8.4.5)$_1$ into Equation (8.4.3), adding the resulting equations and making use of the continuity conditions

$$F_1^{(0,0,1)}(X_1 = 0^-) = -F_1^{(0,0,1)}(X_1 = 0^+),$$

$$u_1(X_1 = 0^-) = u_1(X_1 = 0^+),$$

(8.4.12)

we obtain

$$-8lwh(c_{44}\kappa^2 S_5^{(0,0,0)} - e_{15}\kappa E_1^{(0,0,0)})$$

$$-8\bar{l}\,\overline{w}\,\overline{h}(c_{44}\kappa^2 S_5^{(0,0,0)} - e_{15}\kappa E_3^{(0,0,0)})$$

$$= \frac{8}{3}\rho(lwh^3 + \bar{l}\,\overline{w}\,\overline{h}^3)\ddot{u}_1^{(0,0,1)}.$$

(8.4.13)

Under the known time-harmonic driving voltage $V_1 = \overline{V}_1 \exp(i\omega t)$, for time-harmonic solutions we employ the complex notation and write the unknowns as

$$u_1^{(0,0,1)} = \overline{u}\exp(i\omega t), \quad V_2 = \overline{V}_2\exp(i\omega t),$$

$$I_1 = \overline{I}_1\exp(i\omega t), \quad I_2 = \overline{I}_2\exp(i\omega t).$$

(8.4.14)

The receiving electrodes are connected by an output circuit which, when the motion is time-harmonic, has an impedance Z_L. We have the following circuit condition

$$\overline{I}_2 = \overline{V}_2 / Z_L.$$

(8.4.15)

With successive substitutions, we obtain the following two equations for $\overline{u}$ and $\overline{V}_2$, driven by $\overline{V}_1$

$$-8lwh(c_{44}\kappa^2\overline{u} + e_{15}\kappa\frac{\overline{V}_2}{2l}) - 8\bar{l}\,\overline{w}\,\overline{h}(c_{44}\kappa^2\overline{u} + e_{15}\kappa\frac{\overline{V}_1}{2h})$$

$$= -\frac{8}{3}\rho(lwh^3 + \bar{l}\,\overline{w}\,\overline{h}^3)\omega^2\overline{u},$$

(8.4.16)

$$4whi\omega(e_{15}\kappa\overline{u} - \varepsilon_{11}\frac{\overline{V}_2}{2l}) = \frac{\overline{V}_2}{Z_L}.$$

Once $\bar{u}$ and $\bar{V}_2$ are obtained, the currents are given by

$$\bar{I}_2 = 4whi\omega(e_{15}\kappa\bar{u} - \varepsilon_{11}\frac{\bar{V}_2}{2l}),$$

$$\bar{I}_1 = 4\bar{l}\bar{w}i\omega(e_{15}\kappa\bar{u} - \varepsilon_{11}\frac{\bar{V}_1}{2\bar{h}}). \tag{8.4.17}$$

8.4.2 Forced vibration analysis

We consider the case of $l = \bar{l}$ and $w = \bar{w}$. Solving Equations (8.4.16) for $\bar{u}$ and $\bar{V}_2$, we obtain the transforming ratio and normalized input and output currents as

$$\frac{\bar{V}_2}{\bar{V}_1} = \frac{k_{15}^2}{(1+h/\bar{h})(1+Z_2/Z_L)(\omega^2/\omega_0^2 - 1) - k_{15}^2 h/\bar{h}} \frac{l}{\bar{h}},$$

$$\frac{\bar{I}_2}{(\bar{V}_1/Z_2)} = \frac{k_{15}^2}{(1+h/\bar{h})(1+Z_L/Z_2)(\omega^2/\omega_0^2 - 1) - k_{15}^2 hZ_L/(\bar{h}Z_2)} \frac{l}{\bar{h}},$$

$$-\frac{\bar{I}_1}{(\bar{V}_1/Z_1)} = 1 - \frac{k_{15}^2(1+Z_2/Z_L)}{(1+h/\bar{h})(1+Z_2/Z_L)(\omega^2/\omega_0^2 - 1) - k_{15}^2 h/\bar{h}},$$

$$\tag{8.4.18}$$

where

$$k_{15}^2 = \frac{e_{15}^2}{\varepsilon_{11}c_{44}}, \quad \omega_\infty^2 = \frac{3\kappa^2 c_{44}}{\rho(h^2 - h\bar{h} + \bar{h}^2)},$$

$$Z_1 = \frac{1}{i\omega C_1}, \quad C_1 = \frac{\varepsilon_{11}4\bar{l}\bar{w}}{2\bar{h}}, \tag{8.4.19}$$

$$Z_2 = \frac{1}{i\omega C_2}, \quad C_2 = \frac{\varepsilon_{11}4wh}{2l}.$$

In Equation (8.4.19), ω_∞ is the thickness-shear resonant frequency for shorted receiving electrodes ($Z_L = 0$) as predicted by the zero-dimensional theory, C_1 and C_2 are the static capacitance of the driving and receiving portions, and Z_1 and Z_2 are the impedance of the two portions. As a numerical example, we consider PZT-5H. Material damping is included by allowing c_{pq} to assume complex values. c_{44} is replaced by $c_{44}(1 + iQ^{-1})$, where c_{44} and Q are real and the value of Q is fixed to be 10^2 in the calculation.

The transforming ratio $|V_2/V_1|$ as a function of the driving frequency ω is shown in Figure 8.4.2. When ω is close to the resonant frequency ω_∞, the transforming ratio assumes maximum.

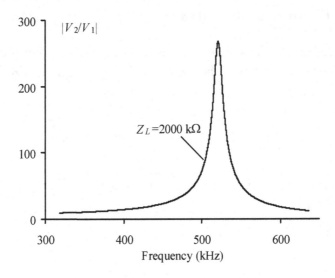

Figure 8.4.2. Transforming ratio versus driving frequency.

$|V_2/V_1|$ versus the aspect ratio $l/\bar{h}$ (the length of the receiving portion over the thickness of the driving portion) is plotted in Figure 8.4.3. An essentially linearly increasing behavior is observed. Figure 8.4.3 exhibits the voltage raising ability of the transformer. For large aspect ratios or long and thin transformers, high voltage output can be achieved. The zero-dimensional equations are particularly suitable for long and thin transformers with almost uniform fields. The dependence of the transforming ration on $l/\bar{h}$ can also be seen from the factor of $l/\bar{h}$ in Equation $(8.4.18)_1$.

It can be concluded from Equation $(8.4.18)_1$ that for small Z_L or almost shorted receiving electrodes, $|V_2/V_1|$ as a function of Z_L is linear

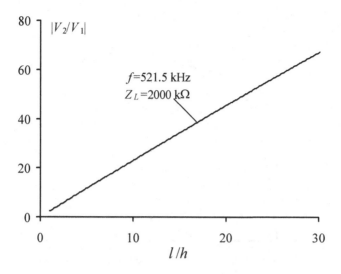

Figure 8.4.3. Transforming ratio versus aspect ratio.

in Z_L. For very large Z_L or almost open receiving electrodes, the transforming ratio approaches a constant (saturation). We note from Equation $(8.4.18)_2$ that the output current $\bar{I}_2$ has such a dependence on Z_L that when Z is small $\bar{I}_2$ has a finite value and when Z_L is large $\bar{I}_2$ approaches zero. This is as expected. $|V_2/V_1|$ as a function of Z_L is shown in Figure 8.4.4.

The input and output powers of the transformer, in terms of the complex notation, are given by

$$P_1 = \frac{1}{4}(\bar{I}_1\bar{V}_1^* + \bar{I}_1^*\bar{V}_1), \quad P_2 = \frac{1}{4}(\bar{I}_2\bar{V}_2^* + \bar{I}_2^*\bar{V}_2). \qquad (8.4.20)$$

Then the efficiency of the transformer is

$$\eta = P_2/P_1. \qquad (8.4.21)$$

It can be concluded from Equations (8.4.18), (8.4.20) and (8.4.21) that the efficiency as a function of Z behaves as follows. For small loads, η depends on the load linearly. For large loads, η decreases to zero. The efficiency as a function of Z_L is shown in Figure 8.4.5.

The behaviors shown in Figures 8.4.2 through 8.4.5 for the thickness-shear transformer are qualitatively very similar to the behaviors of the Rosen extensional transformer discussed in [9].

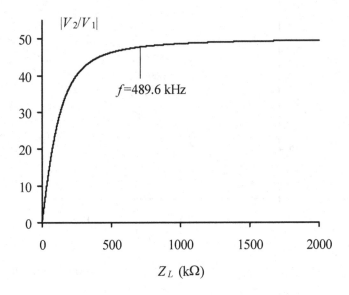

Figure 8.4.4. Transforming ratio versus load.

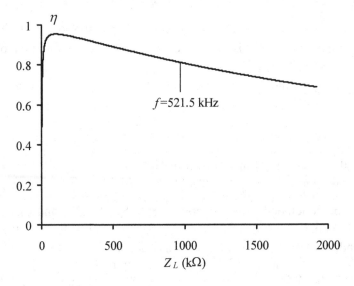

Figure 8.4.5. Efficiency versus load.

References

[1] H. F. Tiersten, On the nonlinear equations of thermoelectroelasticity, Int. J. Engng Sci., 9, 587-604, 1971.

[2] H. F. Tiersten, Electroelastic interactions and the piezoelectric equations, J. Acoust. Soc. Am., 70, 1567-1576, 1981.

[3] A. H. Meitzler, H. F. Tiersten, A. W. Warner, D. Berlincourt, G. A. Couqin and F. S. Welsh, III, IEEE Standard on Piezoelectricity, IEEE, New York, 1988.

[4] H. F. Tiersten, Linear Piezoelectric Plate Vibrations, Plenum, New York, 1969.

[5] J. C. Baumhauer and H. F. Tiersten, Nonlinear electroelastic equations for small fields superposed on a bias, J. Acoust. Soc. Am., 54, 1017-1034, 1973.

[6] H. F. Tiersten, On the accurate description of piezoelectric resonators subject to biasing deformations, Int. J. Engng Sci., 33, 2239-2259, 1995.

[7] H. F. Tiersten, Nonlinear electroelastic equations cubic in the small field variables, J. Acoust. Soc. Am., 57, 660-666, 1975.

[8] H. F. Tiersten, Analysis of intermodulation in thickness-shear and trapped energy resonators, J. Acoust. Soc. Am., 57, 667-681, 1975.

[9] J. S. Yang, An Introduction to the Theory of Piezoelectricity, Springer, New York, 2005.

[10] H. F. Tiersten, Thickness vibrations of piezoelectric plates, J. Acoust. Soc. Am., 35, 53-58, 1963.

[11] H. F. Tiersten, Wave propagation in an infinite piezoelectric plate, J. Acoust. Soc. Am., 35, 234-239, 1963.

[12] J. L. Bleustein, Some simple modes of wave propagation in an infinite piezoelectric plate, J. Acoust. Soc. Am., 45, 614-620, 1969.

[13] R. D. Mindlin, An Introduction to the Mathematical Theory of Vibrations of Plates, the U.S. Army Signal Corps Engineering Laboratories, Fort Monmouth, NJ, 1955.

[14] R. D. Mindlin, Influence of rotatory inertia and shear on flexural motions of isotropic, elastic plates, ASME J. Appl. Mech., 18, 31-38, 1951.

[15] R. D. Mindlin, Thickness-shear and flexural vibrations of crystal plates, J. Appl. Phys., 22, 316-323, 1951.

[16] R. D. Mindlin, High frequency vibrations of crystal plates, Quarterly of Applied Mathematics, 19, 51-61, 1961.

[17] R. D. Mindlin, High frequency vibrations of piezoelectric crystal plates, Int. J. Solids Struct., 8, 895-906, 1972.

[18] H. F. Tiersten, A derivation of two-dimensional equations for the vibration of electroded piezoelectric plates using an unrestricted thickness expansion of the electric potential, Proc. IEEE International Frequency Control Symp. and PDA Exhibition, 571-579, 2001.

[19] P. C. Y. Lee and Z. Nikodem, An approximate theory for high-frequency vibrations of elastic plates, Int. J. Solids Structures, 8, 581-612, 1972.

[20] N. Bugdayci and D. B. Bogy, A two-dimensional theory for piezoelectric layers used in electro-mechanical transducers, Int. J. Solids Struct., 17, 1159-1178, 1981.

[21] P. C. Y. Lee, S. Syngellakis and J. P. Hou, A two-dimensional theory for high-frequency vibrations of piezoelectric crystal plates with or without electrodes, J. Appl. Phys., 61, 1249-1262, 1987.

[22] J. Wang and J. S. Yang, Higher-order theories of piezoelectric plates and applications, Applied Mechanics Reviews, 53, 87-99, 2000.

[23] H. F. Tiersten, Equations for the extension and flexure of relatively thin electrostatic plates undergoing larger electric fields, Mechanics of Electromagnetic Materials and Structures, J. S. Lee, G. A. Maugin and Y. Shindo, ed., AMD Vol. 161, MD Vol. 42, ASME, New York, 21-34, 1993.

[24] J. S. Yang, Equations for the extension and flexure of a piezoelectric beam with rectangular cross section and applications, Int. J. of Appl. Electromagnetics and Mechanics, 9, 409-420, 1998.

[25] J. S. Yang, H. G. Zhou and Z. Y. Wang, Vibrations of an asymmetrically electroded piezoelectric plate, IEEE Trans. on Ultrasonics, Ferroelectrics, and Frequency Control, 52, 2031-2038, 2005.

[26] S. Krishnaswamy and J. S. Yang, On a special class of boundary-value problems in the theory of shells and plates, Thin-Walled Structures, 35, 139-152, 1999.

[27] J. L. Bleustein and H. F. Tiersten, Forced thickness-shear vibrations of discontinuously plated piezoelectric plates, J. Acoust. Soc. Am., 43, 1311-1318, 1968.

[28] J. S. Yang, Analysis of ceramic thickness shear piezoelectric gyroscopes, J. Acoust. Soc. Am., 102, 3542-3548, 1997.

[29] R. D. Mindlin, Bechmann's number for harmonic overtones of thickness/twist vibrations of rotated Y-cut quartz plates, J. Acoust. Soc. Am., 41, 969-973, 1967.

[30] R. D. Mindlin and P. C. Y. Lee, Thickness-shear and flexural vibrations of partially plated, crystal plates, Int. J. Solids Struct., 2, 125-139, 1966.

[31] J. S. Yang and R. C. Batra, Free vibrations of a piezoelectric body, Journal of Elasticity, 34, 239-254, 1994.

[32] J. S. Yang, Variational formulations for the vibration of a piezoelectric plate, Arch. Mech., 45, 639-651, 1993.

[33] T. R. Kane and R. D. Mindlin, High-frequency extensional vibrations of plates, ASME J. Appl. Mech., 23, 277-283, 1956.

[34] R. D. Mindlin and M. A. Medick, Extensional vibrations of elastic plates, ASME J. Appl. Mech., 26, 561-569, 1959.

[35] J. S. Yang and S. Krishnaswamy, An approximate analysis of thickness-stretch waves in an elastic plate, Wave Motion, 30, 291-301, 1999.

[36] R. D. Mindlin, High frequency vibrations of plated, crystal plates, Progress in Applied Mechanics, Macmillan, New York, 73-84, 1963.

[37] H. F. Tiersten, Equations for the control of the flexural vibrations of composite plates by partially electroded piezoelectric actuators, Active Materials and Smart Structures, G. L. Anderson and D. C. Lagoudas, ed., SPIE Proc. Series, Vol. 2427, 326-342, 1994.

[38] J. S. Yang, Equations for elastic plates with partially electroded piezoelectric actuators in flexure with shear deformation and rotary inertia, Journal of Intelligent Material Systems and Structures, 8, 444-451, 1997.

[39] J. S. Yang, H. G. Zhou and S. X. Dong, Analysis of plate piezoelectric unimorphs, IEEE Trans. on Ultrasonics, Ferroelectrics, and Frequency Control, accepted.

[40] J. S. Yang and J. D. Yu, Equations for a laminated piezoelectric plate, Arch. Mech., 45, 653-664, 1993.

[41] J. S. Yang, X. M. Yang and J. A. Turner, Amplification of acoustic waves in laminated piezoelectric semiconductor plates, Archive of Applied Mechanics, 74, 288-298, 2004.

[42] H. F. Tiersten, Surface waves guided by thin films, J. Appl. Phys., 40, 770-789, 1969.

[43] Y. J. Liu, H. Fan and J. S. Yang, Analysis of the shear stress transferred from a partially electroded piezoelectric actuator to an elastic substrate, Smart Materials and Structures, 9, 248-254, 2000.

[44] P. C. Y. Lee, Y. S. Wang and X. Markenscoff, High frequency vibrations of crystal plates under initial stresses, J. Acoust. Soc. Am., 57, 95-105, 1975.

[45] M. C. Dokmeci, Vibration of piezoelectric disks under initial stresses, Proc. 39th Annual Symp. on Frequency Control, 431-435, 1985.

[46] Y. T. Hu, J. S. Yang and Q. Jiang, A model of electroelastic plates under biasing fields with applications in buckling analysis, Int. J. of Solids and Structures, 39, 2629-2642, 2002.

[47] Y. T. Hu, C. Chen, G. Li, J. S. Yang and Q. Jiang, Basic curvilinear coordinate equations of electroelastic plates under biasing fields with applications in buckling analysis, Acta Mechanica Solida Sinica, 15, 189-200, 2002.

[48] J. S. Yang, X. M. Yang, J. A. Turner, J. A. Kosinski and R. A. Pastore, Jr., Two-dimensional equations for electroelastic plates with relatively large shear deformations, IEEE Trans. on Ultrasonics, Ferroelectrics, and Frequency Control, 50, 765-772, 2003.

[49] J. J. Gagnepain and R. Besson, Nonlinear effects in piezoelectric quartz crystals, Physical Acoustics, vol. XI, W. P. Mason and R. N. Thurston, ed., Academic Press, New York, 1975.

[50] H. Kraus, Thin Elastic Shells, John Wiley and Sons, New York, 1967.

[51] M. C. Dokmeci, On the higher order theory of piezoelectric crystal surfaces, J. Mathematical Physics, 15, 2248-2252, 1974.

[52] H. S. Tzou, Piezoelectric Shells, Kluwer, Dordrecht, the Netherlands, 1993.

[53] J. S. Yang, X. M. Yang and J. A. Turner, Amplification of acoustic waves in piezoelectric semiconductor shells, Journal of Intelligent Material Systems and Structures, 16, 613-621, 2005.

[54] M. C. Dokmeci, Shell theory for vibrations of piezoceramics under a bias, IEEE Trans. on Ultrasonics, Ferroelectrics, and Frequency Control, 37, 369-385, 1990.

[55] Y. T. Hu, J. S. Yang and Q. Jiang, On modeling of extension and flexure response of electroelastic shells under biasing fields, Acta Mechanica, 156, 163-178, 2002.

[56] J. S. Yang, X. M. Yang and J. A. Turner, Nonlinear vibrations of electroelastic shells with relatively large shear deformations, Proc. of the International Conference on Heterogeneous Material Mechanics, Chongqing, China, 6/21-26, 314-317, 2004.

[57] J. S. Yang and Y. J. Liu, Boundary formulation and numerical analysis of elastic bodies with surface bounded piezoelectric films, Smart Materials and Structures, 11, 308-311, 2002.

[58] J. S. Yang and S. H. Guo, Frequency shifts in a piezoelectric body due to a surface mass layer with consideration of the layer stiffness, IEEE Trans. on Ultrasonics, Ferroelectrics, and Frequency Control, 52, 1200-1203, 2005.

[59] J. S. Yang, Frequency shifts in a piezoelectric body due to small amounts of additional mass on its surface, IEEE Trans. on Ultrasonics, Ferroelectrics, and Frequency Control, 51, 1199-1202, 2004.

[60] J. S. Yang and S. H. Guo, Torsional waves in a circular cylindrical shell in contact with a viscous fluid, unpublished.

[61] R. L. Panton, Incompressible Flow, John Wiley & Sons, New York, 1984.

[62] R. D. Mindlin and G. Herrmann, A one-dimensional theory of compressional waves in an elastic rod, Proc of the 1st US National Congress in Applied Mechanics, 187-191, 1950.

[63] R. D. Mindlin and H. D. McNiven, Axially symmetric waves in elastic rods, ASME J. Appl. Mech., 27, 145-151, 1960.

[64] J. L. Bleustein and R. M. Stanley, A dynamical theory of torsion, Int. J. Solids Struct., 6, 569-586, 1970.

[65] R. D. Mindlin, Low frequency vibrations of elastic bars, Int. J. Solids Struct., 16, 27-49, 1976.

[66] M. C. Dokmeci, A theory of high frequency vibrations of piezoelectric crystal bars, Int. J. Solids Struct., 10, 401-409, 1974.

[67] J. S. Yang and W. Zhang, A thickness-shear high voltage piezoelectric transformer, Int. J. of Applied Electromagnetics and Mechanics, 10, 105-121, 1999.

[68] J. S. Yang, H. Y. Fang and Q. Jiang, Analysis of a ceramic bimorph piezoelectric gyroscope, Int. J. of Applied Electromagnetics and Mechanics, 10, 459-473, 1999.

[69] J. S. Yang, H. Y. Fang and Q. Jiang, Analysis of a few piezoelectric gyroscopes, Proc. IEEE Frequency Control Symposium, 79-86, 2000.

[70] Y. T. Hu, J. S. Yang and Q. Jiang, Characterization of electroelastic beams under biasing fields with applications in buckling analysis, Archive of Applied Mechanics, 72, 439-450, 2002.

[71] J. S. Yang, H. Y. Fang and Q. Jiang, One-dimensional equations for a piezoelectric ring and applications in a gyroscope, IEEE Trans. on Ultrasonics, Ferroelectrics, and Frequency Control, 48, 1275-1282, 2001.

[72] J. S. Yang, H. Y. Fang and Q. Jiang, Equations for a piezoelectric parallelepiped and applications in a gyroscope, Int. J. of Applied Electromagnetics and Mechanics, 10, 337-350, 1999.

[73] J. S. Yang and X. Zhang, Analysis of a thickness-shear piezoelectric transformer, Int. J. of Applied Electromagnetics and Mechanics, 21, 131-141, 2005.

[74] B. A. Auld, Acoustic Fields and Waves in Solids, vol. 1, John Wiley and Sons, New York, 1973.

[75] H. Jaffe and D. A. Berlincourt, Piezoelectric transducer materials, Proc. of IEEE, 53, 1372-1386, 1965.

[76] R. Bechmann, Elastic and piezoelectric constants of alpha-quartz, Phys. Rev., 110, 1060-1061, 1958.

[77] B. K. Sinha and H. F. Tiersten, First temperature derivatives of the fundamental elastic constants of quartz, J. Appl. Phys., 50, 2732-2739, 1979.

[78] R. N. Thurston, H. J. McSkimin and P. Andreatch, Jr., Third-order elastic constants of quartz, J. Appl. Phys., 37, 267-275, 1966.

[79] D. F. Nelson, Electric, Optic and Acoustic Interactions in Crystals, John Wiley and Sons, New York, 1979.

[80] B. P. Sorokin, P. P. Turchin, S. I. Burkov, D. A. Glushkov and K. S. Alexandrov, Influence of static electric field, mechanical pressure and temperature on the propagation of acoustic waves in $La_3Ga_5SiO_{14}$ piezoelectric single crystals, Proc. IEEE Int. Frequency Control Symp., 161-169, 1996.

[81] A. W. Warner, M. Onoe and G. A. Couqin, Determination of elastic and piezoelectric constants for crystals in class (3m), J. Acoust. Soc. Am., 42, 1223-1231, 1967.

[82] Y. Cho and K. Yamanouchi, Nonlinear, elastic, piezoelectric, electrostrictive, and dielectric constants of lithium niobate, J. Appl. Phys., 61, 875-887, 1987.

Appendix 1
Notation

δ_{ij}, δ_{KL}	—	Kronecker delta
δ_{iK}, δ_{Ki}	—	Shifter
ε_{ijk}, ε_{IJK}	—	Permutation tensor
X_K	—	Reference position of a material point
y_i	—	Present position of a material point
u_K	—	Mechanical displacement vector
J	—	Jacobian
C_{KL}	—	Deformation tensor
S_{KL}	—	Finite strain tensor
S_{kl}	—	Linear strain tensor
v_i	—	Velocity vector
d_{ij}	—	Deformation rate tensor
ω_{ij}	—	Spin tensor
D/Dt	—	Material time derivative
ρ_0	—	Reference mass density (scalar)
ρ	—	Present mass density
ρ_e	—	Free charge density per unit present volume (scalar)
ρ_E	—	Free charge density per unit reference volume (scalar)
σ_e	—	Surface free charge per unit present area (scalar)
σ_E	—	Surface free charge per unit reference area (scalar)
Q_e	—	Free charge (scalar)
I	—	Current
V	—	Voltage
Z	—	Impedance
ε_0	—	Permittivity of free space
ϕ	—	Electrostatic potential
E_i	—	Electric field
P_i	—	Electric polarization per unit present volume
π_i	—	Electric polarization per unit mass

D_i	—	Electric displacement vector
$\mathcal{E}_K$	—	Reference electric field vector
$\mathcal{P}_K$	—	Reference electric polarization vector
$\mathcal{D}_K$	—	Reference electric displacement vector
f_j	—	Mechanical body force per unit mass
σ_{ij}	—	Cauchy stress tensor
σ_{ij}^E	—	Electrostatic stress tensor
$\sigma_{ij}^S, F_{Lj}, T_{KL}^S$	—	Symmetric stress tensor in spatial, two-point, and material form
$\sigma_{ij}^M, M_{Lj}, T_{KL}^M$	—	Symmetric Maxwell stress tensor in spatial, two-point, and material form
$\tau_{ij}, K_{Lj}, \hat{T}_{KL}$	—	Total stress tensor in spatial, two-point, and material form
K_{LM}	—	$K_{Lj}\delta_{jM}$
T_{kl}	—	Linear stress tensor
T_k	—	Mechanical surface traction per unit reference area
t_k	—	Mechanical surface traction per unit present area
ψ	—	Free energy per unit mass
H	—	Electric enthalpy per unit volume
$\gamma_{rs}, \psi_{ks}, \zeta_{kj}$	—	Plate material constants obtained by relaxing T_{3j}.
$\bar{c}_{pq}, \bar{e}_{kp}, \bar{\varepsilon}_{kj}$	—	Plate material constants obtained by relaxing T_{33}.
$\bar{c}_{pq}', \bar{e}_{kp}'$	—	$\bar{c}_{pq}, \bar{e}_{kp}$ modified by shear correction factors.
$\tilde{c}_{pq}, \tilde{e}_{kp}, \tilde{\varepsilon}_{kj}$	—	Beam material constants obtained by relaxing $T_2 - T_6$.
$\hat{c}_{\alpha\beta}, \hat{e}_{k\alpha}, \hat{\varepsilon}_{kj}$	—	Beam material constants obtained by relaxing $T_2 - T_4$.
$\hat{c}_{\alpha\beta}', \hat{e}_{k\alpha}'$	—	$\hat{c}_{\alpha\beta}, \hat{e}_{k\alpha}$ modified by shear correction factors.
ω_∞	—	Thickness-shear frequency of an unbounded plate.
κ	—	Thickness-shear correction factor.

Appendix 2
Electroelastic Material Constants

Material constants for a few common piezoelectrics are summarized below. Numerical results given in this book are calculated from these constants.

Permittivity of free space $\varepsilon_0 = 8.854 \times 10^{-12}$ F/m .

Polarized ceramics

The material matrices for PZT-5H are [74]

$$\rho = 7500 \ kg/m^3,$$

$$[c_{pq}] = \begin{pmatrix} 12.6 & 7.95 & 8.41 & 0 & 0 & 0 \\ 7.95 & 12.6 & 8.41 & 0 & 0 & 0 \\ 8.41 & 8.41 & 11.7 & 0 & 0 & 0 \\ 0 & 0 & 0 & 2.3 & 0 & 0 \\ 0 & 0 & 0 & 0 & 2.3 & 0 \\ 0 & 0 & 0 & 0 & 0 & 2.325 \end{pmatrix} \times 10^{10} \, \text{N/m}^2,$$

$$[e_{ip}] = \begin{pmatrix} 0 & 0 & 0 & 0 & 17 & 0 \\ 0 & 0 & 0 & 17 & 0 & 0 \\ -6.5 & -6.5 & 23.3 & 0 & 0 & 0 \end{pmatrix} \text{C/m}^2,$$

$$[\varepsilon_{ij}] = \begin{pmatrix} 1700\varepsilon_0 & 0 & 0 \\ 0 & 1700\varepsilon_0 & 0 \\ 0 & 0 & 1470\varepsilon_0 \end{pmatrix}$$

$$= \begin{pmatrix} 1.505 & 0 & 0 \\ 0 & 1.505 & 0 \\ 0 & 0 & 1.302 \end{pmatrix} \times 10^{-8}\,\text{C/(V-m)}.$$

For PZT-5H, an equivalent set of material constants are [74]

$$s_{11} = 16.5, \quad s_{33} = 20.7, \quad s_{44} = 43.5,$$

$$s_{12} = -4.78, \quad s_{13} = -8.45 \times 10^{-12}\,\text{m}^2/\text{N},$$

$$d_{31} = -274, \quad d_{15} = 741, \quad d_{33} = 593 \times 10^{-12}\,\text{C/N},$$

$$\varepsilon_{11} = 3130\varepsilon_0, \quad \varepsilon_{33} = 3400\varepsilon_0.$$

When poling is along other directions, the material matrices can be obtained by tensor transformations. For PZT-5H, when poling is along the x_1 axis, we have

$$[c_{pq}] = \begin{pmatrix} 11.7 & 8.41 & 8.41 & 0 & 0 & 0 \\ 8.41 & 12.6 & 7.95 & 0 & 0 & 0 \\ 8.41 & 7.95 & 12.6 & 0 & 0 & 0 \\ 0 & 0 & 0 & 2.325 & 0 & 0 \\ 0 & 0 & 0 & 0 & 2.3 & 0 \\ 0 & 0 & 0 & 0 & 0 & 2.3 \end{pmatrix} \times 10^{10}\,\text{N/m}^2,$$

$$[e_{ip}] = \begin{pmatrix} 23.3 & -6.5 & -6.5 & 0 & 0 & 0 \\ 0 & 0 & 0 & 0 & 0 & 17 \\ 0 & 0 & 0 & 0 & 17 & 0 \end{pmatrix} \text{C/m}^2,$$

$$[\varepsilon_{ij}] = \begin{pmatrix} 1.302 & 0 & 0 \\ 0 & 1.505 & 0 \\ 0 & 0 & 1.505 \end{pmatrix} \times 10^{-8}\,\text{C/Vm}.$$

When poling is along the x_2 axis

$$[c_{pq}] = \begin{pmatrix} 12.6 & 8.41 & 7.95 & 0 & 0 & 0 \\ 8.41 & 11.7 & 8.41 & 0 & 0 & 0 \\ 7.95 & 8.41 & 12.6 & 0 & 0 & 0 \\ 0 & 0 & 0 & 2.3 & 0 & 0 \\ 0 & 0 & 0 & 0 & 2.325 & 0 \\ 0 & 0 & 0 & 0 & 0 & 2.3 \end{pmatrix} \times 10^{10}\,\text{N/m}^2,$$

$$[e_{ip}] = \begin{pmatrix} 0 & 0 & 0 & 0 & 0 & 17 \\ -6.5 & 23.3 & -6.5 & 0 & 0 & 0 \\ 0 & 0 & 0 & 17 & 0 & 0 \end{pmatrix} \text{C/m}^2,$$

$$[\varepsilon_{ij}] = \begin{pmatrix} 1.505 & 0 & 0 \\ 0 & 1.302 & 0 \\ 0 & 0 & 1.505 \end{pmatrix} \times 10^{-8}\,\text{C/Vm}.$$

For PZT-G1195

$$\rho = 7500\text{kg}/\text{m}^3, \quad c_{11}^E = c_{22}^E = 148, \quad c_{33}^E = 131, \quad c_{12}^E = 76.2,$$
$$c_{13}^E = c_{23}^E = 74.2, \quad c_{44}^E = c_{55}^E = 25.4, \quad c_{66}^E = 35.9\text{GPa},$$
$$e_{15} = 9.2, \quad e_{31} = -2.1, \quad e_{33} = 9.5\text{C}/\text{m}^2.$$

Material constants of a few other polarized ceramics are given in the following tables [75]:

Material	c_{11}	c_{12}	c_{13}	c_{33}	c_{44}	c_{66}
PZT-4	13.9	7.78	7.40	11.5	2.56	3.06
PZT-5A	12.1	7.59	7.54	11.1	2.11	2.26
PZT-6B	16.8	8.47	8.42	16.3	3.55	4.17
PZT-5H	12.6	7.91	8.39	11.7	2.30	2.35

Material	c_{11}	c_{12}	c_{13}	c_{33}	c_{44}	c_{66}
PZT-7A	14.8	7.61	8.13	13.1	2.53	3.60
PZT-8	13.7	6.99	7.11	12.3	3.13	3.36
BaTiO$_3$	15.0	6.53	6.62	14.6	4.39	4.24
	$\times 10^{10}$ N/m^2					

Material	e_{31}	e_{33}	e_{15}	ε_{11}	ε_{33}
PZT-4	-5.2	15.1	12.7	0.646	0.562
PZT-5A	-5.4	15.8	12.3	0.811	0.735
PZT-6B	-0.9	7.1	4.6	0.360	0.342
PZT-5H	-6.5	23.3	17.0	1.505	1.302
PZT-7A	-2.1	9.5	9.2	0.407	0.208
PZT-8	-4.0	13.2	10.4	0.797	0.514
BaTiO$_3$	-4.3	17.5	11.4	0.987	1.116
	C/m^2			$\times 10^{-8}$ C/Vm	

Density	PZT-5H	PZT-5A	PZT-6B	PZT-4
kg/m^3	7500	7750	7550	7500

Density	PZT-7A	PZT-8	BaTiO$_3$
kg/m^3	7600	7600	5700

Quartz

When referred to the crystal axes, the second-order material constants for left-hand quartz have the following values [76]:

$$\rho = 2649 \text{ kg/m}^3,$$

$$[c_{pq}] = \begin{pmatrix} 86.74 & 6.99 & 11.91 & -17.91 & 0 & 0 \\ 6.99 & 86.74 & 11.91 & 17.91 & 0 & 0 \\ 11.91 & 11.91 & 107.2 & 0 & 0 & 0 \\ -17.91 & 17.91 & 0 & 57.94 & 0 & 0 \\ 0 & 0 & 0 & 0 & 57.94 & -17.91 \\ 0 & 0 & 0 & 0 & -17.91 & 39.88 \end{pmatrix} \times 10^9 \text{ N/m}^2,$$

$$[e_{ip}] = \begin{pmatrix} 0.171 & -0.171 & 0 & -0.0406 & 0 & 0 \\ 0 & 0 & 0 & 0 & 0.0406 & -0.171 \\ 0 & 0 & 0 & 0 & 0 & 0 \end{pmatrix} \text{C/m}^2,$$

$$[\varepsilon_{ij}] = \begin{pmatrix} 39.21 & 0 & 0 \\ 0 & 39.21 & 0 \\ 0 & 0 & 41.03 \end{pmatrix} \times 10^{-12} \text{ C/Vm}.$$

Temperature derivatives of the elastic constants of quartz at 25 °C are [77]

pq	11	33	12	13
$(1/c_{pq})(dc_{pq}/dT)$ $(10^{-6}/°C)$	18.16	-66.60	-1222	-178.6

pq	44	66	14
$(1/c_{pq})(dc_{pq}/dT)$ $(10^{-6}/°C)$	-89.72	126.7	-49.21

For quartz there are 31 nonzero third-order elastic constants. 14 are given in the following table. These values, at 25 °C, and based on a least-squares fit, are all in 10^{11} N/m^2 [78]

Constant	Value	Standard error
c_{111}	-2.10	0.07
c_{112}	-3.45	0.06
c_{113}	+0.12	0.06
c_{114}	-1.63	0.05
c_{123}	-2.94	0.05
c_{124}	-0.15	0.04
c_{133}	-3.12	0.07
c_{134}	+0.02	0.04
c_{144}	-1.34	0.07
c_{155}	-2.00	0.08
c_{222}	-3.32	0.08
c_{333}	-8.15	0.18
c_{344}	-1.10	0.07
c_{444}	-2.76	0.17

In addition, there are 17 relations among the third-order elastic constants of quartz [79]

$$c_{122} = c_{111} + c_{112} - c_{222}, \quad c_{156} = \frac{1}{2}(c_{114} + 3c_{124}),$$

$$c_{166} = \frac{1}{4}(-2c_{111} - c_{112} + 3c_{222}),$$

$$c_{224} = -c_{114} - 2c_{124}, \quad c_{256} = \frac{1}{2}(c_{114} - c_{124}),$$

$$c_{266} = \frac{1}{4}(2c_{111} - c_{112} - c_{222}),$$

$$c_{366} = \frac{1}{2}(c_{113} - c_{123}), \quad c_{456} = \frac{1}{2}(-c_{144} + c_{155}),$$

$$c_{223} = c_{113}, \quad c_{233} = c_{133}, \quad c_{234} = -c_{134}, \quad c_{244} = c_{155}, \quad c_{255} = c_{144},$$

$$c_{355} = c_{344}, \quad c_{356} = c_{134}, \quad c_{455} = -c_{444}, \quad c_{466} = c_{124}.$$

For the fourth-order elastic constants there are 69 nonzero ones of which 23 are independent [49]

$$c_{1111}, \quad c_{3333}, \quad c_{4444}, \quad c_{6666}, \quad c_{1112}, \quad c_{1113}, \quad c_{1123}, \quad c_{2214}, \quad c_{3331},$$

$$c_{4456}, \quad c_{5524}, \quad c_{4443}, \quad c_{1133}, \quad c_{3344}, \quad c_{1456}, \quad c_{1155}, \quad c_{1134}, \quad c_{2356},$$

$$c_{4423}, \quad c_{4413}, \quad c_{3314}, \quad c_{6614}, \quad c_{6624}.$$

There are 46 relations [49]

$$c_{2222} = c_{1111}, \quad c_{2266} = \frac{1}{6}(c_{1111} - c_{1112}), \quad c_{2223} = c_{1113},$$

$$c_{2221} = c_{1112}, \quad c_{6612} = \frac{1}{6}(c_{1111} - 4c_{6666} - c_{1112}), \quad c_{2213} = c_{1123},$$

$$c_{1166} = c_{2266}, \quad c_{1122} = \frac{1}{3}(-c_{1111} + 4c_{1112} + 8c_{6666}),$$

$$c_{6613} = \frac{1}{4}(c_{1113} - c_{1123}),$$

$$c_{5555} = c_{4444}, \quad c_{4455} = \frac{1}{3}c_{4444}, \quad c_{6623} = c_{6613},$$

$$c_{1124} = -c_{2214} + c_{6614} + c_{6624},$$

$$c_{3312} = -c_{1133}, \quad c_{1114} = 3(-c_{2214} + 2c_{6614} - 2c_{6624}), \quad c_{2233} = c_{1133},$$

$$c_{2256} = \frac{1}{2}(-2c_{2214} + 3c_{6614} - 5c_{6624}), \quad c_{6633} = c_{1133},$$

$$c_{2224} = 3(c_{2214} - 3c_{6614} + c_{6624}),$$

$$c_{3355} = c_{3344}, \quad c_{1156} = \frac{1}{2}(-2c_{2214} + 7c_{6614} - c_{6624}), \quad c_{3332} = c_{3331},$$

$$c_{1256} = \frac{1}{2}(-2c_{2214} + 3c_{6614} - c_{6624}), \quad c_{5534} = -c_{4443},$$

$$c_{6665} = \frac{3}{2}(c_{6614} - c_{6624}),$$

$$c_{4442} = -4c_{4456} - c_{5524}, \quad c_{1234} = c_{1134} - 2c_{2356}, \quad c_{2255} = c_{4412},$$

$$c_{5514} = 2c_{4456} + c_{5524}, \quad c_{1356} = 2c_{1134} - 3c_{2356}, \quad c_{5566} = c_{1456},$$

$$c_{5556} = 3c_{4456}, \quad c_{2234} = 4c_{2356} - 3c_{1134}, \quad c_{3324} = -c_{3314},$$

$$c_{4441} = 2c_{4456} - c_{5524}, \quad c_{6634} = c_{1234}, \quad c_{3356} = c_{3314}, \quad c_{5512} = c_{4412},$$

$$c_{1144} = c_{4412}, \quad c_{5523} = c_{4413}, \quad c_{2456} = c_{1456}, \quad c_{2244} = c_{1155}, \quad c_{5513} = c_{4423},$$

$$c_{4466} = c_{1456}, \quad c_{4412} = c_{1155} - 4c_{1456}, \quad c_{3456} = \frac{1}{2}(c_{4423} - c_{4413}).$$

The fourth-order elastic constants are usually unknown. Some scattered results are [49]

$$c_{1111} = 1.59 \times 10^{13} \, \text{N} / \text{m}^2 \pm 20\%,$$

$$c_{3333} = 1.84 \times 10^{13} \, \text{N} / \text{m}^2 \pm 20\%,$$

and [8]

$$c_{6666}^E = 77 \times 10^{11} \, \text{N} / \text{m}^2.$$

AT-cut quartz is a special case of rotated Y-cut quartz ($\theta = 35.25°$) whose material constants are [4]

$$[c_{pq}] = \begin{pmatrix} 86.74 & -8.25 & 27.15 & -3.66 & 0 & 0 \\ -8.25 & 129.77 & -7.42 & 5.7 & 0 & 0 \\ 27.15 & -7.42 & 102.83 & 9.92 & 0 & 0 \\ -3.66 & 5.7 & 9.92 & 38.61 & 0 & 0 \\ 0 & 0 & 0 & 0 & 68.81 & 2.53 \\ 0 & 0 & 0 & 0 & 2.53 & 29.01 \end{pmatrix} \times 10^9 \, \text{N/m}^2,$$

$$[e_{ip}] = \begin{pmatrix} 0.171 & -0.152 & -0.0187 & 0.067 & 0 & 0 \\ 0 & 0 & 0 & 0 & 0.108 & -0.095 \\ 0 & 0 & 0 & 0 & -0.0761 & 0.067 \end{pmatrix} C/m^2,$$

$$[\varepsilon_{ij}] = \begin{pmatrix} 39.21 & 0 & 0 \\ 0 & 39.82 & 0.86 \\ 0 & 0.86 & 40.42 \end{pmatrix} \times 10^{-12} C/Vm.$$

Langasite

The second-order material constants of $La_3Ga_5SiO_{14}$ are [80]

$$\rho = 5743 \text{ kg/m}^3,$$

$$[c_{pq}] = \begin{pmatrix} 18.875 & 10.475 & 9.589 & -1.412 & 0 & 0 \\ 10.475 & 18.875 & 9.589 & 1.412 & 0 & 0 \\ 9.589 & 9.589 & 26.14 & 0 & 0 & 0 \\ -1.412 & 1.412 & 0 & 5.35 & 0 & 0 \\ 0 & 0 & 0 & 0 & 5.35 & -1.412 \\ 0 & 0 & 0 & 0 & -1.412 & 4.2 \end{pmatrix}$$

$$\times 10^{10} \text{ N/m}^2,$$

$$[e_{ip}] = \begin{pmatrix} -0.44 & 0.44 & 0 & -0.08 & 0 & 0 \\ 0 & 0 & 0 & 0 & 0.08 & 0.44 \\ 0 & 0 & 0 & 0 & 0 & 0 \end{pmatrix} C/m^2,$$

$$[\varepsilon_{ij}] = \begin{pmatrix} 18.92\varepsilon_0 & 0 & 0 \\ 0 & 18.92\varepsilon_0 & 0 \\ 0 & 0 & 50.7\varepsilon_0 \end{pmatrix}$$

$$= \begin{pmatrix} 167.5 & 0 & 0 \\ 0 & 167.5 & 0 \\ 0 & 0 & 448.9 \end{pmatrix} \times 10^{-12} C/Vm.$$

The third-order material constants of $La_3Ga_5SiO_{14}$ at 20°C are given in [80]. The third-order elastic constants c_{pqr} (in 10^{10} N/m^2) are

c_{111}	-97.2	c_{134}	-4.1
c_{112}	0.7	c_{144}	-4.0
c_{113}	-11.6	c_{155}	-19.8
c_{114}	-2.2	c_{222}	-96.5
c_{123}	0.9	c_{333}	-183.4
c_{124}	-2.8	c_{344}	-38.9
c_{133}	-72.1	c_{444}	20.2

The third-order piezoelectric effect constants e_{ipq} (in C/m^2) are

e_{111}	9.3	e_{124}	-4.8
e_{113}	-3.5	e_{134}	6.9
e_{114}	1.0	e_{144}	-1.7
e_{122}	0.7	e_{315}	-4

The third-order electrostriction constants H_{pq} (in 10^{-9}N/V^2) are

H_{11}	-26	H_{31}	-24
H_{12}	65	H_{33}	-40
H_{13}	20	H_{41}	-170
H_{14}	-43	H_{44}	-44

The third-order dielectric permeability ε_{111} (in 10^{-20} F/V) are

ε_{111}	-0.5

Lithium Niobate

The second-order material constants for lithium niobate are [81]
$$\rho = 4700 \text{ kg/m}^3,$$

$$[c_{pq}] = \begin{pmatrix} 2.03 & 0.53 & 0.75 & 0.09 & 0 & 0 \\ 0.53 & 2.03 & 0.75 & -0.09 & 0 & 0 \\ 0.75 & 0.75 & 2.45 & 0 & 0 & 0 \\ 0.09 & -0.09 & 0 & 0.60 & 0 & 0 \\ 0 & 0 & 0 & 0 & 0.60 & 0.09 \\ 0 & 0 & 0 & 0 & 0.09 & 0.75 \end{pmatrix} \times 10^{11} \text{N/m}^2,$$

$$[e_{ip}] = \begin{pmatrix} 0 & 0 & 0 & 0 & 3.70 & -2.50 \\ -2.50 & 2.50 & 0 & 3.70 & 0 & 0 \\ 0.20 & 0.20 & 1.30 & 0 & 0 & 0 \end{pmatrix} \text{C/m}^2,$$

$$[\varepsilon_{ij}] = \begin{pmatrix} 38.9 & 0 & 0 \\ 0 & 38.9 & 0 \\ 0 & 0 & 25.7 \end{pmatrix} \times 10^{-11} \text{C/Vm}.$$

The third-order material constants of lithium niobate are given in [82]. The third-order elastic constants c_{pqr} (in 10^{11} N/m^2) are

Constant	Value	Standard error
c_{111}	-21.2	4.0
c_{112}	-5.3	1.2
c_{113}	-5.7	1.5
c_{114}	2.0	0.8
c_{123}	-2.5	1.0
c_{124}	0.4	0.3
c_{133}	-7.8	1.9

Constant	Value	Standard error
c_{134}	1.5	0.3
c_{144}	-3.0	0.2
c_{155}	-6.7	0.3
c_{222}	-23.3	3.4
c_{333}	-29.6	7.2
c_{344}	-6.8	0.7
c_{444}	-0.3	0.4

The third-order piezoelectric constants e_{ipq} $(= -k_{1\,ipq})$ are

Constant	Value	Standard error
e_{115}	17.1	6.6
e_{116}	-4.7	6.4
e_{125}	19.9	2.1
e_{126}	-15.9	5.3
e_{135}	19.6	2.7
e_{136}	-0.9	2.7
e_{145}	20.3	5.7
e_{311}	14.7	6.0
e_{312}	13.0	11.4
e_{313}	-10.0	8.7
e_{314}	11.0	4.6
e_{333}	-17.3	5.9
e_{344}	-10.2	5.6
	C/m^2	

The third-order electrostirctive constants l_{pq} (compressed from $b_{ijkl} + \varepsilon_0 \delta_{ij} \delta_{kl} - \varepsilon_0 \delta_{ik} \delta_{jl} - \varepsilon_0 \delta_{il} \delta_{kl}$) (in 10^{-9}F/m^2) are

Constant	Value	Standard error
l_{11}	1.11	0.39
l_{12}	2.19	0.56
l_{13}	2.32	0.67
l_{31}	0.19	0.61
l_{33}	-2.76	0.41
l_{14}	1.51	0.17
l_{41}	1.85	0.17
l_{44}	-1.83	0.11

The third-order dielectric constants ε_{ip} (in 10^{-19} F/V) are

Constant	Value	Standard error
ε_{31}	-2.81	0.06
ε_{22}	-2.40	0.09
ε_{33}	-2.91	0.06

Lithium Tantalate

The second-order material constants for lithium tantalate are [81]

$$\rho = 7450 \text{ kg/m}^3,$$

$$[c_{pq}] = \begin{pmatrix} 2.33 & 0.47 & 0.80 & -0.11 & 0 & 0 \\ 0.47 & 2.33 & 0.80 & 0.11 & 0 & 0 \\ 0.80 & 0.80 & 2.45 & 0 & 0 & 0 \\ -0.11 & -0.11 & 0 & 0.94 & 0 & 0 \\ 0 & 0 & 0 & 0 & 0.94 & -0.11 \\ 0 & 0 & 0 & 0 & -0.11 & 0.93 \end{pmatrix} \times 10^{11} \, \text{N/m}^2,$$

$$[e_{ip}] = \begin{pmatrix} 0 & 0 & 0 & 0 & 2.6 & -1.6 \\ -1.6 & 1.6 & 0 & 2.6 & 0 & 0 \\ 0 & 0 & 1.9 & 0 & 0 & 0 \end{pmatrix} \text{C/m}^2,$$

$$[\varepsilon_{ij}] = \begin{pmatrix} 36.3 & 0 & 0 \\ 0 & 36.3 & 0 \\ 0 & 0 & 38.2 \end{pmatrix} \times 10^{-11} \text{C/Vm}.$$

Cadmium Sulfide (CdS)

The second-order material constants [74]:

$\rho = 4820 \, \text{kg/m}^3,$

$c_{11} = 9.07, \quad c_{33} = 9.38, \quad c_{44} = 1.504,$

$c_{12} = 5.81, \quad c_{13} = 5.10 \times 10^{10} \, \text{N/m}^2,$

$e_{15} = -0.21, \quad e_{31} = -0.24, \quad e_{33} = -0.44 \, \text{C/m}^2,$

$\varepsilon_{11} = 9.02\varepsilon_0, \quad \varepsilon_{33} = 9.53\varepsilon_0, \quad \varepsilon_0 = 8.854 \times 10^{-12} \, \text{F/m}.$

Index